GRE
Math Tests

JEFF KOLBY

DERRICK VAUGHN

KUNDA WAMSSHYDAR

GRE is a service mark of
Educational Testing Service.

Additional educational titles from Nova Press (available at novapress.net):

- **GRE Prep Course** (624 pages, includes software)
- **GMAT Prep Course** (624 pages, includes software)
  GMAT Math Prep Course (528 pages)
  GMAT Data Sufficiency Prep Course (422 pages)
  Full Potential GMAT Sentence Correction Intensive (372 pages)
- **Master The LSAT** (608 pages, includes software and 4 official LSAT exams)
- **MCAT Prep Course** (1,340 pages)
- **SAT Prep Course** (640 pages, includes software)
  SAT Math Prep Course (404 pages)
  SAT Critical Reading and Writing Prep Course (350 pages)
- **ACT Math Prep Course** (402 pages)
  ACT Verbal Prep Course (248 pages)
- **Scoring Strategies for the TOEFL® iBT:** (800 pages, includes audio CD)
  Speaking and Writing Strategies for the TOEFL® iBT: (394 pages, includes audio CD)
  500 Words, Phrases, Idioms for the TOEFL® iBT: (238 pages, includes audio CD)
  Practice Tests for the TOEFL® iBT: (292 pages, includes audio CD)
  Business Idioms in America: (220 pages)
  Americanize Your Language and Emotionalize Your Speech! (210 pages)
- **Postal Exam Book** (276 pages)
- **Law School Basics:** A Preview of Law School and Legal Reasoning (224 pages)
- **Vocabulary 4000:** The 4000 Words Essential for an Educated Vocabulary (160 pages)

Copyright © 2014 by Nova Press
All rights reserved.

Duplication, distribution, or data base storage of any part of this work is prohibited without prior written approval from the publisher.

ISBN-10: 1–889057–47–9
ISBN-13: 978-1-889057-47-7

GRE is a service mark of Educational Testing Service, which was not involved in the production of, and does not endorse, this book.

9058 Lloyd Place
West Hollywood, CA 90069

Phone: 1-800-949-6175
E-mail: info@novapress.net
Website: www.novapress.net

ABOUT THIS BOOK

If you don't have a pencil in your hand, get one now! Don't just read this book—write on it, study it, scrutinize it! In short, for the next four weeks, this book should be a part of your life. When you have finished the book, it should be marked-up, dog-eared, tattered and torn.

Although the GRE is a difficult test, it is a *very* learnable test. This is not to say that the GRE is "beatable." There is no bag of tricks that will show you how to master it overnight. You probably have already realized this. Some books, nevertheless, offer "inside stuff" or "tricks" which they claim will enable you to beat the test. These include declaring that answer-choices B, C, or D are more likely to be correct than choices A or E. This tactic, like most of its type, does not work. It is offered to give the student the feeling that he or she is getting the scoop on the test.

The GRE cannot be "beaten." But it can be mastered—through hard work, analytical thought, and by training yourself to think like a test writer.

The GRE math sections are not easy—nor is this book. To improve your GRE math score, you must be willing to work; if you study hard and master the techniques in this book, your score will improve—significantly.

The 23 math tests in this book will introduce you to numerous analytic techniques that will help you immensely, not only on the GRE but in graduate school as well. For this reason, studying for the GRE can be a rewarding and satisfying experience.

CONTENTS

	ORIENTATION	7
Part One	THE TESTS	13
	Test 1	15
	Test 2	33
	Test 3	55
	Test 4	75
	Test 5	93
	Test 6	109
	Test 7	125
	Test 8	143
	Test 9	165
	Test 10	183
	Test 11	205
	Test 12	223
	Test 13	245
	Test 14	265
	Test 15	285
	Test 16	303
	Test 17	323
	Test 18	345
	Test 19	365
	Test 20	383
	Test 21	405
	Test 22	423
	Test 23	443
Part Two:	SUMMARY OF MATH PROPERTIES	463

ORIENTATION

Format of the Math Sections

The math section consists of three types of questions: *Quantitative Comparisons*, *Standard Multiple-Choice*, and *Graphs*. They are designed to test your ability to solve problems, not to test your mathematical knowledge.

There are 2 math sections, each is 35 minutes long, and each contains about 20 questions. The questions can appear in any order.

FORMAT
Quantitative Comparisons
Standard Multiple-Choice
Nonstandard Multiple-Choice
Graphs
Numeric Entry

Level of Difficulty

GRE math is very similar to SAT math. The mathematical skills tested are very basic: only first year high school algebra and geometry (no proofs). However, this does not mean that the math section is easy. The medium of basic mathematics is chosen so that everyone taking the test will be on a fairly even playing field. This way, students who majored in math, engineering, or science don't have an undue advantage over students who majored in humanities. Although the questions require only basic mathematics and **all** have **simple** solutions, it can require considerable ingenuity to find the simple solution. If you have taken a course in calculus or another advanced math topic, don't assume that you will find the math section easy. Other than increasing your mathematical maturity, little you learned in calculus will help on the GRE.

As mentioned above, every GRE math problem has a simple solution, but finding that simple solution may not be easy. The intent of the math section is to test how skilled you are at finding the simple solutions. The premise is that if you spend a lot of time working out long solutions you will not finish as much of the test as students who spot the short, simple solutions. So, if you find yourself performing long calculations or applying advanced mathematics—stop. You're heading in the wrong direction.

Experimental Section

The GRE is a standardized test. Each time it is offered, the test has, as close as possible, the same level of difficulty as every previous test. Maintaining this consistency is very difficult—hence the experimental section. The effectiveness of each question must be assessed before it can be used on the GRE. A problem that one person finds easy another person may find hard, and vice versa. The experimental section measures the relative difficulty of potential questions; if responses to a question do not perform to strict specifications, the question is rejected.

The experimental section can be a verbal section or a math section. You won't know which section is experimental. You will know which type of section it is, though, since there will be an extra one of that type.

Because the "bugs" have not been worked out of the experimental section—or, to put it more directly, because you are being used as a guinea pig to work out the "bugs"—this portion of the test is often more difficult and confusing than the other parts.

This brings up an ethical issue: How many students have run into the experimental section early in the test and have been confused and discouraged by it? Crestfallen by having done poorly on, say, the first—though experimental—section, they lose confidence and perform below their ability on the rest of the test. Some testing companies are becoming more enlightened in this regard and are administering experimental sections as separate practice tests. Unfortunately, ETS has yet to see the light.

Knowing that the experimental section can be disproportionately difficult, if you do poorly on a particular section you can take some solace in the hope that it may have been the experimental section. In other words, do not allow one difficult section to discourage your performance on the rest of the test.

Research Section

You may also see a research section. This section, if it appears, will be identified and will be last. The research section will not be scored and will not affect your score on other parts of the test.

Multiple-Choice Questions (select one or more choices)

In addition to the classic multiple-choice question with only one correct answer, some multiple-choice questions ask you to select one or more answers. We'll discuss this type of multiple-choice question in the problems that follow. For now, here are the official directions:

Directions: Select one or more answer choices according to the specific question directions.

If the question does not specify how many answer choices to select, select all that apply.

- The correct answer may be just one of the choices or may be as many as all of the choices, depending on the question.

- No credit is given unless you select all of the correct choices and no others.

If the question specifies how many answer choices to select, select exactly that number of choices.

Numeric Entry Questions

This type of question requires you to enter the answer as an integer or a decimal. If the answer is a fraction, then there will be two answer boxes—one for the numerator and one for the denominator. Entering the answers is quite natural, but read the following official directions to make sure there are no surprises. We'll discuss this type of question in the problems that follow.

Directions: Enter your answer as an integer or a decimal if there is a single answer box OR as a fraction if there are two separate boxes—one for the numerator and one for the denominator.

To enter an integer or a decimal, either type the number in the answer box using the keyboard or use the Transfer Display button on the calculator.

- First, click on the answer box—a cursor will appear in the box—and then type the number.
- To erase a number, use the Backspace key.
- For a negative sign, type a hyphen. For a decimal point, type a period.
- To remove a negative sign, type the hyphen again and it will disappear; the number will remain.
- The Transfer Display button on the calculator will transfer the calculator display to the answer box.
- Equivalent forms of the correct answer, such as 2.5 and 2.50, are all correct.
- Enter the exact answer unless the question asks you to round your answer.

To enter a fraction, type the numerator and the denominator in the respective boxes using the keyboard.

- For a negative sign, type a hyphen. A decimal point cannot be used in a fraction.
- The Transfer Display button on the calculator cannot be used for a fraction.
- Fractions do **not** need to be reduced to lowest terms, though you may need to reduce your fraction to fit in the boxes.

Calculator

An on-screen calculator is provided during the test, but use it sparingly. Here's what it looks like:

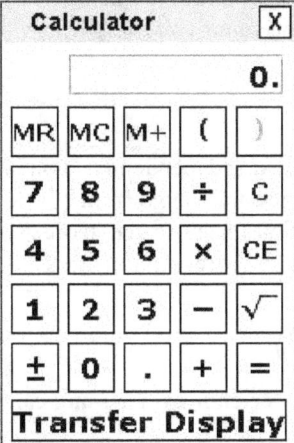

As you can see, there are a limited number of basic functions available. One neat feature is that you can transfer a particular calculation from the calculator to the answer box by clicking the Transfer Display button.

Note: The calculator uses the standard order of operations when performing multiple operations, so it does not necessarily perform operations from left to right. The order is **P**arentheses, **E**xponentiation, **M**ultiplication and **D**ivision (from left to right), **A**ddition and **S**ubtraction (from left to right). This is often remembered by the acronym **PEMDAS**. So, if you enter $2 - 3 \times 4$ and click the equal symbol (=), the calculator will multiply the 3 and 4 and subtract the result from 2, giving the answer -10.

The Computer Based Test & the Paper-&-Pencil Test

The computer based GRE uses the same type of questions as the paper-&-pencil test. The only difference is the medium, that is the way the questions are presented.

There are advantages and disadvantages to the computer based test. Probably the biggest advantages are that you can take the computer based test just about any time and you can take it in a small room with just a few other people—instead of in a large auditorium with hundreds of other stressed people. One the other hand, it is easier to misread a computer screen than it is to misread printed material, and it can be distracting looking back and forth from the computer screen to your scratch paper.

Pacing

Although time is limited on the GRE, working too quickly can damage your score. Many problems hinge on subtle points, and most require careful reading of the setup. Because undergraduate school puts such heavy reading loads on students, many will follow their academic conditioning and read the questions quickly, looking only for the gist of what the question is asking. Once they have found it, they mark their answer and move on, confident they have answered it correctly. Later, many are startled to discover that they missed questions because they either misread the problems or overlooked subtle points.

To do well in your undergraduate classes, you had to attempt to solve every, or nearly every, problem on a test. Not so with the GRE. In fact, if you try to solve every problem on the test, you will probably damage your score. For the vast majority of people, the key to performing well on the GRE is not the number of questions they solve, within reason, but the percentage they solve correctly.

The level of difficulty of the second verbal or math section you see will depend on how well you perform on the first verbal or math section. If you do well on the first verbal section, then the second section will be a little harder. And if you do poorly on the first math section, then the second section will be a little easier. Within each section, you can change answers and skip questions and later return to them ("mark and review" feature). But once you leave a section, you cannot return to it.

Scoring the GRE

The three major parts of the test are scored independently. You will receive a verbal score, a math score, and a writing score. The verbal and math scores range from 130 to 170, in 1-point increments. The writing score is on a scale from 0 to 6. In addition to the scaled score, you will be assigned a percentile ranking, which gives the percentage of students with scores below yours.

Skipping and Guessing

If you can eliminate even one of the answer-choices, guessing can be advantageous. We'll talk more about this later.

Often students become obsessed with a particular problem and waste time trying to solve it. To get a top score, learn to cut your losses and move on.

If you are running out of time, randomly guess on the remaining questions. This is unlikely to harm your score. In fact, if you do not obsess about particular questions, you probably will have plenty of time to solve a sufficient number of questions.

Because the total number of questions answered contributes to the calculation of your score, you should answer ALL the questions—even if this means guessing randomly before time runs out.

Directions and Reference Material

Be sure you understand the directions below so that you do not need to read or interpret them during the test.

Directions

Solve each problem and decide which one of the choices given is best.

Notes

1. All numbers used are real numbers.
2. Figures are intended to provide information useful in answering the questions. However, unless a note states that a figure is drawn to scale, you should not solve these problems by estimating sizes by sight or by measurement.
3. All figures lie in a plane unless otherwise indicated. Position of points, angles, regions, etc. can be assumed to be in the order shown; and angle measures can be assumed to be positive.

Note 1 indicates that complex numbers, $i = \sqrt{-1}$, do not appear on the test.

Note 2 indicates that figures are not drawn accurately. Hence, an angle that appears to be 90° may not be or an object that appears congruent to another object may not be.

Note 3 indicates that two-dimensional figures do not represent three-dimensional objects. That is, the drawing of a circle is not representing a sphere, and the drawing of a square is not representing a cube.

Reference Information

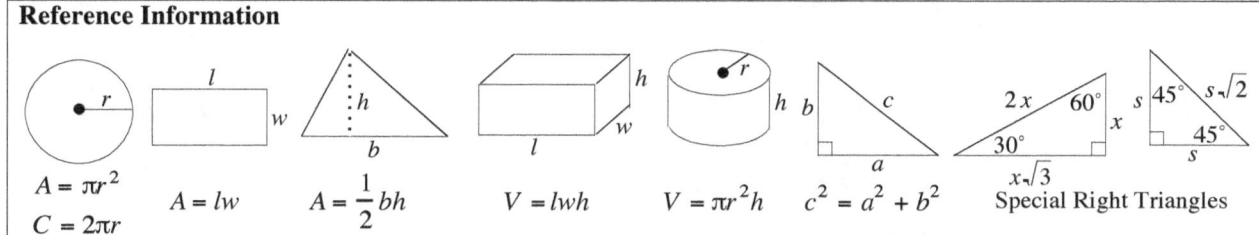

$A = \pi r^2$
$C = 2\pi r$
$A = lw$
$A = \frac{1}{2}bh$
$V = lwh$
$V = \pi r^2 h$
$c^2 = a^2 + b^2$
Special Right Triangles

The number of degrees of arc in a circle is 360.
The sum of the measures in degrees of the angles of a triangle is 180.

Although this reference material can be handy, be sure you know it well so that you do not waste time looking it up during the test.

Part One
THE TESTS

Test 1

Questions: 24
Time: 45 minutes

[Multiple-choice Question – Select One Answer Choice Only]
1. If $1 < p < 3$, then which of the following could be true?

    I. $p^2 < 2p$
    II. $p^2 = 2p$
    III. $p^2 > 2p$

    (A) I only
    (B) II only
    (C) III only
    (D) I and II only
    (E) I, II, and III

[Multiple-choice Question – Select One or More Answer Choices]
2.
$$x + 2y = 7$$

Which of the following could the value of x be in the given equation?

(A) $x = 0$
(B) $x = 2$
(C) $x = 3$
(D) $x = 5$
(E) $x = 7$

3.
Column A	Column B
The least positive integer divisible by 2, 3, 4, 5, and 6	The least positive integer that is a multiple of 2, 3, 4, 5 and 6

[Multiple-choice Question – Select One or More Answer Choices]
4. The product of which three of the following numbers is the greatest?

    (A)   −9
    (B)   −5
    (C)   −3
    (D)   0
    (E)   4
    (F)   7
    (G)   10

5.
Column A	Column B
The count of the numbers between 100 and 300 that are divisible by both 5 and 6	The count of the numbers between 100 and 300 that are divisible by either 5 or 6

6.
Column A	a and b are integers greater than zero.	Column B
a/b		a^2

[Multiple-choice Question – Select One or More Answer Choice Only]
7. From the figure, which of the following must be true?

(A) $y = z$
(B) $y < z$
(C) $y \le z$
(D) $y > z$
(E) $y + z = 2x$

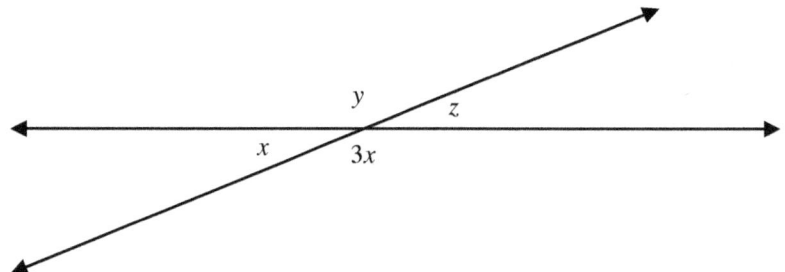

[Numeric Entry Question]
8. In the figure, ABCD is a square, and OB is a radius of the circle. If BC is a tangent to the circle and PC = 2, then what is the area of the square?

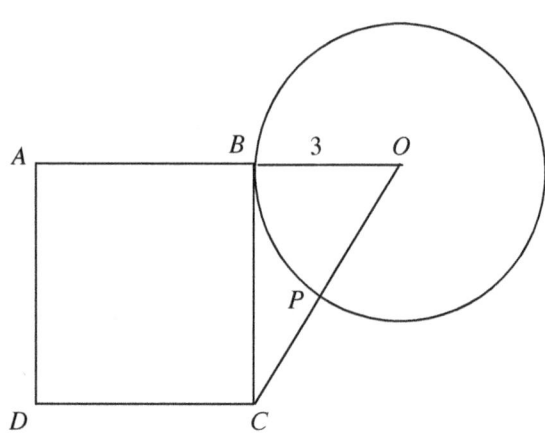

[Multiple-choice Question – Select One Answer Choice Only]

9. In the figure, ABCD is a rectangle, and F and E are points on AB and BC, respectively. The area of △DFB is 9 and the area of △BED is 24. What is the perimeter of the rectangle?

(A) 18
(B) 23
(C) 30
(D) 42
(E) 48

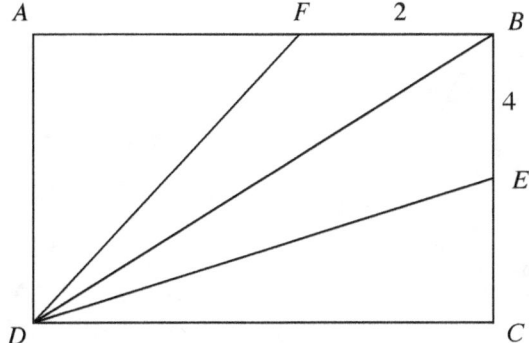

[Multiple-choice Question – Select One Answer Choice Only]

10. In the figure, ABCD is a square and BCP is an equilateral triangle. What is the measure of x?

(A) 7.5
(B) 15
(C) 30
(D) 45
(E) 60

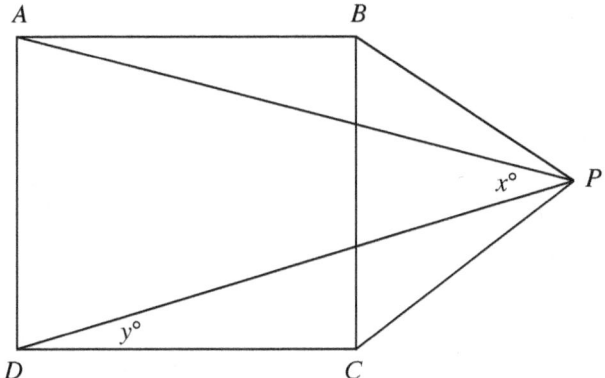

[Multiple-choice Question – Select One Answer Choice Only]
11. After being marked down 20 percent, a calculator sells for $10. The original selling price was

- (A) $20
- (B) $12.5
- (C) $12
- (D) $9
- (E) $7

12. | Column A | $a^2 + 7a < 0$ | Column B |
|---|---|---|
| a | | 0 |

13. | Column A | | Column B |
|---|---|---|
| $\frac{30}{31}$ of $\frac{31}{32}$ | | $\frac{30}{31}$ |

[Multiple-choice Question – Select One or More Answer Choice Only]

14. If $(x+5)\left(\dfrac{1}{x}+\dfrac{1}{5}\right)=4$, then $x=$

    (A) 1/5
    (B) 1/2
    (C) 1
    (D) 5
    (E) 10

[Multiple-choice Question – Select One or More Answer Choices]

15.

$$3x + 4y = c$$
$$kx + 12y = 36$$

In the system, if $k=9$, $c=12$, and x and y are integers, then which of the following could y be?

(A) –6
(B) 0
(C) 4
(D) 5
(E) 12

[Multiple-choice Question – Select One Answer Choice Only]

16. The ratio of the number of chickens to the number of pigs to the number of horses on Richard's farm is 33 : 17 : 21. What fraction of the animals are either pigs or horses?

    (A) 16/53
    (B) 17/54
    (C) 38/71
    (D) 25/31
    (E) 38/33

[Multiple-choice Question – Select One or More Answer Choices]

17. The ratio of the angles in △ABC is 2 : 3 : 4. Which of the following triangles is similar to △ABC ?

    (A) △DEF has angles in the ratio 4 : 3 : 2.
    (B) △PQR has angles in the ratio 1 : 2 : 3.
    (C) △LMN has angles in the ratio 1 : 1 : 1.
    (D) △STW has sides in the ratio 1 : 1 : 1.
    (E) △XYZ has sides in the ratio 4 : 3 : 2.
    (F) △MRQ has angles in the ratio 3 : 2 : 4.

[Multiple-choice Question – Select One Answer Choice Only]

18. If $p = \dfrac{\sqrt{3}-2}{\sqrt{2}+1}$, then which one of the following equals $p - 4$?

    (A) $\sqrt{3} - 2$
    (B) $\sqrt{3} + 2$
    (C) 2
    (D) $-2\sqrt{2} + \sqrt{6} - \sqrt{3} - 2$
    (E) $-2\sqrt{2} + \sqrt{6} - \sqrt{3} + 2$

19. Column A   The population of a country increases at a fixed percentage each year.   Column B

Increase in population in the first decade 1980–1990

Increase in population in the second decade 1990–2000

20. | Column A | Patrick purchased 80 pencils and sold them at a loss that is equal to the selling price of 20 pencils. | Column B |

Cost of 80 pencils | | Selling price of 100 pencils

[Multiple-choice Question – Select One or More Answer Choices]
21. An off-season discount of 10% is being offered at a store for any purchase with list price above $500. No other discounts are offered at the store. John purchased a computer from the store for $459. Which of the following could be the list price (in dollars)?

   (A) 459
   (B) 465
   (C) 479
   (D) 510
   (E) 525

---

22. | Column A | In a jar, 60% of the marbles are red and the rest are green. | Column B |

40% of the red marbles in the jar | | 60% of the green marbles in the jar

[Multiple-choice Question – Select One Answer Choice Only]
23. How many possible combinations can a 3-digit safe code have?

    (A)   84
    (B)   504
    (C) 3^9
    (D) 9^3
    (E) 10^3

[Multiple-choice Question – Select One Answer Choice Only]
24. A coin is tossed five times. What is the probability that the fourth toss would turn a head?

    (A)   1/8
    (B)   1/4
    (C)   1/2
    (D)   3/4
    (E)   4/5

Answers and Solutions Test 1:

Question	Answer
1.	E
2.	A, B, C, D, E
3.	C
4.	A, B, G
5.	B
6.	D
7.	D
8.	16
9.	D
10.	C
11.	B
12.	B
13.	B
14.	D
15.	A, B, E
16.	C
17.	A, F
18.	D
19.	B
20.	C
21.	A, D
22.	C
23.	E
24.	C

1. Let's chose a number to substitute into the problem such as $p = 3/2$. Then $p^2 = (3/2)^2 = 9/4 = 2.25$ and $2p = 2 \cdot 3/2 = 3$. Hence, $p^2 < 2p$, I is true, and clearly II ($p^2 = 2p$) and III ($p^2 > 2p$) are both false. This is true for all $1 < p < 2$.

Next, if $p = 2$, then $p^2 = 2^2 = 4$ and $2p = 2 \cdot 2 = 4$. Hence, $p^2 = 2p$, II is true, and clearly I ($p^2 < 2p$) and III ($p^2 > 2p$) are both false.

Finally, if $p = 5/2$, then $p^2 = (5/2)^2 = 25/4 = 6.25$ and $2p = 2 \cdot 5/2 = 5$. Hence, $p^2 > 2p$, III is true, and clearly I ($p^2 < 2p$) and II ($p^2 = 2p$) are both false. This is true for all $2 < p < 3$.

Hence, each of the three choices I, II, and III can be true (for a given value of p). The answer is (E).

2. Each value of x yields one value of y. For example, substituting $x = 1$ in the system $x + 2y = 7$ yields 3 for y. Similarly, other values of x yield other values for y. Hence, the system is feasible for all values of x. Hence, choose all the answer choices.

3. Any number that is divisible by the five numbers 2, 3, 4, 5, and 6 must also be a multiple of all five numbers. So, Column A (the set of numbers that are divisible by 2, 3, 4, 5, and 6) and Column B (the set of numbers that are a multiple of 2, 3, 4, 5 and 6) refer to the same set of integers. Hence, the least values of the two columns must be the same. So, Column A equals Column B, and the answer is (C).

4. The product of *n* numbers is greatest when

   a) The individual *n* numbers are as far as possible from zero on the number line and the resultant product is positive.

   Or

   b) The individual *n* numbers are as close as possible to zero on the number line and the product is negative.

With this as the criterion, choose the greater of Choice (A) –9, Choice (B) –5, and Choice (G) 10 so that the product is positive and the choices are as far as possible from zero on the number line.

Or

Choice (E) 4, Choice (F) 7, and Choice (G) 10 so that the product is positive and the choices are as far as possible from zero on the number line.

$$(-9) \cdot (-5) \cdot 10 = 450$$
$$4 \cdot 7 \cdot 10 = 280$$

Select (A), (B), and (G).

---

5. If a number is divisible by 5 and 6, then the number must be multiple of the least common multiple of 5 and 6, which is the product of the two (since 5 and 6 have no common factors): $5 \cdot 6 = 30$. Such numbers occur every 30 integers on the number line, while numbers divisible by *either* 5 or 6, occur every 5 or 6 integers on the number line. Hence, the frequency of later is more than former. The answer is (B).

---

6. Intuitively, we expect a^2 to be larger than the fraction *a/b*. So, that probably will not be the answer. It's an eye-catcher. Now, if $a = b = 1$, then both columns equal 1. However, if $a = b = 2$, then Column B is larger. Hence, the answer is (D): Not enough information to decide.

7.

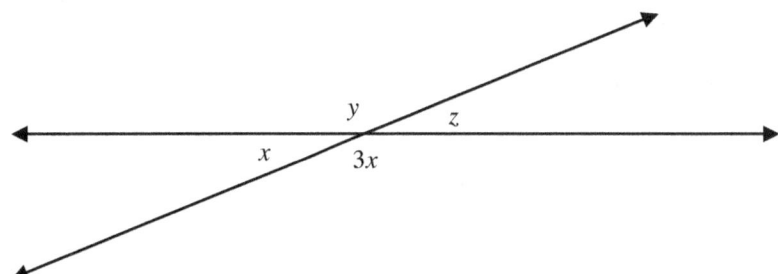

Equating the two pairs of vertical angles in the figure yields $y = 3x$ and $x = z$. Replacing *x* in first equation with *z* yields $y = 3z$. This equation says that *y* is 3 times as large as *z*. Hence, $y > z$. The answer is (D).

8.

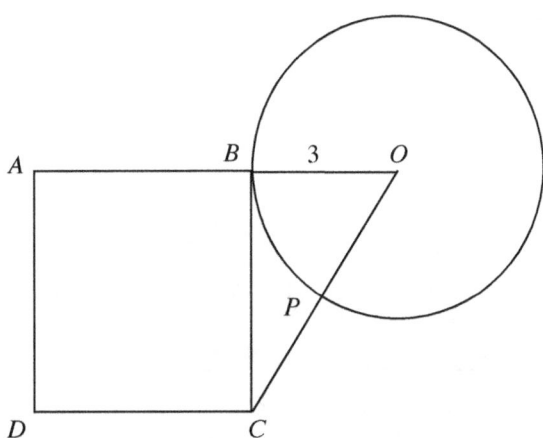

Side *BC* in the square is also a tangent to the circle shown. Since, at the point of tangency, a tangent is perpendicular to the radius of a circle, ∠*CBO* is a right angle. Hence, △*CBO* is right angled; and by The Pythagorean Theorem, we have $OC^2 = OB^2 + BC^2$.

Now, we are given $PC = 2$. Hence, $OC = OP + PC = Radius + PC = 3 + 2 = 5$. Putting the results in the known equation $OB^2 + BC^2 = OC^2$ yields $3^2 + BC^2 = 5^2$, or $BC^2 = 5^2 - 3^2 = 25 - 9 = 16$.

Now, the area of the square is $side^2 = BC^2 = 16$. Enter in the grid.

9.

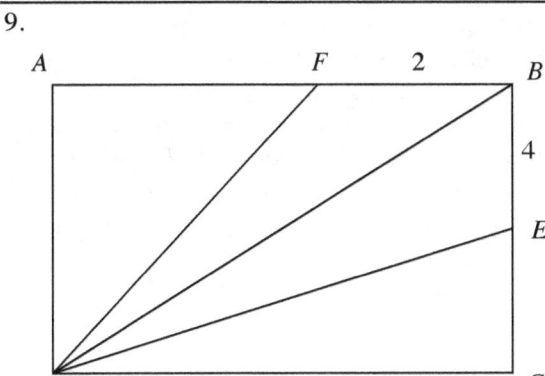

The formula for the area of a triangle is $1/2 \times base \times height$. Hence, the area of △*DFB* is $1/2 \times FB \times AD$. We are given that the area of △*DFB* is 9. Hence, we have

$1/2 \times FB \times AD = 9$

$1/2 \times 2 \times AD = 9$

$AD = 9$

Similarly, the area of △*BED* is $1/2 \times BE \times DC$, and we are given that the area of the triangle is 24. Hence, we have

$1/2 \times BE \times DC = 24$

$1/2 \times 4 \times DC = 24$ from the figure, $BE = 4$

$DC = 12$

Now, the formula for the perimeter of a rectangle is

2 × (the sum of the lengths of any two adjacent sides of the rectangle)

Hence, the perimeter of the rectangle $ABCD = 2(AD + DC) = 2(9 + 12) = 2 \times 21 = 42$. The answer is (D).

10.

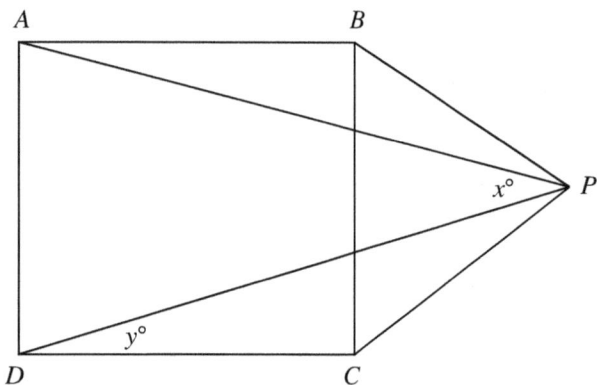

Through point P, draw a line QP parallel to the side AB of the square. Since QP cuts the figure symmetrically, it bisects $\angle BPC$. Hence, $\angle QPC$ = (Angle in equilateral triangle)/2 = 60°/2 = 30°. Now, $\angle QPD = \angle CDP$ (alternate interior angles are equal) = $y°$ (given). Since sides in a square are equal, $BC = CD$; and since sides in an equilateral triangle are equal, $PC = BC$. Hence, we have $CD = PC$ (= BC). Since angles opposite equal sides in a triangle are equal, in $\triangle CDP$ we have that $\angle DPC$ equals $\angle CDP$, which equals $y°$. Now, from the figure, $\angle QPD + \angle DPC = \angle QPC$ which equals 30° (we know from earlier work). Hence, $y° + y° = 30°$, or $y = 30/2 = 15$. Similarly, by symmetry across the line QP, $\angle APQ = \angle QPD = 15°$. Hence, $x = \angle APD = \angle APQ + \angle QPD = 15° + 15° = 30°$. The answer is (C).

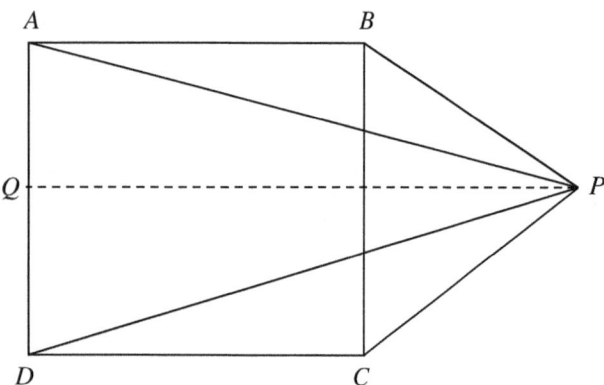

Test 1—Answers and Solutions

11. Twenty dollars is too large. The discount was only 20 percent—eliminate (A). Both (D) and (E) are impossible since they are less than the selling price—eliminate. Choice (C), 12, is the eye-catcher: 20% of 10 is 2 and 10 + 2 = 12. This is too easy for a hard problem like this—eliminate. Thus, by process of elimination, the answer is (B).

Method II: Let's calculate the answer directly. Let x be the original price. Since the discount is 20%, the markdown is $20\%x$. Hence, we get the equation

$$\textit{Original Price} - \textit{Markdown} = \$10$$

Or

$$x - 20\%x = 10$$
$$x - 0.2x = 10$$
$$(1 - 0.2)x = 10$$
$$0.8x = 10$$
$$x = 10/0.8$$
$$x = 12.5$$

12. Factoring out the common factor a on the left-hand side of the inequality $a^2 + 7a < 0$ yields $a(a + 7) < 0$. The product of two numbers (here, a and $a + 7$) is negative when one is negative and the other is positive. Hence, we have two cases:

   1) $a < 0$ and $a + 7 > 0$
   2) $a > 0$ and $a + 7 < 0$

Case 2) is impossible since if a is positive then $a + 7$ cannot be negative.

Case 1) is valid for all values of a between 0 and –7. Any number in the range is less than 0. So, Column A is less than Column B, and the answer is (B).

Method II: Since a square, a^2, is positive or zero, the inequality $a^2 + 7a < 0$ is possible only if $7a$ is negative, which means a is negative. Hence, $a < 0$, and Column B is larger. The answer is (B).

13. Column A: $\dfrac{30}{31}$ of $\dfrac{31}{32} = \dfrac{30}{31} \cdot \dfrac{31}{32} = \dfrac{30}{32}$.

Now, $\dfrac{30}{32} < \dfrac{30}{31}$ because the fractions have the same numerators and the denominator of 30/32 is larger than the denominator of 30/31.

Hence, Column B is larger than Column A, and the answer is (B).

14. We are given the equation

$$(x+5)\left(\frac{1}{x} + \frac{1}{5}\right) = 4$$

$$(x+5)\left(\frac{x+5}{5x}\right) = 4$$

$(x + 5)^2 = 4x(5)$
$x^2 + 5^2 + 2x(5) = 4x(5)$ since $(a + b)^2 = a^2 + b^2 + 2ab$
$x^2 + 5^2 - 2x(5) = 0$
$(x - 5)^2 = 0$ since $(a - b)^2 = a^2 + b^2 - 2ab$
$x - 5 = 0$ by taking the square root of both sides
$x = 5$

The answer is (D) only.

15. Substituting the given values for k and c yields

$$3x + 4y = 12$$
$$9x + 12y = 36$$

Dividing the second equation by 3 yields

$$3x + 4y = 12$$

This equation is the same as the first equation. Effectively, we have a system of just one equation in 2 variables. So, there are an infinite number of possible solutions. Dividing the equation by 3 yields

$$x + (4/3)y = 4$$

$$x = 4 - (4/3)y \text{ is an integer}$$

$$= 4 - (4/3)(y \text{ is divisible by 3})$$

Thus, select choices that are divisible by 3. Choices (A), (B), and (E) correspond the type. Select them.

---

16. Let the number of chickens, pigs and horses on Richard's farm be c, p, and h. Then forming the given ratio yields

$$c : p : h = 33 : 17 : 21$$

Let c equal $33k$, p equal $17k$, and h equal $21k$, where k is a positive integer (such that $c : p : h = 33 : 17 : 21$). Then the total number of pigs and horses is $17k + 21k = 38k$; and the total number of pigs, horses and chickens is $17k + 21k + 33k = 71k$. Hence, the required fraction equals $38k/71k = 38/71$. The answer is (C).

---

17. Two triangles are similar when their corresponding angle ratios or corresponding side ratios are the same.

Name the first triangle ABC and the second triangle DEF. Now, the angle ratio of the first triangle is $\angle A : \angle B : \angle C = 2 : 3 : 4$, and angle ratio for the second triangle is $\angle D : \angle E : \angle F = 4 : 3 : 2$, while the sum of the angles is 180 degrees.

The corresponding ordering ratio can be safely reversed and doing this for the second ratio equation yields $\angle F : \angle E : \angle D = 2 : 3 : 4$, while the sum of the angles is 180 degrees.

Since the angles in triangle ABC equal the corresponding angles in triangle FED, triangles ABC and FED are similar. Hence, (A) is an answer.

Since the angles in triangle ABC equal the corresponding angles in triangle RMQ, both being $2 : 3 : 4$, the triangles ABC and RMQ are similar. Hence, (F) is an answer.

The answer is (A) and (F).

Test 1—Answers and Solutions

18. Since none of the answers are fractions, let's rationalize the denominator of the given fraction by multiplying top and bottom by the conjugate of the bottom of the fraction:

$$p = \frac{\sqrt{3}-2}{\sqrt{2}+1} \cdot \frac{\sqrt{2}-1}{\sqrt{2}-1} \qquad \text{the conjugate of } \sqrt{2}+1 \text{ is } \sqrt{2}-1$$

$$= \frac{\sqrt{3}\sqrt{2}+\sqrt{3}(-1)+(-2)\sqrt{2}+(-2)(-1)}{\left(\sqrt{2}\right)^2 - 1^2}$$

$$= \frac{\sqrt{6}-\sqrt{3}-2\sqrt{2}+2}{2-1}$$

$$= \sqrt{6}-\sqrt{3}-2\sqrt{2}+2$$

Now, $p - 4 = \left(\sqrt{6}-\sqrt{3}-2\sqrt{2}+2\right) - 4 = \sqrt{6}-\sqrt{3}-2\sqrt{2}-2$. The answer is (D).

---

19. The population of the country grows at a fixed percentage each year. Hence, just as we have a fixed percentage for each year, we will have a different and fixed percentage for each 10-year period (a decade). Let the later rate be $R\%$. Hence, if P is the population in 1980, the population in 1990 (after a decade) would grow by $PR/100$ (= Column A). So, the population in 1990 would be

Original Plus Increase = $P + PR/100$

Similarly, in the next decade (1990–2000) and at the same rate $R\%$ it would increase by

$(P + PR/100)(R/100)$ (= Column B)

Substituting the results in the columns yields

$PR/100$ $\qquad\qquad\qquad\qquad\qquad\qquad (P + PR/100)(R/100)$

Canceling $R/100$ from both sides yields

P $\qquad\qquad\qquad\qquad\qquad\qquad P + PR/100$

Now, Column A = P and Column B = $P + PR/100$ = Column A + $PR/100$. Hence, Column B > Column A, and the answer is (B).

---

20. Let c be the cost of each pencil and s be the selling price of each pencil. Then the loss incurred by Patrick on each pencil is $c - s$. The net loss on 80 pencils is $80(c - s)$. Since we are given that the loss incurred on the 80 pencils equaled the selling price of 20 pencils which is $20s$, we have the equation:

$80(c - s) = 20s$
$80c - 80s = 20s$
$80c = 100s$

Column A: Cost of 80 pencils equals $80c$.

Column B: Selling price of 100 pencils equals $100s$.

Column A = Column B.

Hence, the answer is (C).

GRE Math Tests

21. We do not know whether the $459 price that John paid for the computer was with the discount offer or without the discount offer. We have two cases.

Case I:
If he did not get the discount offer, the list price of the computer should be $459 and John paid the exact amount for the computer. Select choice (A).

Case II:
If the price corresponds to the price after the discount offer, then $459 should equal a 10% discount on the list price. Hence, if l represents the list price, then we have

$$\$459 = l(1 - 10/100) =$$
$$l(1 - 1/10) =$$
$$(9/10)l$$

Solving the equation for l yields $l = (10/9)459 = 510$ dollars (a case when discount was offered because the list price is greater than $500). Hence, it is also possible that John got the 10% discount on the computer originally list priced at $510. Select choice (D).

The answers to choose are (A) and (D).

---

22. Let j be the total number of marbles in the jar. Then $60\%j$ must be red (given), and the remaining $40\%j$ must be green (given). Now,

Column A equals 40% of the red marbles = $40\%(60\%j) = .40(.60j) = .24j$.

Column B equals 60% of the green marbles = $60\%(40\%j) = .60(.40j) = .24j$.

Since both columns equal $.24j$, the answer is (C).

---

23. A possible safe combination could be 433 or 334; the combinations are the same, but their ordering is different. Since order is important for the safe combinations, this is a permutation problem.

A safe code can be made of any of the numbers $\{0, 1, 2, 3, 4, 5, 6, 7, 8, 9\}$.

Repetitions of numbers in the safe code are possible. For example, 334 is a possible safe code.

Safe codes allow 0 to be first digit (just as other digits). So, there are the same number of options for each of the 3 digits, which we will call slots, in the code. You can choose any one of the 10 digits for each of the slots. Thus, we have 10 ways of selecting a number for each of the 3 slots. Therefore, the number of ways of selecting a code is

$$10 \cdot 10 \cdot 10 = 1000$$

The answer is (E).

---

24. The fourth toss is independent of any other toss. The probability of a toss turning heads is 1 in 2, or simply 1/2. Hence, the probability of the fourth toss being a head is 1/2. The answer is (C).

Test 2

GRE Math Tests

Questions: 24
Time: 45 minutes

[Multiple-choice Question – Select One or More Answer Choices]

1. Mr. Smith's average annual income in each of the years 2006 and 2007 is positive x dollars. His average annual income in each of the years 2008, 2009, and 2010 is positive y dollars. He did earn at least some money each year. Which of the following could be his average annual income for the four continuous years 2006 through 2009?

    (A) $x/2$
    (B) $x/2 + y/2$
    (C) $x/2 + y$
    (D) $x/2 + 3y/4$
    (E) $x/2 + 3y/5$
    (F) $3x/5 + y/2$

2. | Column A | | Column B |
   |---|---|---|
   | | For any positive integer n, $\pi(n)$ represents the number of factors of n, inclusive of 1 and itself. a and b are unequal prime numbers. | |
   | $\pi(a) + \pi(b)$ | | $\pi(a \times b)$ |

[Multiple-choice Question – Select One or More Answer Choices]

3. a and b are multiples of 20, b and c are multiples of 30, and a and c are multiples of 40. Which of the following must be individually divisible by a, b, and c?

    (A) 10
    (B) 20
    (C) 30
    (D) 40
    (E) 120
    (F) 240

[Multiple-choice Question – Select One or More Answer Choices]
4. Which of the following could be the fraction of numbers that are divisible by both 7 and 10 in a set of 71 consecutive integers?

    (A) 1/71
    (B) 2/71
    (C) 3/71
    (D) 4/71
    (E) 5/71

[Multiple-choice Question – Select One or More Answer Choices]
5. In the triangle, what is the value of x?

    (A) 25
    (B) $25 + y$
    (C) $25 - y$
    (D) 55
    (E) 60

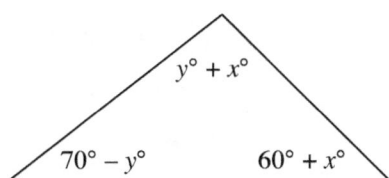

[Multiple-choice Question – Select One or More Answer Choices]

6. In the figure, *O* is the center of the circle. Which of the following must be true about the perimeter of the triangle shown?

(A) Less than 10
(B) Greater than 20
(C) Greater than 30
(D) Less than 40
(E) Less than 40 and greater than 20

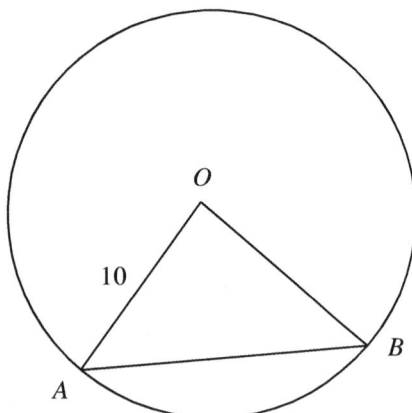

[Multiple-choice Question – Select One Answer Choice Only]
7. In the figure, *ABCD* is a rectangle inscribed in the circle shown. What is the length of the smaller arc *DC* ?

    (A)   π/4
    (B)   π/3
    (C)   2π/3
    (D)   3π/4
    (E)   4π/3

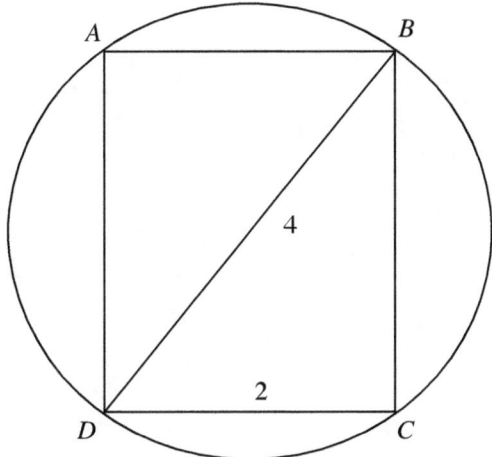

[Multiple-choice Question – Select One Answer Choice Only]
8. In the figure, $y =$

    (A) -12
    (B) -3
    (C) 1
    (D) $5\sqrt{3}$
    (E) 12

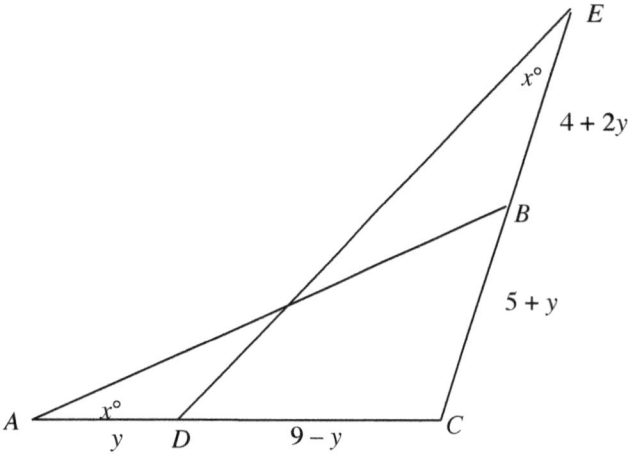

[Multiple-choice Question – Select One or More Answer Choices]
9. In triangle ABC, $CA = 4$, and $CB = 6$. Which of the following could be the area of triangle ABC ?

    (A) 7
    (B) 8
    (C) 9
    (D) 12
    (E) 15

[Multiple-choice Question – Select One Answer Choice Only]
10. Suppose five circles, each 4 inches in diameter, are cut from a rectangular strip of paper 12 inches long. If the least amount of paper is to be wasted, what is the width of the paper strip?

   (A) 5
   (B) $4 + 2\sqrt{3}$
   (C) 8
   (D) $4(1+\sqrt{3})$
   (E) not enough information

---

11. Column A $\qquad\qquad (4-5x)^2 = 1 \qquad\qquad$ Column B
     $(4-5x) + (4-5x)^2$ $\qquad\qquad\qquad\qquad\qquad\qquad$ 1

---

12. Column A $\qquad\qquad 0 < x < y \qquad\qquad$ Column B
     $x + 1/x$ $\qquad\qquad\qquad\qquad\qquad\qquad$ $y + 1/y$

[Multiple-choice Question – Select One or More Answer Choices]
13. If $x = a + 2$ and $b = x + 1$, then which of the following must be false?

   (A) $a > b$
   (B) $a < b$
   (C) $a = b$
   (D) $a = b^2$
   (E) $b = a^2$

[Multiple-choice Question – Select One or More Answer Choices]
14. The difference between two angles of a triangle is 24°. If all three angles are numerically double-digit integers, then which of the following could be the measure of the third angle?

   (A) 10°
   (B) 12°
   (C) 59°
   (D) 74°
   (E) 98°

15.   Column A   The monthly incomes of A and B Column B
                                      are in the ratio 3 : 2, and the
                                      monthly expenditures of the two
                                      are in the ratio 4 : 3.

                                      The income of both is greater than
                                      the expenditures, and the *Savings*
                                      from income is defined as the
                                      Income – Expenditure.

             Savings of A Savings of B

[Multiple-choice Question – Select One or More Answer Choices]
16. If $x - 3 = 10/x$ and $x > 0$, then x could equal which of the following?

 (A) -2
 (B) -1
 (C) 3
 (D) 5
 (E) 10

[Multiple-choice Question – Select One Answer Choice Only]
17. If r and s are two positive numbers and r is 25% greater than s, what is the value of the ratio r/s ?

 (A) 0.75
 (B) 0.8
 (C) 1
 (D) 1.2
 (E) 1.25

[Multiple-choice Question – Select One Answer Choice Only]
18. If $1/x + 1/y = 1/3$, then $\dfrac{xy}{x+y} =$

 (A) 1/5
 (B) 1/3
 (C) 1
 (D) 3
 (E) 5

19. Column A — A stores sells Brand A items and Brand B items at 75% and 80%, respectively, of list prices. During festival season, discounts of 20% and 25%, on Brand A and Brand B, respectively, are offered. Spinach tins of Brand A and Brand B have the same list price. — Column B

Cost of spinach tins of Brand A during festival season

Cost of spinach tins of Brand B during festival season

[Numeric Entry Question]

20. $2ab5$ is a four-digit number divisible by 25. If the number ab formed from the two digits a and b is a multiple of 13, then the three-digit number $2ab =$

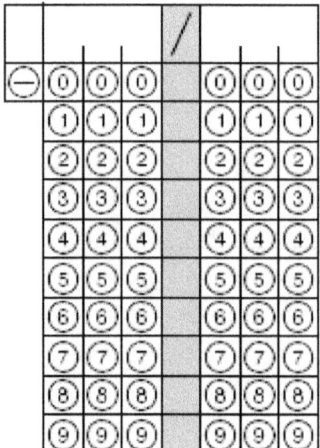

[Multiple-choice Question – Select One or More Answer Choices]
21. Two data sets S and R are defined as follows:

 Data set S: 28, 30, 25, 28, 27
 Data set R: 22, 19, 15, 17, 21, 25

Which of the following are the medians of the sets S and R or the difference between the medians?

(A) 8
(B) 10
(C) 20
(D) 25
(E) 28

---

22. Column A $x > y > 0$ Column B

 $x - y$ $\dfrac{x}{3} + \dfrac{y}{3}$

[Multiple-choice Question – Select One Answer Choice Only]
23. Seven years ago, Scott was 3 times as old as Kathy was at that time. If Scott is now 5 years older than Kathy, how old is Scott?

(A) 12½
(B) 13
(C) 13½
(D) 14
(E) 14½

24. The University of Maryland, University of Vermont, and Emory University have each 4 soccer players. If a team of 9 is to be formed with an equal number of players from each university, how many possible teams are there?

    (A) 3
    (B) 4
    (C) 12
    (D) 16
    (E) 64

USE THIS SPACE FOR SCRATCHWORK.

Answers and Solutions Test 2:

Question	Answer
1.	E, F
2.	C
3.	E, F
4.	A, B
5.	A
6.	B, D, E
7.	C
8.	C
9.	D, E
10.	B
11.	D
12.	D
13.	A, C, D
14.	B, D
15.	A
16.	D
17.	D
18.	D
19.	C
20.	252
21.	A, C, E
22.	D
23.	E
24.	E

1. Since Mr. Smith's average annual income in each of the two years 2006 and 2007 is x dollars, his total income in the two years is $2 \cdot x = 2x$.

Since Mr. Smith's average annual income in each of the three years 2008 and 2010 is y dollars, his total income in the three years is $3 \cdot y = 3y$.

Now, average income from 2006 through 2009 is

[(2 x amount earned between 2006, 2007) + (Amount earned between 2008, 2009, which could be between 0 and $3y$)]/4 years =

Between Min = $(2x + 0)/4 = x/2$ and Max = $(2x + 3y)/4 = x/2 + 3y/4$

The possible incomes in the range NOT including both Min and Max values above are Choices. Select such choices.

Since, the range is $x/2$ through $x/2 + 3y/4$ NOT including both, Reject choices (A) and (D), which are the extremes Min and Max.

Since $y/2 < 3y/4$, $x/2 + y/2 < x/2 + 3y/4$; Choice (B) < Max.
$x/2 + y/2$ is clearly greater than just $x/2$. Choice (B) > Min.
Accept.

Since $y > 3y/4$, Choice(C) = $x/2 + y > x/2 + 3y/4$; Choice (C) > Max. Reject.

Choice (E): $x/2 + 3y/5$ is similarly > Min and < Max. Accept.

GRE Math Tests

Choice (F): $3x/5 + y/2$; $3x/5$ could be in the range and $y/2$ could be in the range and their sum is not necessarily out of range. Hence, select Choice (F) as well.

Hence, the answer is (E) and (F).

---

2. The only factors of a prime number are 1 and itself. Hence, π(any prime number) = 2. So, $\pi(a) = 2$ and $\pi(b) = 2$ and therefore Column A equals $\pi(a) + \pi(b) = 2 + 2 = 4$.

Now, the factors of ab are $1, a, b$, and ab itself. Since a and b are different, the total number of factors of ab is 4. In other words, $\pi(a \times b) = 4$. Hence, Column B also equals 4.

Since the columns are equal, the answer is (C).

---

3. We are given that "a and b are multiples of 20, b and c are multiples of 30, and a and c are multiples of 40."

We can summarize this information as follows:

a is multiple of 20; a is multiple of 40. Hence, $a = 40l$ [40 is the LCM of 20 and 40]
b is multiple of 20; b is multiple of 30. Hence, $b = 60m$ [60 is the LCM of 20 and 30]
c is multiple of 30; c is multiple of 40. Hence, $a = 120n$ [120 is the LCM of 30 and 40]

where l, m, n are integers. Now, a number divisible by $a, b,$ and c individually must be a multiple of the least common multiple of 40, 60, and 120, which is 120. Hence, the answers are (E) 120 and (F) 240.

---

4. The numbers that are divisible by both 7 and 10 must be divisible by their least common multiple $7 \cdot 10 = 70$ also. The numbers divisible by 70 occur once every 70 consecutive integers.

Now, if for example, the set is 70 through 140, then two numbers, 70 and 140, in the set are divisible by 70. Here, the fraction is 2/71. Select (B).

But if for example, the set is 71 through 141, then only one number, 140, is divisible by 70. Here the fraction is 1/71. Select (A).

The possible choices are (A) and (B).

---

5. The angle sum of a triangle is 180°. Hence,

$$(y + x) + (60 + x) + (70 - y) = 180$$

Simplifying the equation yields $2x + 130 = 180$. Solving for x yields $x = 25$. The answer is (A)—Only one choice must be chosen.

Note: Sometimes when a problem says, "Select One or More Answer Choices," there will be only one answer. When you find just one answer, double-check your work.

6. In △AOB, OA and OB are radii of the circle. Hence, both equal 10 (since OA = 10 in the figure). Now, the perimeter of a triangle equals the sum of the lengths of the sides of the triangle. Hence,

Perimeter of △AOB =

OA + OB + AB =

10 + 10 + AB =

20 + AB

In a triangle, the length of any side is less than the sum of the lengths of the other two sides and is greater than their difference. This makes AB less than AO + OB (= 10 + 10 = 20) and AB greater than 10 − 10 = 0. So, the range of AB is 0 through 20, not including either number. Hence, the range of 20 + AB is evaluated as the inequality

20 < 20 + AB < 20 + 20

20 < Perimeter < 40

Logically, the choices that apply are (B), (D), and (E).

---

7. In the figure, BD is a diagonal of the rectangle inscribed in the circle. Hence, BD is a diameter of the circle. So, the midpoint of the diagonal must be the center of the circle, and the radius must equal half the length of the diameter: BD/2 = 4/2 = 2. Now, joining the center of the circle, say, O to the point C yields the following figure:

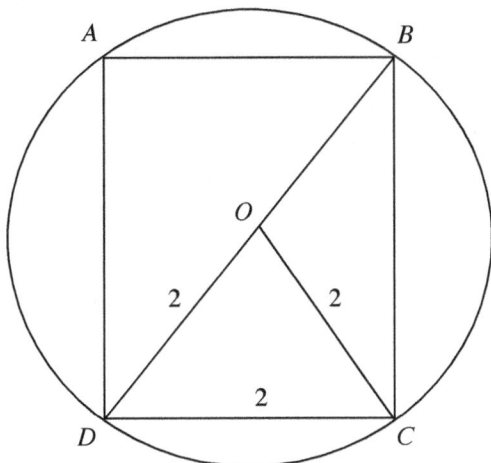

Now, in △DOC since OD and OC are radii of the circle, both equal 2. So, OD = OC = 2. Since DC also equals 2 (in figure given), OD = OC = DC = 2 (all three sides are equal). Hence, the triangle is equilateral and each angle must equal 60°, including ∠DOC. Now, the circumference of the given circle equals $2\pi \times radius = 2\pi(2) = 4\pi$. The fraction of the complete angle that the arc DC makes in the circle is 60°/360° = 1/6. The arc length would also be the same fraction of the circumference of the circle. Hence, the arc length equals $1/6 \times 4\pi = 2\pi/3$. The answer is (C).

8. In the figure, $y =$

(A) -12
(B) -3
(C) 1
(D) $5\sqrt{3}$
(E) 12

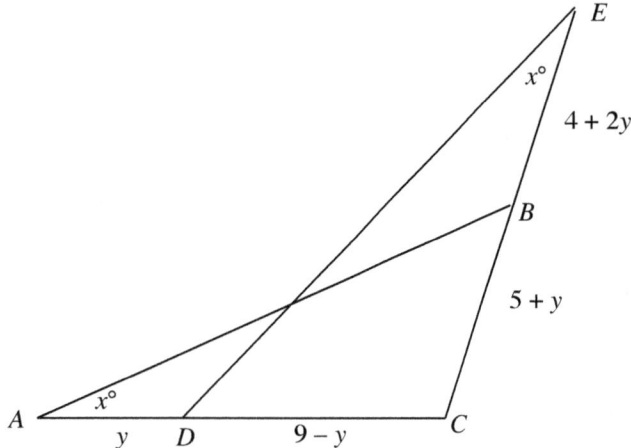

In the figure, in the triangles *ABC* and *EDC*, we have

$\angle ACB = \angle DCE$ Common Angle
$\angle BAC = \angle DEC = x$

Therefore, triangles *ABC* and *EDC* are similar. Equating corresponding side ratios yields

$AC/BC = EC/DC$
$(y + 9 - y)/(5 + y) = (5 + y + 4 + 2y)/(9 - y)$
$9/(5 + y) = (9 + 3y)/(9 - y)$
$9(9 - y) = (9 + 3y)(5 + y)$ cross-multiplication
$81 - 9y = 45 + 9y + 15y + 3y^2$
$81 = 45 + 9y + 15y + 9y + 3y^2$
$27 = 15 + 3y + 5y + 3y + y^2$ dividing by 3
$12 = 11y + y^2$
$0 = -12 + 11y + y^2$
$y^2 + 11y - 12 = 0$
$y^2 + 12y - y - 12 = 0$
$y(y + 12) - (y + 12) = 0$
$(y - 1)(y + 12) = 0$
$y = 1$ or -12
$AD = y = 1$. Eliminate -12 because the side $AD = y$ cannot have a negative length.

The answer is (C) only.

9. We are given that in triangle *ABC*, *CA* = 4 and *CB* = 6. The area of the triangle is greatest when the sides form a right angle, and least when the sides align in a line. When the sides form a right-angle, the area is $1/2 \times base \times height = 1/2 \times 4 \times 6 = 12$; and when they are aligned in a line, the area is 0 because the height is zero. Hence, the permissible range is 0 through 12. Select choices in this range. The answer is (D) and (E).

10. Since this is a hard problem, we can eliminate (E), "not enough information." And because it is too easily derived, we can eliminate (C), (8 = 4 + 4). Further, we can eliminate (A), 5, because answer-choices (B) and (D) are a more complicated. At this stage we cannot apply any more elimination rules; so if we could not solve the problem, we would guess either (B) or (D).

Let's solve the problem directly. The drawing below shows the position of the circles so that the paper width is a minimum.

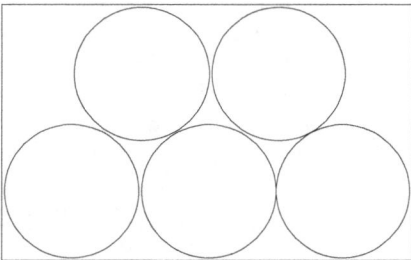

Now, take three of the circles in isolation, and connect the centers of these circles to form a triangle:

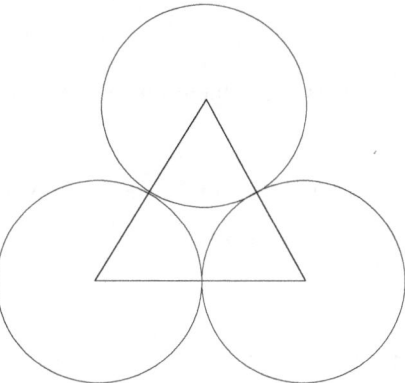

Since the triangle connects the centers of circles of diameter 4, the triangle is equilateral with sides of length 4.

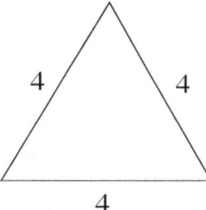

Drawing an altitude gives

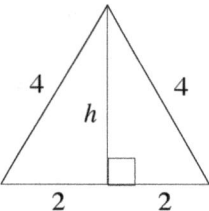

Applying the Pythagorean theorem to either right triangle gives	$h^2 + 2^2 = 4^2$
Squaring yields	$h^2 + 4 = 16$
Subtracting 4 from both sides of this equation yields	$h^2 = 12$
Taking the square root of both sides yields	$h = \sqrt{12} = \sqrt{4 \cdot 3}$
Removing the perfect square 4 from the radical yields	$h = 2\sqrt{3}$
Summarizing gives	

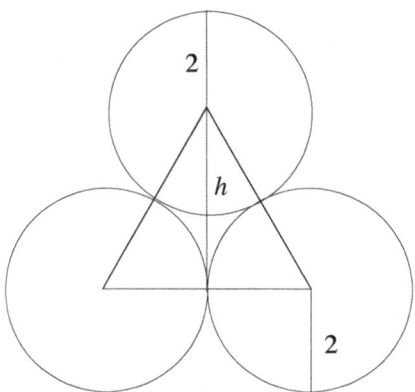

Adding to the height, $h = 2\sqrt{3}$, the distance above the triangle and the distance below the triangle to the edges of the paper strip gives

$$width = (2 + 2) + 2\sqrt{3} = 4 + 2\sqrt{3}$$

The answer is (B).

---

11. $(4 - 5x)^2$ equals 1 when either $4 - 5x = -1$ or $+1$. If $4 - 5x$ equals -1, then

$$(4 - 5x) + (4 - 5x)^2 =$$
$$-1 + 1 =$$
$$0$$

And if $4 - 5x$ equals 1, then

$$(4 - 5x) + (4 - 5x)^2 =$$
$$1 + 1 =$$
$$2$$

In the first case, Column A is less than Column B; and in the second case, Column B is less than Column A. Hence, we have a double case, and the answer is (D).

Test 2—Answers and Solutions

12. If $x = 1$ and $y = 2$, then the columns become

$1 + 1/1$ $2 + 1/2$

2 $2\ 1/2$

In this case, Column B is greater.

If $x = 1/2$ and $y = 1$, the columns become

$\dfrac{1}{2} + \dfrac{1}{\frac{1}{2}}$ $1 + \dfrac{1}{1}$

$2\dfrac{1}{2}$ 2

In this case, Column A is greater.

Hence, we have a double case, and the answer is (D).

---

13. We have the equations $x = a + 2$ and $b = x + 1$. Substituting the first equation into the second equation yields $b = (a + 2) + 1 = a + 3$. This equation shows that b is 3 units greater than a. Hence, b must be greater than a. So, (A) and (C) are false and therefore correct answers (remember, we are looking for statements that must be false).

Next, assume $a = b^2$ [Choice (D)]. Then a must be positive. If a is between 0 and 1, b is also between 0 and 1, which violates $b = a + 3$. Here, our assumption fails. If a were greater than 1, then b equals $\sqrt{a}$, which must be less than a—but we know that b is greater than a. Again, our assumption fails. Accept choice (D).

Next, assume $b = a^2$ [Choice (E)]. This could be true. For example, consider $b = a + 3 = a^2$. Subtracting a and 3 from both sides yields $a^2 - a - 3 = 0$. The quadratic equation is in the form $ax^2 + bx + c = 0$. Since the discriminant of the quadratic equation which is $b^2 - 4ac$ equals $(-1)^2 - 4(1)(-3) = 1 + 12 = 13$, is positive, we will have valid real solutions (Note, the GRE is only interested in Real Numbers), which means that $b = a^2$ is possible with given information. Hence, reject this choice (remember, we are looking for answers that must be false).

The answer consists of (A), (C), and (D).

GRE Math Tests

14. Let a, b, and c be the angles. Their sum is 180: $a + b + c = 180$. Let a and b be the angles differing by 24°, with a being the greater angle. So, we have $a - b = 24$. Subtracting the equations yields

$$a + b + c - (a - b) = 180 - 24$$
$$2b + c = 156$$
$$2b = 156 - c$$
$$b = (156 - c)/2$$
$$= 78 - c/2$$

Since b is at least a 2-digit number, c must be in the range between 10 and 98, including both. So,

$78 - 10/2 \geq b \geq 78 - 98/2$; $73 \geq b \geq 29$; $72 \geq b \geq 30$ (EVEN integers in the range).

Since $a = b + 24$, the range of a is

$72 + 24 \geq a \geq 30 + 24$; $96 \geq a \geq 54$ (EVEN integers in the range);

Hence, the range of $a + b$ is

$96 + 72 \geq a + b \geq 54 + 30$; $168 \geq a + b \geq 84$ (EVEN integers in the range);

Now, $c = 180 - (a + b)$ yields the range for c as

$180 - 168 \leq c \leq 180 - 84$; $12 \leq c \leq 96$ (EVEN integers in the range);

Thus, the third-angle could be any EVEN integer from 12 through 96.

Select (B) and (D).

---

15. Let the income of A and B be $3s$ and $2s$, respectively, and let their expenditures be $4t$ and $3t$. Then since savings is defined as income minus expenditure, Column A = the saving of A = $3s - 4t$, and Column B = saving of B = $2s - 3t$.

Column A is greater than Column B when $3s - 4t > 2s - 3t$, or $s > t$.

Since money is positive, s and t are positive. Since income is greater than expenditure (given), the Income of B = $2s$ > Expenditure of B = $3t$. Hence, $s > 3t/2$. Clearly, $s > t$. Since we know that Column A > Column B when $s > t$, the answer is (A).

---

16. We have the equation $x - 3 = 10/x$. Multiplying the equation by x yields $x^2 - 3x = 10$. Subtracting 10 from both sides yields $x^2 - 3x - 10 = 0$. Factoring the equation yields $(x - 5)(x + 2) = 0$. The possible solutions are 5 and –2. The only solution that also satisfies the given inequality $x > 0$ is $x = 5$. The answer is (D).

---

17. We are given that r is 25% greater than s. Hence, $r = s + 25\%s = s + 0.25s = (1 + 0.25)s = 1.25s$. So, $r/s = 1.25s/s = 1.25$. The answer is (E).

Test 2—Answers and Solutions

18. Multiplying the given equation $1/x + 1/y = 1/3$ by xy yields

$$y + x = xy/3, \text{ or } x + y = xy/3$$

Multiplying both sides of the equation $x + y = xy/3$ by $3/(x + y)$ yields

$$\frac{xy}{x + y} = 3$$

The answer is (D).

19. Since the spinach tins of either brand have the same list price, let each be x dollars. Now, 75% of x is $(75/100)x = 3x/4$, and 80% of x is $(80/100)x = 4x/5$. So, spinach tins of brands A and B are sold at $3x/4$ and $4x/5$, respectively. After festival discounts of 20% and 25% on the respective brand items, the cost of spinach tins drop to

$$\left(1 - \frac{20}{100}\right)\left(\frac{3x}{4}\right) = \left(\frac{4}{5}\right)\left(\frac{3x}{4}\right) = \frac{3x}{5} \text{ and } \left(1 - \frac{25}{100}\right)\left(\frac{4x}{5}\right) = \left(\frac{3}{4}\right)\left(\frac{4x}{5}\right) = \frac{3x}{5}$$

Hence, the columns are equal, and the answer is (C).

20. We have that the number $2ab5$ is divisible by 25. Any number divisible by 25 ends with the last two digits as 00, 25, 50, or 75. So, $b5$ should equal 25 or 75. Hence, $b = 2$ or 7.

We also have that ab is divisible by 13. The multiples of 13 are 13, 26, 39, 52, 65, 78, and 91. Among these, the only number ending with 2 or 7 is 52. Hence, $ab = 52$.

Then, $2ab = 252$. Enter 252 in the grid.

21. The definition of *median* is "When a set of numbers is arranged in order of size, the *median* is the middle number. If a set contains an even number of elements, then the median is the average of the two middle elements."

Data set S (arranged in increasing order of size) is 25, 27, 28, 28, 30. The median of the set is the third number 28.

Data set R (arranged in increasing order of size) is 15, 17, 19, 21, 22, 25. The median is the average of the two middle numbers (the 3rd and 4th numbers): $(19 + 21)/2 = 40/2 = 20$.

The difference of 28 and 20 is 8. The correct answers are (A), the difference between the medians, (C), the median of the set R, and (E), the median of the set S.

22. If $x = 2$ and $y = 1$, then $x - y = 2 - 1 = 1 = \frac{3}{3} = \frac{2}{3} + \frac{1}{3} = \frac{x}{3} + \frac{y}{3}$. In this case, the columns are equal.

If $x = 3$ and $y = 1$, then $x - y = 3 - 1 = 2 \neq \frac{3}{3} + \frac{1}{3} = \frac{4}{3} = \frac{x}{3} + \frac{y}{3}$. In this case, the columns are not equal and therefore the answer is (D).

GRE Math Tests

23. Let S be Scott's age and K be Kathy's age. Then translating the sentence *"If Scott is now 5 years older than Kathy, how old is Scott"* into an equation yields

$$S = K + 5$$

Now, Scott's age 7 years ago can be represented as $S = -7$, and Kathy's age can be represented as $K = -7$. Then translating the sentence *"Seven years ago, Scott was 3 times as old as Kathy was at that time"* into an equation yields $S - 7 = 3(K - 7)$.

Combining this equation with $S = K + 5$ yields the system:

$$S - 7 = 3(K - 7)$$
$$S = K + 5$$

Solving this system gives $S = 14½$. The answer is (E).

---

24. The selection from the 3 universities can be done in $4 \times 4 \times 4 = 4^3 = 64$ ways.

The answer is (E).

Test 3

Questions: 24
Time: 45 minutes

[Multiple-choice Question – Select One Answer Choice Only]

1. Which of the following equals the product of the smallest prime number greater than 21 and the largest prime number less than 16?

(A) $13 \cdot 16$
(B) $13 \cdot 29$
(C) $11 \cdot 23$
(D) $15 \cdot 23$
(E) $16 \cdot 21$

USE THIS SPACE FOR SCRATCHWORK.

[Multiple-choice Question – Select One Answer Choice Only]

2. How many 3-digit numbers do not have an even digit or a zero?

(A) 50
(B) 60
(C) 75
(D) 100
(E) 125

USE THIS SPACE FOR SCRATCHWORK.

3.	Column A		Column B
		ABC and *DEF* are right triangles. Each side of ΔABC is twice the length of the corresponding side of ΔDEF.	
	$\dfrac{\text{The area of } \Delta DEF}{\text{The area of } \Delta ABC}$		1/2

USE THIS SPACE FOR SCRATCHWORK.

[Multiple-choice Question – Select One Answer Choice Only]
4. In the figure, $ABCD$ is a rectangle and AF is parallel to BE. If $x = 5$, and $y = 10$, then what is the area of $\triangle AFD$?

   (A) 2.5
   (B) 5
   (C) 12.5
   (D) 50
   (E) $50 + 5y$

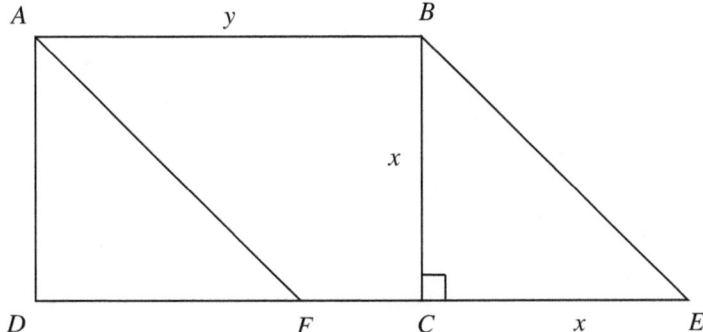

[Multiple-choice Question – Select One Answer Choice Only]
5. In the figure, if $AB = 8$, $BC = 6$, $AC = 10$ and $CD = 9$, then $AD =$

   (A) 12
   (B) 13
   (C) 15
   (D) 17
   (E) 24

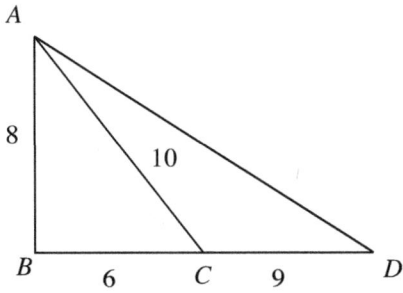

[Multiple-choice Question – Select One or More Answer Choices]
6. In the figure, which of the following are the two largest angles?

   (A) ∠A
   (B) ∠B
   (C) ∠C
   (D) ∠D
   (E) ∠CDB

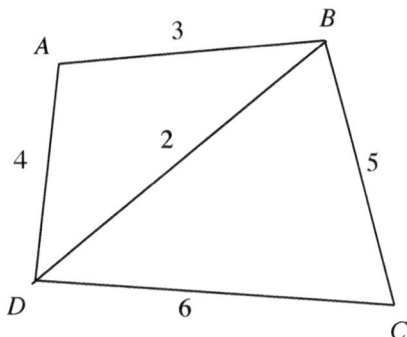

The figure is not drawn to scale.

[Numeric Entry Question]
7. In the figure, if ∠A = ∠C, then x =

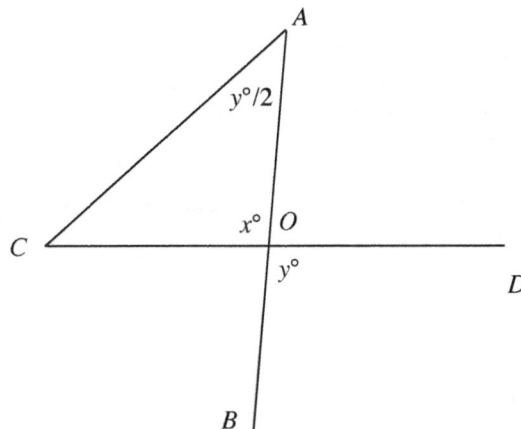

The dimensions in the figure may be different from what they appear to be.

GRE Math Tests

[Multiple-choice Question – Select One or More Answer Choices]
8. If $x^2 + 4x + 3$ is odd, then which of the following could be the value of x?

   (A) 3
   (B) 5
   (C) 8
   (D) 13
   (E) 16

9. | Column A | | Column B |
    |---|---|---|
    | | Mike and Fritz ran a 30-mile Marathon. Mike ran 10 miles at 10 mph and the remaining 20 miles at 5 mph. Fritz ran one-third (by time) of the Marathon at 10 mph and the remaining two-thirds at 5 mph. | |
    | Average speed of Mike | | Average speed of Fritz |

[Multiple-choice Question – Select One Answer Choice Only]
10. In the coordinate system shown, if (b, a) lies in Quadrant III, then in which quadrant can the point (a, b) lie?

   (A) I only
   (B) II only
   (C) III only
   (D) IV only
   (E) Origin

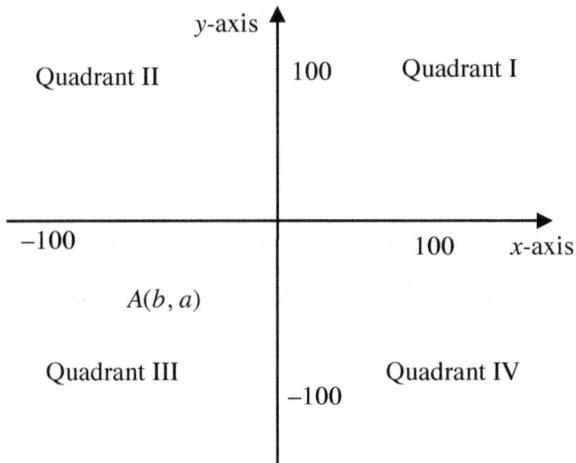

[Multiple-choice Question – Select One Answer Choice Only]
11. If $x > 2$ and $x < 3$, then which of the following must be positive?

   (I) $(x - 2)(x - 3)$
   (II) $(2 - x)(x - 3)$
   (III) $(2 - x)(3 - x)$

   (A) I only
   (B) II only
   (C) III only
   (D) I and II only
   (E) I and III only

[Multiple-choice Question – Select One Answer Choice Only]

12. Water is poured into an empty cylindrical tank at a constant rate. In 10 minutes, the height of the water increased by 7 feet. The radius of the tank is 10 feet. What is the rate at which the water is poured?

    (A) 11/8 π cubic feet per minute.
    (B) 11/3 π cubic feet per minute.
    (C) 7/60 π cubic feet per minute.
    (D) 11π cubic feet per minute.
    (E) 70π cubic feet per minute.

[Multiple-choice Question – Select One Answer Choice Only]

13. The selling price of 15 items equals the cost of 20 items. What is the percentage profit earned by the seller?

    (A) 15
    (B) 20
    (C) 25
    (D) 33.3
    (E) 40

[Multiple-choice Question – Select One Answer Choice Only]

14. If $4p$ is equal to $6q$, then $2p - 3q$ equals which one of the following?

    (A) 0
    (B) 2
    (C) 3
    (D) 4
    (E) 6

15.
Column A	Let <x> denote the greatest integer less than or equal to x.	Column B
<3.1> + <–3.1>		0

[Multiple-choice Question – Select One Answer Choice Only]

16. In what proportion must rice at $0.8 per pound be mixed with rice at $0.9 per pound so that the mixture costs $0.825 per pound?

   (A) 1 : 3
   (B) 1 : 2
   (C) 1 : 1
   (D) 2 : 1
   (E) 3 : 1

17.
Column A	Column B
$\sqrt{12.5} + \sqrt{12.5}$	$\sqrt{25}$

[Multiple-choice Question – Select One or More Answer Choices]

18. Removing which of the following numbers from the set S = {1, 2, 3, 4, 5, 6} would actually increase the average of the set?

   (A) 1
   (B) 2
   (C) 3
   (D) 4
   (E) 5
   (F) 6

19.
Column A	Column B
45% of 90	90% of 45

Questions 20–23 refer to the following graph.

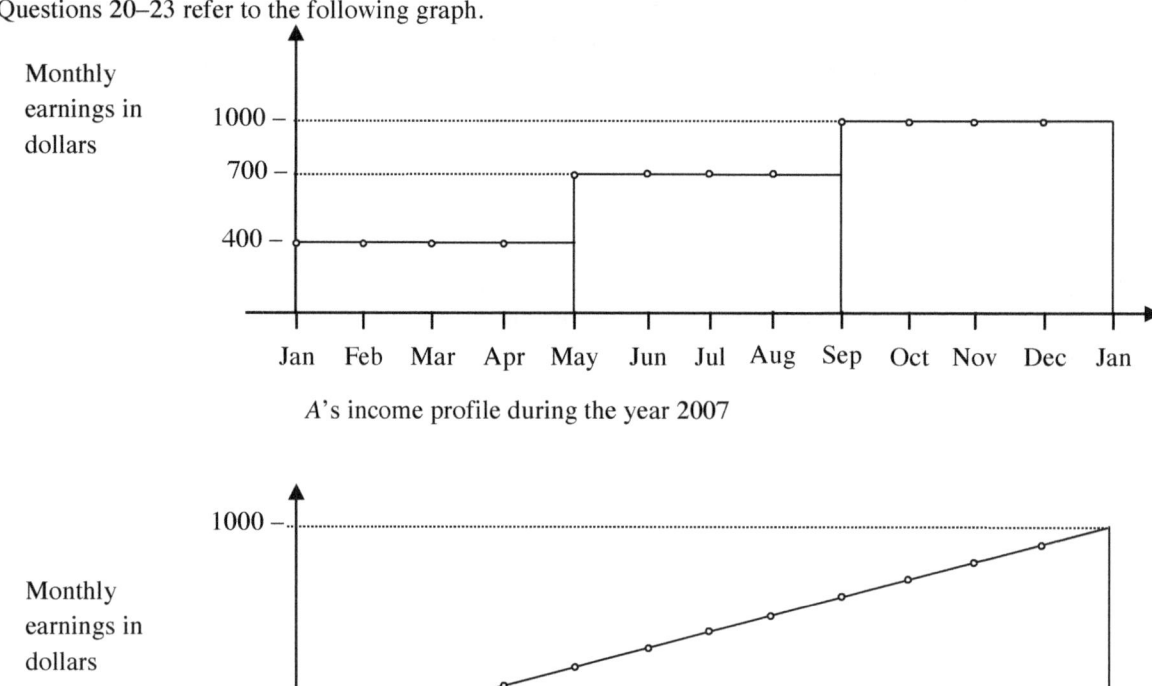

A's income profile during the year 2007

B's income profile during the year 2007

[Multiple-choice Question – Select One Answer Choice Only]

20. A launched 3 products in the year 2007 and earns income from the sales of the products only. The top graph shows his monthly earnings for the year. B's earnings consist of continuously growing salary, growing by same amount each month as shown in the figure. Which one of the following equals the total earnings of A and B in the year 2007?

(A) 7500, 8100
(B) 7850, 8300
(C) 8150, 8400
(D) 8400, 8100
(E) 8400, 8700

[Multiple-choice Question – Select One or More Answer Choices]
21. In which months was A's income equal to B's ?

   (A) January 2007
   (B) June 2007
   (C) July 2007
   (D) December 2007
   (E) January 2008

[Multiple-choice Question – Select One or More Answer Choices]
22. In which months was A's income less than B's ?

   (A) Apr 2007
   (B) May 2007
   (C) July 2007
   (D) August 2007
   (E) January 2008

[Multiple-choice Question – Select One or More Answer Choices]
23. In which months was the net income earned so far by A is less than that of B ?

   (A) Apr 2007
   (B) May 2007
   (C) July 2007
   (D) October 2007
   (E) January 2008

GRE Math Tests

[Multiple-choice Question – Select One Answer Choice Only]
24. The following values represent the number of cars owned by the 20 families on Pearl Street.

   1, 1, 2, 3, 2, 5, 4, 3, 2, 4, 5, 2, 6, 2, 1, 2, 4, 2, 1, 1

What is the probability that a family randomly selected from Pearl Street has at least 3 cars?

(A)   1/6
(B)   2/5
(C)   9/20
(D)   13/20
(E)   4/5

Answers and Solutions Test 3:

Question	Answer
1.	C
2.	E
3.	B
4.	C
5.	D
6.	B, D
7.	90
8.	C, E
9.	B
10.	C
11.	B
12.	E
13.	D
14.	A
15.	B
16.	E
17.	A
18.	A, B, C
19.	C
20.	D
21.	A, C, E
22.	A, D
23.	A, B, C
24.	B

1. The smallest prime number greater than 21 is 23, and the largest prime number less than 16 is 13. The product of the two is $13 \cdot 23$, which is listed in choice (C). The answer is (C).

2. There are 5 digits that are not even or zero: 1, 3, 5, 7, and 9. Now, let's count all the three-digit numbers that can be formed from these five digits. The first digit of the number can be filled in 5 ways with any one of the mentioned digits. Similarly, since repetition of a number is allowed, the second and third digits of the number can also be filled in 5 ways. Hence, the total number of ways of forming the three-digit number is $125 (= 5 \times 5 \times 5)$. The answer is (E).

3. We are given that each side of triangle *ABC* is twice the length of the corresponding side of triangle *DEF*. Hence, each leg of triangle *ABC* must be twice the length of the corresponding leg in triangle *DEF*. The formula for the area of a right triangle is 1/2 • (*product of the measures of the two legs*). Hence,

$$\text{Column A} = \frac{\text{The area of } \Delta DEF}{\text{The area of } \Delta ABC} =$$

$$\frac{\frac{1}{2}(\text{leg 1 of } \Delta DEF)(\text{leg 2 of } \Delta DEF)}{\frac{1}{2}(\text{leg 1 of } \Delta ABC)(\text{leg 2 of } \Delta ABC)} =$$

$$\frac{\frac{1}{2}(\text{leg 1 of } \Delta DEF)(\text{leg 2 of } \Delta DEF)}{\frac{1}{2}(2 \cdot \text{leg 1 of } \Delta DEF)(2 \cdot \text{leg 2 of } \Delta DEF)} =$$

$$\frac{\frac{1}{2}}{\frac{1}{2}(2)(2)} =$$

$$\frac{1}{2 \cdot 2} =$$

$$\frac{1}{4} \text{ and this is less than } \frac{1}{2} \text{ (= Column B)}$$

Hence, the answer is (B).

Method II:

Since we are not given the sizes of the two triangles, we can choose any measurements such that each side of Δ*ABC* is twice the length of the corresponding side of Δ*DEF*. Suppose each leg of Δ*ABC* is 2 units long and each leg of Δ*DEF* is 1 unit long. Then the area of Δ*DEF* is (1/2)(1)(1) = 1/2, and the area of Δ*ABC* is (1/2)(2)(2) = 2. Forming the ratio yields.

$$\frac{\text{The area of } \Delta DEF}{\text{The area of } \Delta ABC} =$$

$$\frac{\frac{1}{2}}{2} =$$

$$\frac{1}{2 \cdot 2} =$$

$$\frac{1}{4}$$

Since 1/4 is less than 1/2, Column A is less than Column B. The answer is (B).

4.

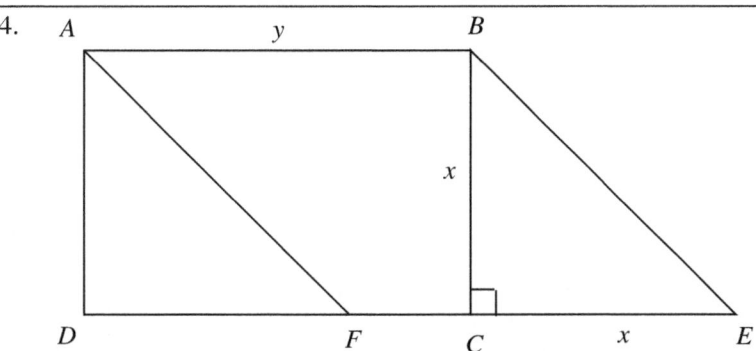

Since opposite sides of a rectangle are equal and parallel, *AD* is parallel to *BC* and *AD* = *BC*. We are also given that *AF* is parallel to *BE*. Further, points *D*, *C*, and *E* must be on the same line because the angle made by points *D* and *E* at *C* is 180° (∠*DCE* = ∠*DCB* + ∠*BCE* = angle in a rectangle + a right angle [given] = 90° + 90° = 180°). So, *DC* and *CE* can be considered parallel.

Any three pairs of parallel lines (here *AF* and *BE*, *DF* and *CE*, and *AD* and *BC*) make two similar triangles (*AFD* and *BEC*). And if one pair of corresponding sides of two similar triangles are equal (here, *AD* = *BC*), then the triangles are congruent (equal).

The areas of congruent triangles are equal. So, area of

$$\begin{aligned}\triangle AFD &= \text{area of } \triangle BEC \\ &= 1/2 \cdot base \cdot height \\ &= 1/2 \cdot BC \cdot CE \\ &= 1/2 \cdot 5 \cdot 5 \\ &= 1/2 \cdot 25 \\ &= 12.5\end{aligned}$$

The answer is (C).

5.

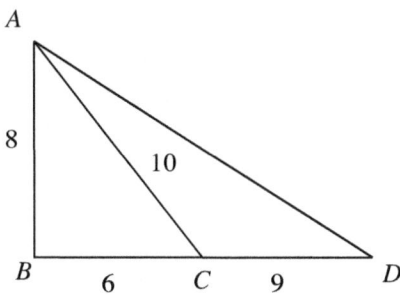

The lengths of the three sides of $\triangle ABC$ are $AB = 8$, $BC = 6$, and $AC = 10$. The three sides satisfy the Pythagorean Theorem: $AC^2 = BC^2 + AB^2$ ($10^2 = 6^2 + 8^2$). Hence, triangle *ABC* is right angled and ∠*B*, the angle opposite the longest side *AC* (hypotenuse), is a right angle. Now, from the figure, this angle is part of $\triangle ADB$, so $\triangle ADB$ is also right angled. Applying The Pythagorean Theorem to the triangle yields

$$\begin{aligned}AD^2 &= AB^2 + BD^2 \\ &= AB^2 + (BC + CD)^2 \quad \text{from the figure, } BD = BC + CD \\ &= 8^2 + (6 + 9)^2 \\ &= 8^2 + 15^2 \\ &= 289 \\ &= 17^2 \\ AD &= 17 \quad \text{by square rooting}\end{aligned}$$

The answer is (D).

6.

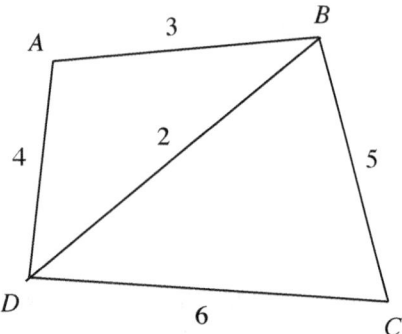

The figure is not drawn to scale.

In $\triangle ABD$, $AD = 4$, $AB = 3$, and $BD = 2$ (from the figure). Forming the inequality relation for the side lengths yields $AD > AB > BD$. Since in a triangle, the angle opposite the longer side is greater, we have a similar inequality for the angles opposite the corresponding sides: $\angle ABD > \angle BDA > \angle A$.

Similarly, in $\triangle BCD$, $DC = 6$, $BC = 5$, and $BD = 2$. Forming the inequality for the side lengths yields $DC > BC > BD$. Also the angles opposite the corresponding sides follow the relation $\angle DBC > \angle CDB > \angle C$.

Now, summing the two known inequalities $\angle ABD > \angle BDA > \angle A$ and $\angle DBC > \angle CDB > \angle C$ yields $\angle ABD + \angle DBC > \angle BDA + \angle CDB > \angle A + \angle C$; $\angle B > \angle D > \angle A + \angle C$. From this inequality, clearly $\angle B$ is the greatest angle, and $\angle D$ is the next greatest angle in the quadrilateral. Hence, the answer is (B) and (D).

7.

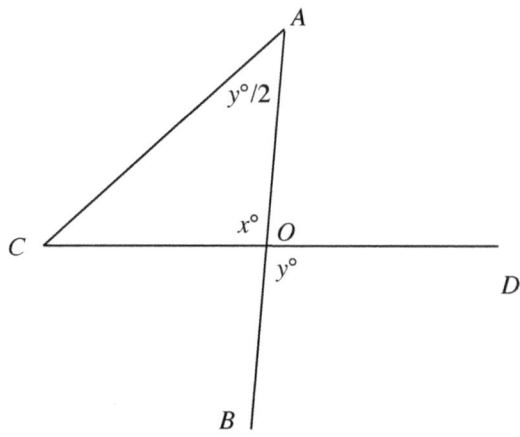

The dimensions in the figure may be different from what they appear to be.

From the figure, $\angle A = y/2$. Since we are given that $\angle A = \angle C$, $\angle C$ also equals $y/2$. Also, in figure, $\angle O = x = y$ (vertical angles are equal). Summing the angles of $\triangle AOC$ to 180 yields

$\angle A + \angle O + \angle C = 180$
$y/2 + x + y/2 = 180$
$x/2 + x + x/2 = 180$ (since $x = y$)
$2x = 180$
$x = 180/2 = 90$

Enter in the grid.

Test 3—Solutions

8. Let's substitute the given choices for x in the expression $x^2 + 4x + 3$ and find out which one results in an odd number.

   Choice (A): $x = 3$. $x^2 + 4x + 3 = 3^2 + 4(3) + 3 = 9 + 12 + 3 = 24$, an even number. Reject.
   Choice (B): $x = 5$. $x^2 + 4x + 3 = 5^2 + 4(5) + 3 = 25 + 20 + 3 = 48$, an even number. Reject.
   Choice (C): $x = 8$. $x^2 + 4x + 3 = 8^2 + 4(8) + 3 = 64 + 32 + 3 = 99$, an odd number. Accept the choice.
   Choice (D): $x = 13$. $x^2 + 4x + 3 = 13^2 + 4(13) + 3 = 169 + 52 + 3 = 224$, an even number. Reject.
   Choice (E): $x = 16$. $x^2 + 4x + 3 = 16^2 + 4(16) + 3 = 256 + 64 + 3 = 323$, an odd number. Accept the choice.

The answer is (C) and (E).

Method II (without substitution):

$x^2 + 4x + 3 =$ An Odd Number
$x^2 + 4x =$ An Odd Number $- 3$
$x^2 + 4x =$ An Even Number
$x(x + 4) =$ An Even Number. This happens only when x is even. If x is odd, $x(x + 4)$ is not even. Hence, x must be even. Since 8 and 16 are the only even answer-choices, the answer is (C) and (E).

---

9. Mike ran 10 miles at 10 mph (for the *Time = Distance/Rate* = 10 miles/10 mph = 1 hour). He ran the remaining 20 miles at 5 mph (for the *Time = Distance/Rate* = 20 miles/5 mph = 4 hrs). The length of the Marathon track is 30 miles, and the total time taken to cover the track is 5 hours.

Now, let the time taken by Fritz to travel the 30-mile Marathon track be t hours. Then as given, Fritz ran at 10 mph for $t/3$ hours and at 5 mph for the rest of $2t/3$ hours. Hence, by formula, *Distance = Rate×time*, the distance covered is $(10 \text{ mph}) \cdot \frac{t}{3} + (5 \text{ mph}) \cdot \frac{2t}{3} = \left(\frac{10}{3} + \frac{10}{3}\right) \cdot t = \frac{20}{3} t = 30 \text{ miles}$. Solving the equation for t yields $t = 90/20 = 4.5$ hours.

Since Fritz took less time to cover the Marathon than Mike, the average speed of Fritz is greater. Hence, Column B is greater than Column A, and the answer is (B).

---

10. We are given that the point (b, a) lies in Quadrant III. In this quadrant, both x- and y-coordinates are negative. So, both b and a are negative. So, the point (a, b) also lies in the same quadrant (since both x- and y-coordinates are again negative). Hence, the answer is (C).

---

11. Combining the given inequalities $x > 2$ and $x < 3$ yields $2 < x < 3$. So, x lies between 2 and 3. Hence,

   $x - 2$ is positive and $x - 3$ is negative. Hence, the product $(x - 2)(x - 3)$ is negative. I is false.

   $2 - x$ is negative and $x - 3$ is negative. Hence, the product $(x - 2)(x - 3)$ is positive. II is true.

   $2 - x$ is negative and $3 - x$ is positive. Hence, the product $(2 - x)(3 - x)$ is negative. III is false.

Hence, the answer is (B), II only is correct.

12. The formula for the volume of a cylindrical tank is (*Area of the base*) × *height*. Since the base is circular (in a cylinder), the *Area of the base* = $\pi(radius)^2$. Also, The rate of filling the tank equals

The volume filled ÷ Time taken =

(*Area of the base* × *height*) ÷ *Time taken* =

[$\pi (radius)^2$ × height] ÷ *Time taken* =

$\pi(10 \text{ feet})^2 \times 7 \text{ feet} \div 10 \text{ minutes}$ =

70π cubic feet per minute

The answer is (E).

---

13. Let c and s be the cost and the selling price, respectively, for the seller on each item.

We are given that the selling price of 15 items equals the cost of 20 items. Hence, we have $15s = 20c$, or $s = (20/15)c = 4c/3$. Now, the profit equals selling price – cost = $s - c = 4c/3 - c = c/3$. The percentage profit on each item is

$$\frac{\text{Profit}}{\text{Cost}} \cdot 100 = \frac{\frac{c}{3}}{c} \cdot 100 = \frac{100}{3} = 33.3\%$$

The answer is (D).

---

14. We are given that $4p = 6q$. Dividing both sides by 2 yields $2p = 3q$. Subtracting $3q$ from both sides yields $2p - 3q = 0$. The answer is (A).

Method II: If $2p - 3q$ were not 0, it would be some expression in p or q. But there is no such choice.

---

15. The eye-catcher is that the columns are equal: $3.1 - 3.1 = 0$. But that won't be the answer to this hard problem. Now, <x> denotes the greatest integer less than or equal to x. That is, <x> is the first integer smaller than x. Further, if x is an integer, then <x> is equal to x itself. Therefore, <3.1> = 3, and <–3.1> = –4 (not –3). Hence, <3.1> + <–3.1> = 3 + (–4) = –1. Therefore, Column B is larger. The answer is (B).

---

16. Let 1 pound of the rice of the first type ($0.8 per pound) be mixed with p pounds of the rice of the second type ($0.9 per pound). Then the total cost of the 1 + p pounds of the rice is

($0.8 per pound × 1 pound) + ($0.9 per pound × p pounds) =

$0.8 + 0.9p$

Hence, the cost of the mixture per pound is

$$\frac{\text{Cost}}{\text{Weight}} = \frac{0.8 + 0.9p}{1 + p}$$

If this equals $0.825 per pound (given), then we have the equation

$\frac{0.8 + 0.9p}{1 + p} = 0.825$
$0.8 + 0.9p = 0.825(1 + p)$
$0.8 + 0.9p = 0.825 + 0.825p$
$0.9p - 0.825p = 0.825 - 0.8$
$900p - 825p = 825 - 800$
$75p = 25$
$p = 25/75 = 1/3$

Hence, the proportion of the two rice types is 1 : 1/3, which also equals 3 : 1. Hence, the answer is (E).

Test 3—Solutions

17. Since the columns are positive, we can square both columns without affecting the inequality relation between the columns. Doing this yields

Column A	Column B
$\left(\sqrt{12.5}+\sqrt{12.5}\right)^2$	$\left(\sqrt{25}\right)^2$
$= 12.5 + 12.5 + 2\sqrt{12.5}\sqrt{12.5}$	$= 25$
$= 25 + 2\sqrt{12.5}\sqrt{12.5}$	

Subtracting 25 from both columns yields

Column A	Column B
$2\sqrt{12.5}\sqrt{12.5}$	0

Since $2\sqrt{12.5}\sqrt{12.5}$ is greater than 0, Column A is greater than Column B. The answer is (A).

Method II:
Whatever the value of $\sqrt{12.5}$ is, it is greater than 3 since $\sqrt{12.5} > \sqrt{9} = 3$. Hence, $\sqrt{12.5} + \sqrt{12.5} > 3 + 3 = 6 > 5 =$ Column B.

---

18. The average of the numbers in the set is

$$(1 + 2 + 3 + 4 + 5 + 6)/6 = 21/6 = 3.5$$

Now, removing the numbers less than 3.5 would increase the average and removing the numbers greater than the average would decrease the average. Hence, choose (A), (B), and (C).

---

19. Column A: 45% of 90 = $\dfrac{45}{100} \times 90 = \dfrac{45 \times 90}{100}$.

Column B: 90% of 45 = $\dfrac{90}{100} \times 45 = \dfrac{90 \times 45}{100}$.

Since the columns are equal, the answer is (C).

---

20. From the figure, the monthly income of *A* for the first four months is $400. Hence, the net earnings in the 4 months is 4 × 400 = 1600 dollars.

From the figure, the monthly income of *A* for the second four months is $700. Hence, the net earnings in the 4 months is 4 × 700 = 2800 dollars.

From the figure, the monthly income of *A* for the last four months is $1000. Hence, the net earnings in the 4 months is 4 × 1000 = 4000 dollars.

Hence, the total income in the year is 1600 + 2800 + 4000 = 8400 dollars.

The monthly income of *B* grew regularly from 400 in January to 950 in December. Hence, the net income is 400 + 450 + 500 + 550 + 600 + 650 + 700 + 750 + 800 + 850 + 900 + 950 = 8100.

The answer is (D) 8400, 8100.

21. If we superimpose the two graphs, we can get that the two curves coincide at the months January 2007, July 2007, and again at January 2008. The answer is (A), (C), and (E).

22. If we superimpose the two graphs, we can see that curve A is below curve C for the following months: February 2007, March 2007, April 2007, and August 2007. Select the months available in the choices. Select (A) and (D).

23.

Month	Income of the month		Net Income in the year	
	A	B	A	B
Jan-2007	400	400	400	400
Feb-2007	400	800	450	850
Mar-2007	400	1200	500	1350
Apr-2007	400	1600	550	1900
May-2007	700	2300	600	2500
June-2007	700	3000	650	3150
July-2007	700	3700	700	3850
Aug-2007	700	4600	750	4600
Sep-2007	1000	5600	800	5400
Oct-2007	1000	6600	850	6250
Nov-2007	1000	7600	900	7150
Dec-2007	1000	8600	950	8100
Jan-2008	1000	1000 (Year start)	1000	1000 (Year start)

The values in the third column are less than those in fifth column in the following months: February 2007, March 2007, April 2007, May 2007, June 2007, July 2007 only. Select choices (A), (B), and (C).

24. From the distribution given, the 4th, 6th, 7th, 8th, 10th, 11th, 13th, and 17th families, a total of 8, have at least 3 cars. Hence, the probability of selecting a family having at least 3 cars out of the available 20 families is 8/20, which reduces to 2/5. The answer is (B).

Test 4

GRE Math Tests

Questions: 24
Time: 45 minutes

[Multiple-choice Question – Select One Answer Choice Only]
1. Which of the following could be an integer?

    (A) The average of two consecutive integers
    (B) The average of three consecutive integers
    (C) The average of four consecutive integers
    (D) The average of six consecutive integers
    (E) The average of 6 and 9

2. | Column A | m and n are two positive integers. $5m + 7n = 46$. | Column B |
   |---|---|---|
   | m | | n |

3. | Column A | The nth term of the sequence $a_1, a_2, a_3, ..., a_n$ is defined as $a_n = -(a_{n-1})$. The first term a_1 equals -1. | Column B |
   |---|---|---|
   | a_5 | | 1 |

[Multiple-choice Question – Select One Answer Choice Only]
4. ABCD is a square and one of its sides AB is also a chord of the circle as shown in the figure. What is the area of the square?

(A) 3
(B) 9
(C) 12
(D) $12\sqrt{2}$
(E) 18

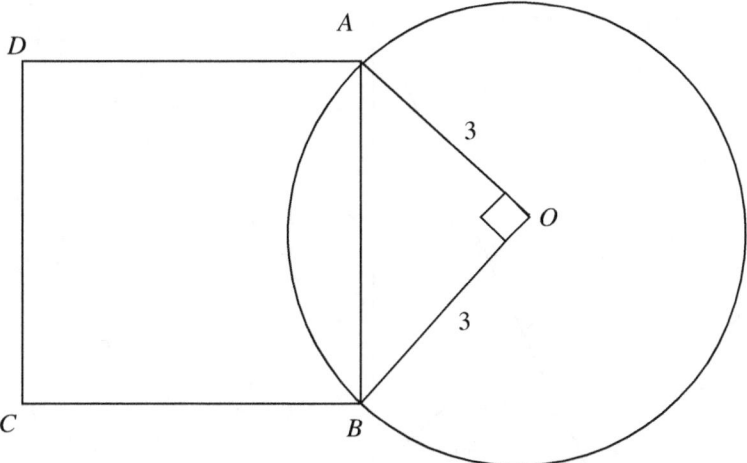

5.   Column A   Column B
    The perimeter of $\triangle ABC$ The circumference of the circle

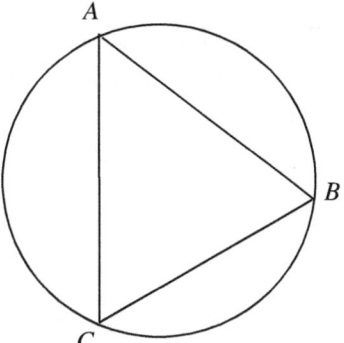

[Multiple-choice Question – Select One Answer Choice Only]
6. In the figure, the areas of parallelograms *EBFD* and *AECF* are 3 and 2, respectively. What is the area of rectangle *ABCD* ?

   (A)   3
   (B)   4
   (C)   5
   (D) $4\sqrt{3}$
   (E)   7

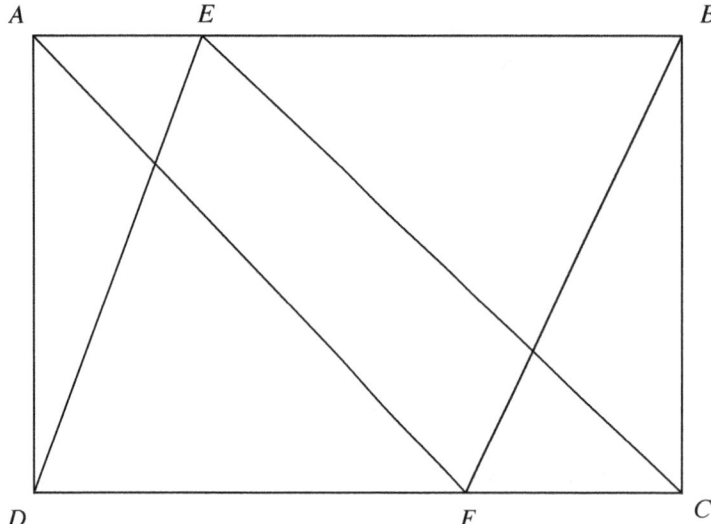

[Multiple-choice Question – Select One Answer Choice Only]
7. The diagonal length of a square is 14.1 sq. units. What is the area of the square, rounded to the nearest integer? ($\sqrt{2}$ is approximately 1.41.)

   (A)   96
   (B)   97
   (C)   98
   (D)   99
   (E)   100

[Multiple-choice Question – Select One Answer Choice Only]
8. *AB* and *CD* are chords of the circle, and *E* and *F* are the midpoints of the chords, respectively. The line *EF* passes through the center *O* of the circle. If *EF* = 17, then what is radius of the circle?

   (A) 10
   (B) 12
   (C) 13
   (D) 15
   (E) 25

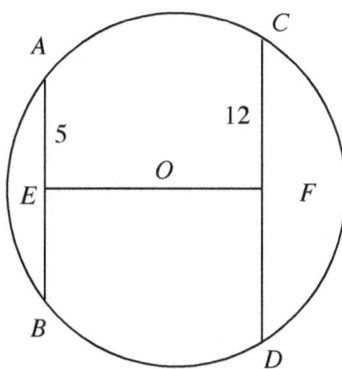

[Multiple-choice Question – Select One Answer Choice Only]

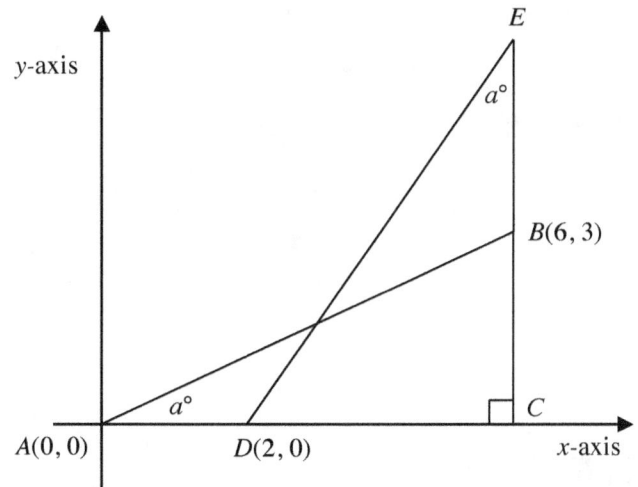

9. In the rectangular coordinate plane shown, what are the coordinates of point *E* ?

   (A) (2, 0)
   (B) (2, 3)
   (C) (6, 2)
   (D) (6, 6)
   (E) (6, 8)

GRE Math Tests

[Multiple-choice Question – Select One Answer Choice Only]
10. In the game of chess, the Knight can make any of the moves displayed in the diagram. If a Knight is the only piece on the board, what is the greatest number of spaces from which not all 8 moves are possible?

   (A) 8
   (B) 24
   (C) 38
   (D) 48
   (E) 56

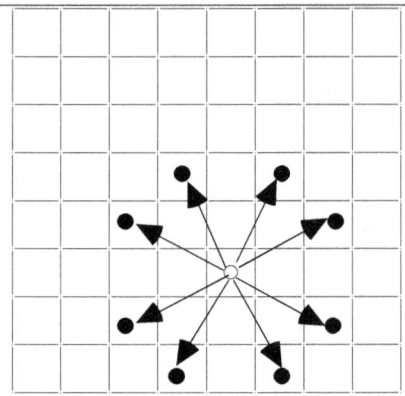

[Multiple-choice Question – Select One Answer Choice Only]
11. Which one of the following is nearest to 0.313233 ?

   (A) 3/10
   (B) 31/100
   (C) 313/1000
   (D) 3132/10000
   (E) 31323/100000

[Multiple-choice Question – Select One Answer Choice Only]
12. If $x^2 - 4x + 3$ equals 0, then what is the value of $(x - 2)^2$?

   (A) 0
   (B) 1
   (C) 2
   (D) 3
   (E) 4

[Multiple-choice Question – Select One Or More Answer Choices]
13. If $2x = 2y + 1$, then which of the following could be true?

   (A) $x > y$
   (B) $x < y$
   (C) $x = y$
   (D) $x = y + 1$
   (E) $9x + 12y = 36$

[Multiple-choice Question – Select One Answer Choice Only]

14. The average length of all the sides of a rectangle equals twice the width of the rectangle. If the area of the rectangle is 18, what is its perimeter?

   (A) $6\sqrt{6}$
   (B) $8\sqrt{6}$
   (C) 24
   (D) 32
   (E) 48

[Multiple-choice Question – Select One Or More Answer Choices]

15. If Robert can assemble a model car in 30 minutes and Craig can assemble the same model car in 20 minutes, how long would it take them, working together, to assemble the model car?

   (A) 12 minutes
   (B) 13 minutes
   (C) 14 minutes
   (D) 15 minutes
   (E) 16 minutes

[Multiple-choice Question – Select One Or More Answer Choices]

16. The ratio of the number of chickens to the number of pigs to the number of horses on Richard's farm is 33 : 17 : 21. What fraction of the animals are either pigs or horses?

   (A) 16/53
   (B) 17/54
   (C) 38/71
   (D) 25/31
   (E) 38/33

17. Column A — n equals $10^4 + (2 \times 10^2)$ — Column B
    Number of zeros in n Number of zeros in n^2

18. Column A   Column B
    $(1111.0)^2 - (999.0)^2$ $(1111.5)^2 - (999.5)^2$

[Multiple-choice Question – Select One Answer Choice Only]
19. Each person in a group of 110 investors has investments in either equities or securities or both. Exactly 25% of the investors in equities have investments in securities, and exactly 40% of the investors in securities have investments in equities. How many have investments in equities?

    (A) 65
    (B) 80
    (C) 120
    (D) 135
    (E) 150

[Multiple-choice Question – Select One Answer Choice Only]
20. Katrina has a wheat business. She purchases wheat from a local wholesaler at a particular cost per pound. The price of the wheat at her stores is $3 per pound. Her faulty spring balance reads 0.9 pounds for a pound. Also, in the festival season, she gives a 10% discount on the wheat. She found that she made neither a profit nor a loss in the festival season. At what price did Katrina purchase the wheat from the wholesaler?

    (A) 2.43
    (B) 2.5
    (C) 2.7
    (D) 3
    (E) 3.3

[Multiple-choice Question – Select One Answer Choice Only]
21. In a zoo, each pigeon has 2 legs, and each rabbit has 4 legs. The head count of the two species together is 12, and the leg count is 32. How many pigeons and how many rabbits are there in the zoo?

    (A)   4, 8
    (B)   6, 6
    (C)   6, 8
    (D)   8, 4
    (E)   8, 6

[Multiple-choice Question – Select One Or More Answer Choices]
22. The table shows the distribution of a team of 16 engineers by gender and level.

	Junior Engineers	Senior Engineers	Lead Engineers
Male	3		2
Female	2	4	

The number of Male senior Engineers and the number of Female Lead Engineers is not yet known. If one engineer is selected from the team, which of the following could be the probability that the engineer is a male senior engineer when the numbers are known?

    (A)   1/32
    (B)   1/16
    (C)   3/16
    (D)   1/8
    (E)   3/4

[Multiple-choice Question – Select One Answer Choice Only]

23. The following values represent the exact number of cars owned by the 20 families on Pearl Street.

    1, 1, 2, 3, 2, 5, 4, 3, 2, 4, 5, 2, 6, 2, 1, 2, 4, 2, 1, 1

    This can be expressed in frequency distribution format as follows:

x	The number of families having x number of cars
1	5
2	7
3	a
4	3
5	b
6	1

    What are the values of a and b, respectively?

    (A) 1 and 1
    (B) 1 and 2
    (C) 2 and 1
    (D) 2 and 2
    (E) 2 and 3

[Multiple-choice Question – Select One Answer Choice Only]

24. A school has a total enrollment of 150 students. There are 63 students taking French, 48 taking chemistry, and 21 taking both. How many students are taking NEITHER French nor chemistry?

    (A) 60
    (B) 65
    (C) 71
    (D) 75
    (E) 97

Test 4—Solutions

Answers and Solutions Test 4:

Question	Answer
1.	B
2.	A
3.	B
4.	E
5.	B
6.	C
7.	D
8.	C
9.	E
10.	D
11.	E
12.	B
13.	A, E
14.	B
15.	A
16.	C
17.	B
18.	B
19.	B
20.	A
21.	D
22.	B, C, D
23.	D
24.	A

1. For choice (A), choose any two consecutive integers, say, 1 and 2. Forming their average* yields

$$\frac{1+2}{2} = \frac{3}{2}$$

Since 3/2 is not an integer, let's move on to choice (B). Note: We cannot eliminate choice (A) because it could be an integer some other pair of consecutive integers (though that will not actually happen).

For choice (B), choose any three consecutive integers, say, 1, 2, and 3. Forming their average yields

$$\frac{1+2+3}{3} = \frac{6}{3} = 2$$

Since 2 is an integer, the answer is (B).

---

* Recall that the average of N numbers is their sum divided by N, that is, $Average = \frac{sum}{N}$.

2. Usually, a system having a single constraint in two variables such as $5m + 7n = 46$ will not have a unique solution.

But the given system has two constraints:

    1) $5m + 7n = 46$
    2) m and n are positive integers

Hence, we might have a unique solution. Let's see:

Let $p = 5m$ (p is a multiple of 5) and $q = 7n$ (q is a multiple of 7) such that and $p + q = 46$. Subtracting q from both sides yields $p = 46 - q$ [(a positive multiple of 5) equals 46 − (a positive multiple of 7)]. Let's look at how many such solutions exist:

    If $q = 7, p = 46 - 7 = 39$, not a multiple of 5. Reject.

    If $q = 14, p = 46 - 14 = 32$, not a multiple of 5. Reject.

    If $q = 21, p = 46 - 21 = 25$, a multiple of 5. Acceptable. So, $m = 25/5 = 5$ and $n = q/7 = 21/7 = 3$.

    If $q = 28, p = 46 - 28 = 18$, not a multiple of 5. Reject.

    If $q = 35, p = 46 - 35 = 11$, not a multiple of 5. Reject.

    If $q = 42, p = 46 - 42 = 4$, not a multiple of 5. Reject.

    If $q \geq 49, p \leq 46 - 49 = -3$, not positive. Reject.

The solution is Column A = m = 5, and Column B = n = 3. Hence, Column A is greater and the answer is (A).

---

3. The rule for the sequence is $a_n = -(a_{n-1})$. Putting $n = 2$ and 3 in the rule yields

$$a_2 = -(a_{2-1}) = -a_1 = -(-1) = 1 \quad\quad \text{(given that } a_1 = -1\text{)}$$

$$a_3 = -(a_{3-1}) = -a_2 = -1$$

Similarly, we get that *each* even numbered term (when n is even) equals 1 and *each* odd numbered term (when n is odd) equals −1. Since a_5 is an odd numbered term, it equals −1. Hence, Column A equals −1 and is less than 1 (= Column B). The answer is (B).

---

4. Side AB is the hypotenuse of $\triangle AOB$. Hence, by The Pythagorean Theorem, we have

$$AB^2 = AO^2 + BO^2 = 3^2 + 3^2 = 18$$

Hence, the area of the square $ABCD$ equals $side^2 = AB^2 = 18$. The answer is (E).

5. Since the shortest distance between two points is a straight line, the length of a chord is always less than the length of the arc it makes (subtends) on the circle. Hence, from the figure, we have

$$AB < \text{arc } AB$$
$$BC < \text{arc } BC$$
$$CA < \text{arc } CA$$

Adding the three inequalities yields $AB + BC + CA < \text{arc } AB + \text{arc } BC + \text{arc } CA$. The left side of the inequality is the perimeter of triangle ABC (which Column A equals), and the right side is the circumference of the circle (which Column B equals). Hence, Column A is less than Column B, and the answer is (B).

---

6. The area of the rectangle $ABCD$ is $length \times width = DC \times AD$.

The area of the parallelogram $AECF$ is $FC \times AD$ ($base \times height$). Also, the area of parallelogram $EBFD$ is $DF \times AD$. Now, summing the areas of the two parallelograms $AECF$ and $EBFD$ yields $FC \times AD + DF \times AD = (FC + DF)(AD) = DC \times AD =$ the area of the rectangle $ABCD$. Hence, the area of rectangle $ABCD$ equals the sum of areas of the two parallelograms, which is $2 + 3 = 5$. The answer is (C).

---

7. If a is the length of a side of the square, then a diagonal divides the square into two congruent (equal) right triangles. Applying The Pythagorean Theorem to either triangle yields $diagonal^2 = side^2 + side^2 = a^2 + a^2 = 2a^2$. Taking the square root of both sides of this equation yields the $diagonal$ $a\sqrt{2}$. We are given that the diagonal length is 14.1. Hence, $a\sqrt{2} = 14.1$ or $a = \dfrac{14.1}{\sqrt{2}}$. Now, the area, a^2, equals

$$\left(\dfrac{14.1}{\sqrt{2}}\right)^2 = \dfrac{14.1^2}{\left(\sqrt{2}\right)^2} = \dfrac{14.1^2}{2} = \dfrac{198.81}{2} = 99.4.$$ The number 99.4 is nearest to 99. Hence, the answer is (D).

Note: If you had approximated $\dfrac{14.1}{\sqrt{2}}$ with 10, you would have mistakenly gotten 100 and would have answered (E). Approximation is the culprit. Save it for last.

---

8. If a line joining the midpoints of two chords of a circle passes through the center of the circle, then it cuts both chords perpendicularly.

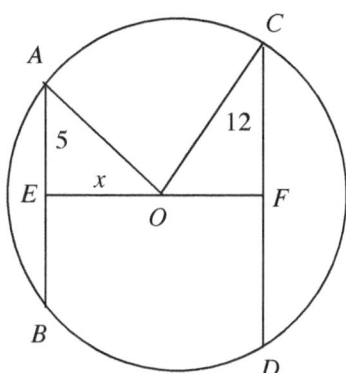

Let x be the length of the line segment from center of the circle to chord AB (at E). Since the length of EF is 17, $OF = 17 - x$.

Now, let r be the radius of the circle. Applying The Pythagorean Theorem to the right triangle AEO yields

$$AE^2 + EO^2 = AO^2$$
$$5^2 + x^2 = r^2 \qquad (1)$$

Also, applying The Pythagorean Theorem to $\triangle CFO$ yields

$$CF^2 + OF^2 = OC^2$$
$$12^2 + (17 - x)^2 = r^2 \qquad (2)$$

Equating the left-hand sides of equations (1) and (2), since their right-hand sides are the same, yields

$$x^2 + 5^2 = 12^2 + (17 - x)^2$$
$$x^2 + 5^2 = 12^2 + 17^2 + x^2 - 34x$$
$$34x = 12^2 + 17^2 - 5^2$$
$$34x = 144 + 289 - 25$$
$$34x = 408$$
$$x = 408/34$$
$$x = 12$$

Now, substituting this value of x in equation (1) yields

$$5^2 + 12^2 = r^2$$
$$25 + 144 = r^2$$
$$r = \sqrt{169} = 13$$

Hence, the radius is 13, and the answer is (C).

9. From the figure, since EC is perpendicular to the x-axis, C is a point vertically below point B. Hence, both have the same x-coordinate.

The line AC is horizontal and therefore its length equals the x-coordinate difference of A and C, which equals $6 - 0 = 6$.

The line DC is horizontal and therefore its length equals the x-coordinate difference of D and C, which is $6 - 2 = 4$.

The length of the vertical line BC equals the y-coordinate difference of B and C, which is $3 - 0 = 3$.

The length of the vertical line EC equals the y-coordinate difference of E and C.

Now, in $\triangle ABC$, $\angle A = a°$, $\angle B = 90° - a°$, and $\angle C = 90°$. The sides opposite angles A and B are in the ratio $BC/AC = 3/6 = 1/2$.

Similarly, in $\triangle DEC$, $\angle E = a°$, $\angle D = 90° - a°$, and $\angle C = 90°$ (So, ABC and DEC are similar triangles and their corresponding sides are proportional). Hence, the sides opposite angles E and D are in the ratio $DC/EC = 1/2$. Hence, we have

$$DC/EC = 1/2$$
$$EC = 2DC$$
$$EC = 2 \cdot 4 = 8$$

Hence, the y-coordinate of point E is 8, and the answer is (E).

Test 4—Solutions

10. Since we are looking for the greatest number of spaces from which not all 8 moves are possible, we can eliminate the greatest number, 56. Now, clearly not all 8 moves are possible from the outer squares, and there are 28 outer squares—not 32. Also, not all 8 moves are possible from the next to outer squares, and there are 20 of them—not 24. All 8 moves are possible from the remaining squares. Hence, the answer is 28 + 20 = 48. The answer is (D). Notice that 56, (32 + 24), is given as an answer-choice to catch those who don't add carefully.

11. The given decimal 0.313233 rounded to first, second, third, fourth and fifth digits after the decimal respectively equal 0.3 [= Choice (A)], 0.31 [= Choice (B)], 0.313 [= Choice (C)], 0.3132 [= Choice (D)], and 0.31323 [= Choice (E)]. Accuracy can be maintained by rounding the decimals only to later digits. So, choice (E) is the most accurate and hence the nearest. The answer is (E).

12. We have $x^2 - 4x + 3 = 0$. By the Perfect Square Trinomial formula, $(x - 2)^2 = x^2 - 4x + 4$; and this can be rewritten as $(x^2 - 4x + 3) + 1 = 0 + 1 = 1$. Hence, the answer is (B).

13. Dividing the equation $2x = 2y + 1$ by 2 yields $x = y + 1/2$. This equation says that "x is 1/2 unit greater than y." Or more simply, $x > y$. The answer is (A)—automatically making (B) and (C) false.

Choice (D) $x = y + 1$ clearly violates the equation $x = y + 1/2$. Hence, choice (D) is not possible. Reject.

Choice (E): Solving the equation $9x + 12y = 36$ for x yields $x = -4y/3 + 4$, which is an equation with slope different from the slope of given equation $x = y + 1/2$. Hence, the two lines intersect at some point, and at the point, choice (E) must be true.

The answer is (A) and (E).

14. The perimeter of a rectangle is twice the sum of its length and width. Hence, if l and w are length and width, respectively, of the given rectangle, then the perimeter of the rectangle is $2(l + w)$. Also, the average side length of the rectangle is 1/4 times the sum. So, the average side length is $2(l + w)/4 = l/2 + w/2$.

Now, we are given that the average equals twice the width. Hence, we have $l/2 + w/2 = 2w$. Multiplying the equation by 2 yields $l + w = 4w$ and solving for l yields $l = 3w$.

Now, the area of the rectangle equals $length \times width = l \times w = 18$ (given). Plugging $3w$ for l in the equation yields $3w \times w = 18$. Dividing the equation by 3 yields $w^2 = 6$, and square rooting both sides yields $w = \sqrt{6}$. Finally, the perimeter equals $2(l + w) = 2(3w + w) = 8w = 8\sqrt{6}$. The answer is (B).

15. Let t be the time it takes the boys, working together, to assemble the model car. Then their combined rate is $1/t$, and their individual rates are $1/30$ and $1/20$. Now, their combined rate is merely the sum of their individual rates:

$$\frac{1}{t} = \frac{1}{30} + \frac{1}{20}$$

Solving this equation for t yields $t = 12$. The answer is (A).

GRE Math Tests

16. Let the number of chickens, pigs and horses on Richard's farm be c, p, and h. Then forming the given ratio yields

$$c : p : h = 33 : 17 : 21$$

Let c equal $33k$, p equal $17k$, and h equal $21k$, where k is a positive integer (such that $c : p : h = 33 : 17 : 21$). Then the total number of pigs and horses is

$$17k + 21k = 38k$$

And the total number of pigs, horses and chickens is

$$17k + 21k + 33k = 71k$$

Hence, the required fraction equals $38k/71k = 38/71$. The answer is (C).

---

17. Since n equals $10^4 + (2 \times 10^2) = 10200$ and has 3 zeros, Column A equals 3.

By the Perfect Square Trinomial formula,

$$n^2 =$$

$$[10^4 + (2 \times 10^2)]^2 =$$

$$10^{4 \cdot 2} + (2 \times 10^2)^2 + 2(2 \times 10^2)10^4 =$$

$$10^8 + 4 \times 10^4 + 4 \times 10^6 =$$

$$104040000$$

There are 6 zero digits. So, Column B equals 6.

Column B is greater, and the answer is (B).

---

18. Column B = $(1111.5)^2 - (999.5)^2 =$
    = $(1111.5 - 999.5)(1111.5 + 999.5)$ by the Difference of Squares formula
                                                  $a^2 - b^2 = (a - b)(a + b)$
    = $(1111 + 0.5 - 999 - 0.5)(1111 + 0.5 + 999 + 0.5)$
    = $(1111 - 999)(1111 + 999 + 1)$
    = $(1111 - 999)(1111 + 999) + (1111 - 999)$
    = $1111^2 - 999^2 + (1111 - 999)$ by the Difference of Squares formula
                                                  $(a - b)(a + b) = a^2 - b^2$
    = Column A + $(1111 - 999)$
    = Column A + (a positive number)

From this equation, it is clear that Column B is greater than Column A, by $1111 - 999$, a positive number. Hence, the answer is (B).

Test 4—Solutions

19. The investors can be categorized into three groups:

(1) Those who have investments in equities only.
(2) Those who have investments in securities only.
(3) Those who have investments in both equities and securities.

Let x, y, and z denote the number of people in the respective categories. Since the total number of investors is 110, we have

$$x + y + z = 110 \qquad (1)$$

Also,
The number of people with investments in equities is $x + z$ and
The number of people with investments in securities is $y + z$.

Since exactly 25% of the investors in equities have investments in securities, we have the equation

$$25/100 \bullet (x + z) = z$$
$$25/100 \bullet x + 25/100 \bullet z = z$$
$$25/100 \bullet x = 75/100 \bullet z$$
$$x = 3z \qquad (2)$$

Since exactly 40% of the investors in securities have investments in equities, we have the equation

$$40/100 \bullet (y + z) = z$$
$$2/5 \bullet (y + z) = z$$
$$y + z = 5z/2$$
$$y = 3z/2 \qquad (3)$$

Substituting equations (2) and (3) into equation (1) yields

$$3z + 3z/2 + z = 110$$
$$11z/2 = 110$$
$$z = 110 \bullet 2/11 = 20$$

Hence, the number of people with investments in equities is $x + z = 3z + z = 3 \bullet 20 + 20 = 60 + 20 = 80$. The answer is (B).

20. The cost of wheat at Katrina's store is $3 per Katrina's pound. After the 10% discount (festival season discount), the cost of the wheat would be

$$\text{(the price)}(1 - 10\%) =$$
$$3(1 - .10) =$$
$$3(.90) =$$
2.7 dollars per Katrina's pound

Since her faulty balance was reading 0.9 pounds (0.9 Katrina Pound's) for a real pound, she was unknowingly selling 1 real pound in the name of 0.9 Katrina's pounds. Equating yields 0.9 Katrina Pounds = 1 real pound. Therefore, a Katrina pound equals 1/0.9 = 10/9 real pounds. Now, for each of Katrina's pounds, she received the price 2.7 dollars. Therefore, 10/9 real pounds yields 2.7 dollars. This means, each real pound yields (9/10)(2.7 dollars) = 2.43 dollars per real pound. Since she earned neither a profit nor a loss, she must have purchased the wheat at this same cost per real pound from the wholesaler. Hence, the answer is (A).

21. Let the number of pigeons be p and the number of rabbits be r. Since the head count together is 12,

$$p + r = 12 \quad (1)$$

Since each pigeon has 2 legs and each rabbit has 4 legs, the total leg count is

$$2p + 4r = 32 \quad (2)$$

Dividing equation (2) by 2 yields $p + 2r = 16$. Subtracting this equation from equation (1) yields

$$(p + r) - (p + 2r) = 12 - 16$$
$$p + r - p - 2r = -4$$
$$r = 4$$

Substituting this into equation (1) yields $p + 4 = 12$, which reduces to $p = 8$.

Hence, the number of pigeons is $p = 8$, and the number of rabbits is $r = 4$. The answer is (D).

22. The total number of engineers available is 16.

The number of engineers shown in the table is $3 + 2 + 4 + 2 = 11$.

Hence, the number of engineers not shown in the table must be $16 - 11 = 5$.

Hence, the number of Male Senior Engineers must be 1, 2, 3, 4, or 5.

Hence, the probability must be $1/16$, $2/16 = 1/8$, $3/16$, $4/16 = 1/4$, or $5/16$ depending upon the number of male senior engineers available.

Select choices (B), (C), and (D).

23. In the frequency distribution table, the first column represents the number of cars and the second column represents the number of families having the particular number of cars. Now, from the data given, the number of families having exactly 3 cars is 2, and the number of families having exactly 5 cars is 2. Hence, $a = 2$ and $b = 2$. The answer is (D).

24. Adding the number of students taking French and the number of students taking chemistry and then subtracting the number of students taking both yields $(63 + 48) - 21 = 90$. This is the number of students enrolled in *either* French or chemistry or both. Since the total school enrollment is 150, there are $150 - 90 = 60$ students enrolled in *neither* French nor chemistry. The answer is (A).

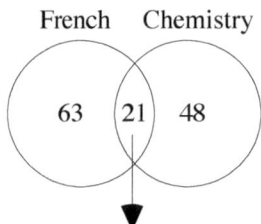

Both French and Chemistry

Test 5

GRE Math Tests

Questions: 24
Time: 45 minutes

---

1. Column A Column B
    ab^2 a^2b

USE THIS SPACE FOR SCRATCHWORK.

Multiple-choice Question – Select One or More Answer Choices]
2. Which of the following are divisible by both 2 and 3? [Note: If an integer is divisible by 3, then the sum of its digits is also divisible by 3.]

   (A) 1005
   (B) 1296
   (C) 1351
   (D) 1406
   (E) 1414
   (F) 3456

USE THIS SPACE FOR SCRATCHWORK.

---

3. Column A   The digits of a two-digit number X Column B
                                          differ by 4.
    The positive difference between   15
    the squares of the digits of X

USE THIS SPACE FOR SCRATCHWORK.

94

[Multiple-choice Question – Select One Answer Choice Only]
4.   What is the remainder when 3^7 is divided by 8?

   (A)   1
   (B)   2
   (C)   3
   (D)   5
   (E)   7

USE THIS SPACE FOR SCRATCHWORK.

5.   Column A $0 < x < 1$ Column B
             x^2 x

USE THIS SPACE FOR SCRATCHWORK.

[Multiple-choice Question – Select One Answer Choice Only]
6. From the figure, which of the following must be true?

(I) $x + y = 90$
(II) x is 35 units greater than y
(III) x is 35 units less than y

(A) I only
(B) II only
(C) III only
(D) I and II only
(E) I and III only

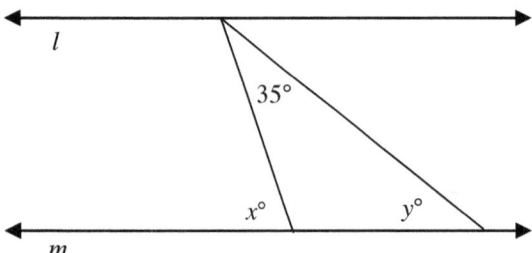

[Multiple-choice Question – Select One Answer Choice Only]
7. Which one of the following is true regarding the triangle shown in figure?

(A) $x > y > z$
(B) $x < y < z$
(C) $x = y = z$
(D) $2x = 3y/2 = z$
(E) $x/2 = 2y/3 = z$

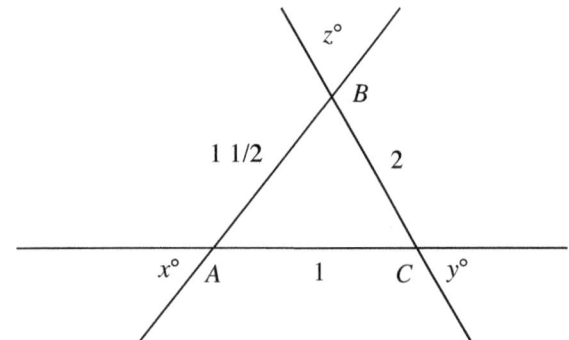

8. | Column A | $|a|$ is the distance point a is from the origin on the number line. $x \neq 0$. | Column B |
   |---|---|---|
   | $|x| + |-2|$ | | $|x - 2|$ |

USE THIS SPACE FOR SCRATCHWORK.

[Multiple-choice Question – Select One Answer Choice Only]

9. Let u represent the sum of the integers from 1 through 20, and let v represent the sum of the integers from 21 through 40. What is the value of $v - u$?

   (A) 21
   (B) 39
   (C) 200
   (D) 320
   (E) 400

USE THIS SPACE FOR SCRATCHWORK.

[Multiple-choice Question – Select One Answer Choice Only]

10. Richard leaves to visit his friend who lives 200 miles down Interstate 10. One hour later his friend Steve leaves to visit Richard via Interstate 10. If Richard drives at 60 mph and Steve drives at 40 mph, how many miles will Steve have driven when they cross paths?

    (A) 56
    (B) 58
    (C) 60
    (D) 65
    (E) 80

USE THIS SPACE FOR SCRATCHWORK.

[Multiple-choice Question – Select One Answer Choice Only]
11. How many different ways can 3 cubes be painted if each cube is painted one color and only the 3 colors red, blue, and green are available? (Order is not considered, for example, green, green, blue is considered the same as green, blue, green.)

    (A)   2
    (B)   3
    (C)   9
    (D)   10
    (E)   27

12.
Column A	Column B
2/3 − 3/4	3/4 − 4/5

[Multiple-choice Question – Select One Answer Choice Only]
13. If x is not equal to 1 and $y = \dfrac{1}{x-1}$, then which one of the following cannot be the value of y?

    (A)   0
    (B)   1
    (C)   2
    (D)   3
    (E)   4

Test 5—Questions

[Multiple-choice Question – Select One Answer Choice Only]
14. Which one of the following numbers is the greatest positive integer x such that 3^x is a factor of 27^5 ?

(A) 5
(B) 8
(C) 10
(D) 15
(E) 19

15. Column A

The arithmetic mean (average) of the numbers a and b is 17. The geometric mean of the numbers a and b is 8. The *geometric mean* of two numbers is defined to be the square root of their product.

Column B

a

b

[Multiple-choice Question – Select One Answer Choice Only]
16. If $x = 10a, y = 3b, z = 7c$, and $x : y : z = 10 : 3 : 7$, then $\dfrac{7x + 2y + 5z}{8a + b + 3c} =$

(A) 111/12
(B) 7/6
(C) 8/15
(D) 108/123
(E) 135/202

[Multiple-choice Question – Select One Answer Choice Only]

17. A perfect square is a positive integer that is the result of squaring a positive integer. If $N = 3^4 \cdot 5^3 \cdot 7$, then what is the biggest perfect square that is a factor of N?

    (A) 3^2
    (B) 5^2
    (C) 9^2
    (D) $(9 \cdot 5)^2$
    (E) $(3 \cdot 5 \cdot 7)^2$

[Multiple-choice Question – Select One Answer Choice Only]

18. If 50% of x equals the sum of y and 20, then what is the value of $x - 2y$?

    (A) 20
    (B) 40
    (C) 60
    (D) 80
    (E) 100

19. Column A   1 Pound = 16 Ounces   Column B

Weight of 16,000 ounces of rice   Weight of 1,000 pounds of coal

[Multiple-choice Question – Select One Answer Choice Only]
20. An old man distributed all the gold coins he had to his two sons into two different numbers such that the difference between the squares of the two numbers is 36 times the difference between the two numbers. How many coins did the old man have?

   (A) 24
   (B) 26
   (C) 30
   (D) 36
   (E) 40

21.   Column A   In a jar, 60% of the marbles are   Column B
                                red and the rest are green.

   40% of the red marbles in the   60% of the green marbles in the
   jar   jar

[Multiple-choice Question – Select One Answer Choice Only]
22. A prize of $200 is given to anyone who solves a hacker puzzle independently. The probability that Tom will win the prize is 0.6, and the probability that John will win the prize is 0.7. What is the probability that both will win the prize?

   (A) 0.35
   (B) 0.36
   (C) 0.42
   (D) 0.58
   (E) 0.88

GRE Math Tests

[Multiple-choice Question – Select One Answer Choice Only]
23. The buyer of a particular car must choose 2 of 3 optional colors and 3 of 4 optional luxury features. In how many different ways can the buyer select the colors and luxury features?

    (A)   3
    (B)   6
    (C)   9
    (D)   12
    (E)   20

[Multiple-choice Question – Select One Answer Choice Only]
24. In the two-digit number x, both the sum and the difference of its digits is 4. What is the value of x?

    (A)   13
    (B)   31
    (C)   40
    (D)   48
    (E)   59

Answers and Solutions Test 5:

Question	Answer
1.	D
2.	B, F
3.	A
4.	C
5.	B
6.	B
7.	A
8.	D
9.	E
10.	A
11.	D
12.	B
13.	A
14.	D
15.	D
16.	A
17.	D
18.	B
19.	C
20.	D
21.	C
22.	D
23.	D
24.	C

If you got 18/24 correct on this test, you are likely to get 750+ on the actual GRE **by the time you complete all the tests in the book.**

1. If $a = 0$, both columns equal zero. If $a = 1$ and $b = 2$, the two columns are unequal. This is a double case and the answer is (D).

2. A number divisible by 2 ends with one of the digits 0, 2, 4, 6, or 8.

If a number is divisible by 3, then the sum of its digits is also divisible by 3.

Hence, a number divisible by both 2 and 3 will follow both of the above rules.

Choices (A) and (C) do not end with an even digit. Hence, eliminate them.

The sum of digits of Choice (B) is $1 + 2 + 9 + 6 = 18$, which is divisible by 3. Also, the last digit is 6. Hence, choice (B) is correct.

This is also true with choice (F). The sum of digits of Choice (F) is $3 + 4 + 5 + 6 = 18$, which is divisible by 3. Also, the last digit is 6. Hence, choice (F) is correct.

Next, the sum of the digits of choices (D) and (E) are $1 + 4 + 0 + 6$ (= 11) and $1 + 4 + 1 + 4$ (= 10), respectively, and neither is divisible by 3. Hence, reject the two choices.

Hence, the answer is (B) and (F).

3. Since the digits of the number differ by 4, the number can be any one of the numbers 15, 26, 37, 40, 48, and 59 or any one of their reverse numbers, except 40. Hence, the positive difference between the digits varies from $4^2 - 0^2 = 16$ (for the number 40) to $9^2 - 5^2 = 56$ (for the numbers 59 or 95). Hence, Column A is a number between 16 and 56, inclusive, and therefore is always greater than 15, the value of Column B. Hence, the answer is (A).

4. We know that $3^7 = 3 \cdot 3^3 \cdot 3^3 = 3 \cdot 27 \cdot 27 = 3(27^2)$. The number immediately before 27 that is divisible by 8 is 24. Hence, replace 27 with $24 + 3$. Then we have

$$3^7 = 3(27^2) = 3(24 + 3)^2 = 3(24^2 + 2 \cdot 24 \cdot 3 + 3^2)$$
$$= 3 \cdot 24^2 + 3 \cdot 2 \cdot 24 \cdot 3 + 3 \cdot 9$$

Now,

$$\frac{3^7}{8} = \frac{3 \cdot 24^2 + 3 \cdot 2 \cdot 24 \cdot 3 + 3 \cdot 9}{8}$$
$$= \frac{3 \cdot 24^2}{8} + \frac{3 \cdot 2 \cdot 24 \cdot 3}{8} + \frac{3 \cdot 9}{8}$$
$$= \text{Integer} + \text{Integer} + \frac{27}{8}$$
$$= \text{Integer} + \text{Integer} + \frac{24 + 3}{8}$$
$$= \text{Integer} + \text{Integer} + 3 + \frac{3}{8}$$

Hence, the remainder is 3, and the answer is (C).

5. Since $0 < x < 1$, we know that x is positive. Now, multiplying both sides of the inequality $0 < x < 1$ by x yields

$$0 < x^2 < x$$

Hence, Column B is greater than Column A, and the answer is (B).

6.

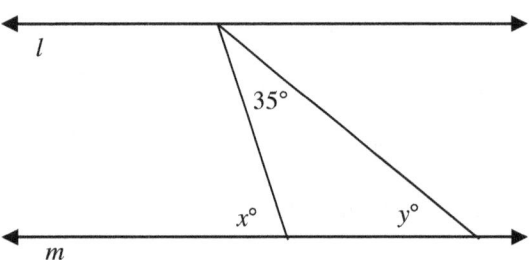

Angle x is an exterior angle of the triangle and therefore equals the sum of the remote interior angles, 35 and y. That is, $x = y + 35$. This equation says that x is 35 units greater than y. So, (II) is true and (III) is false. Now, if x is an obtuse angle ($x > 90$), then $x + y$ is greater than 90. Hence, $x + y$ need not equal 90. So, (I) is not necessarily true. The answer is (B).

7. From the figure, we have the following inequality between the sides of the triangle: BC (= 2) > AB (= 1 1/2) > AC (= 1). In a triangle, the longer the side, the bigger the angle opposite it. Hence, we have the following inequality between the angles of the triangle: $\angle A > \angle C > \angle B$. Replacing the angles in the inequality with their respective vertical angles in the figure yields $x > y > z$. Select Choice (A).

Test 5—Solutions

8. Suppose $x = -2$. Then Column A equals $|x| + |-2| = |-2| + |-2| = 2 + 2 = 4$, and Column B equals $|x - 2| = |-2 - 2| = |-4| = 4$. Here, Column A equals Column B.

Now, suppose $x = 2$. Then Column A equals $|x| + |-2| = |2| + |-2| = 2 + 2 = 4$, and Column B equals $|x - 2| = |2 - 2| = |0| = 0$. Here, Column A is greater than Column B.

This is a double case, and therefore the answer is (D).

---

9. Forming the series for u and v yields

$$u = 1 + 2 + \cdots + 19 + 20$$
$$v = 21 + 22 + \cdots + 39 + 40$$

Subtracting the series for u from the series for v yields

$$v - u = \underbrace{20 + 20 + \cdots + 20 + 20}_{20 \text{ times}} = 20 \cdot 20 = 400$$

The answer is (E).

---

10. Let t be time that Steve has been driving. Then $t + 1$ is time that Richard has been driving. Now, the distance traveled by Steve is $D = rt = 40t$, and Richard's distance is $60(t + 1)$. At the moment they cross paths, they will have traveled a combined distance of 200 miles. Hence,

$$40t + 60(t + 1) = 200$$
$$40t + 60t + 60 = 200$$
$$100t + 60 = 200$$
$$100t = 140$$
$$t = 1.4$$

Therefore, Steve will have traveled $D = rt = 40(1.4) = 56$ miles. The answer is (A).

---

11. Clearly, there are more than 3 color combinations possible. This eliminates (A) and (B). We can also eliminate (C) and (E) because they are both multiples of 3, and that would be too ordinary, too easy, to be the answer. Hence, by process of elimination, the answer is (D).

Let's also solve this problem directly. The following list displays all 27 (= 3 · 3 · 3) color combinations possible (without restriction):

RRR	BBB	GGG
RRB	BBR	GGR
RRG	BBG	GGB
RBR	BRB	GRG
RBB	BRR	GRR
RBG	BRG	GRB
RGR	BGB	GBG
RGB	BGR	GBR
RGG	BGG	GBB

If order is not considered, then there are 10 distinct color combinations in this list. You should count them.

GRE Math Tests

12. The least common multiple of the denominators 3, 4, and 5 of all the fractions is 60. Multiplying both columns by 60 to clear the fractions yields

$$40 - 45 \qquad\qquad 45 - 48$$

Subtracting the numbers yields

$$-5 \qquad\qquad -3$$

Since $-5 < -3$, Column B is greater than Column A and the answer is (B).

---

13. Since the numerator of the fraction $\dfrac{1}{x-1}$ does not contain a variable, it can never equal 0. Hence, the fraction can never equal 0. The answer is (A).

---

14. $27^5 = \left(3^3\right)^5 = 3^{15}$. Hence, $x = 15$ and the answer is (D).

---

15. The arithmetic mean of the numbers a and b is 17. Hence, $\dfrac{a+b}{2} = 17$, or $a + b = 34$.

The geometric mean of the numbers a and b is 8. Hence, $\sqrt{ab} = 8$, or $ab = 8^2 = 64$.

Solving the equation $a + b = 34$ for a yields $a = 34 - b$. Substituting this into the equation $ab = 64$ yields

$$(34 - b)b = 64$$
$$34b - b^2 = 64$$
$$34b - b^2 - 64 = 0$$
$$b^2 - 34b + 64 = 0$$
$$(b - 32)(b - 2) = 0$$
$$b - 32 = 0 \quad \text{or} \quad b - 2 = 0$$
$$b = 32 \quad \text{or} \quad b = 2$$

Now, if $b = 32$, then $a = 34 - b = 34 - 32 = 2$. In this case, Column B is larger.

Now, if $b = 2$, then $a = 34 - b = 34 - 2 = 32$. In this case, Column A is larger.

Hence, we have a double case, and the answer is (D).

16. We are given the equations $x = 10a$, $y = 3b$, $z = 7c$ and the proportion $x : y : z = 10 : 3 : 7$.

Substituting the first three equations into the last equation (ratio equation) yields $10a : 3b : 7c = 10 : 3 : 7$. Forming the resultant ratio yields $10a/10 = 3b/3 = 7c/7$. Simplifying the equation yields $a = b = c$. Hence, both a and b equal c. Substituting the result in the given equations $x = 10a$, $y = 3b$, $z = 7c$ yields $x = 10a = 10c$, $y = 3b = 3c$, and $z = 7c$. Now, we have

$$\frac{7x + 2y + 5z}{8a + b + 3c} =$$

$$\frac{7 \cdot 10c + 2 \cdot 3c + 5 \cdot 7c}{8c + c + 3c} = \quad \text{because } x = 10c, y = 3c, z = 7c, \text{ and } a = b = c$$

$$\frac{111c}{12c} =$$

$$\frac{111}{12}$$

The answer is (A).

---

17. Every positive integer can be uniquely factored into powers of primes. When the integer is squared, all powers of these primes are doubled. Hence, a perfect square has only even powers of the primes in its factorization, and clearly any positive integer whose prime factorization has only even powers of primes is a perfect square. Any factor of N is the product of some or all of the primes contained in $3^4\ 5^3\ 7$; the largest such product containing only even powers of is $3^4\ 5^2 = (9\ 5)^2$. The answer is (D).

---

18. 50% of x equals the sum of y and 20. Expressing this as an equation yields

$(50/100)x = y + 20$
$x/2 = y + 20$
$x = 2y + 40$
$x - 2y = 40$

The answer is (B).

---

19. Column B has 1,000 pounds of coal, and there are 16 ounces in 1 pound. So, Column B has 1,000 pounds = 1,000(16 ounces) = 16,000 ounces. Hence, each column weighs 16,000 ounces. The answer is (C).

---

20. Let x and y be the numbers of gold coins the two sons received. Since we are given that the difference between the squares of the two numbers is 36 times the difference between the two numbers, we have the equation

$x^2 - y^2 = 36(x - y)$
$(x - y)(x + y) = 36(x - y)$ by the Difference of Squares formula $a^2 - b^2 = (a - b)(a + b)$
$x + y = 36$ by canceling $(x - y)$ from both sides

Hence, the total number of gold coins the old man had, namely $x + y$, equals 36. The answer is (D).

21. Let j be the total number of marbles in the jar. Then $60\%j$ must be red (given), and the remaining $40\%j$ must be green (given). Now,

Column A equals 40% of the red marbles = $40\%(60\%j) = .40(.60j) = .24j$.

Column B equals 60% of the green marbles = $60\%(40\%j) = .60(.40j) = .24j$.

Since both columns equal $.24j$, the answer is (C).

---

22. The number of attendees at the meeting is 750 of which 450 are female. Hence, the number of male attendees is $750 - 450 = 300$. Half of the female attendees are less than 30 years old. One half of 450 is $450/2 = 225$. Also, one-fourth of the male attendees are less than 30 years old. One-fourth of 300 is $300/4 = 75$.

Now, the total number of (male and female) attendees who are less than 30 years old is $225 + 75 = 300$.

So, out of the total 750 attendees 300 attendees are less than 30 years old. Hence, the probability of randomly selecting an attendee less than 30 years old (equals the fraction of all the attendees who are less than 30 years old) is $300/750 = 2/5$. The answer is (D).

---

23. Let A, B, C stand for the three colors, and let W, X, Y, Z stand for the four luxury features. There are three ways of selecting the colors:

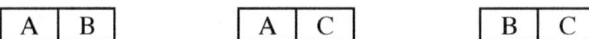

There are four ways of selecting the luxury features:

| W | X | Y | | W | Y | Z | | W | X | Z | | X | Y | Z |

Hence, there are $3 \times 4 = 12$ ways of selecting all the features. The answer is (D).

---

24. Since the sum of the digits is 4, x must be 13, 22, 31, or 40. Further, since the difference of the digits is 4, x must be 40, 51, 15, 62, 26, 73, 37, 84, 48, 95, or 59. We see that 40 and only 40 is common to the two sets of choices for x. Hence, x must be 40. The answer is (C).

Test 6

GRE Math Tests

Questions: 24
Time: 45 minutes

[Multiple-choice Question – Select One or More Answer Choices]

1. If *p* and *q* are both positive integers such that $p/9 + q/10$ is also an integer, then which of the following numbers could *p* equal?

    (A) 3
    (B) 9
    (C) 15
    (D) 27
    (E) 35

[Multiple-choice Question – Select One or More Answer Choices]

2. If *a*, *b*, and *c* are positive prime integers such that $(a + 1)(b + 2)(c + 3)$ equals the product of exactly three prime numbers, then which of the following must be equal?

    (A) *a*
    (B) *b*
    (C) *a* + 1
    (D) *b* + 1
    (E) *c* + 1

3. | Column A | | Column B |
    |---|---|---|
    | | *x* is a two-digit number. The digits of the number differ by 6, and the squares of the digits differ by 60. | |
    | *x* | | 60 |

110

4.

Column A	The average of a set of six positive numbers is 30.	Column B
The average of the numbers in the set after replacing the smallest number in the set with 0		The average of the remaining numbers in the set after removing the smallest number from the set

USE THIS SPACE FOR SCRATCHWORK.

5.

Column A	q is an integer greater than 1. Let $<q>$ stand for the smallest positive integer factor of q that is greater than 1.	Column B
$<q>$		$<q^3>$

USE THIS SPACE FOR SCRATCHWORK.

[Numeric Entry Question]
6. In the figure, what is the value of a?

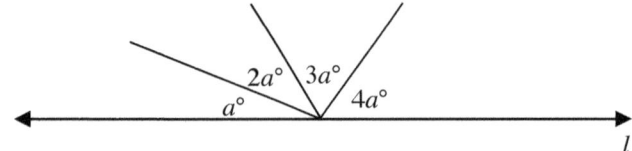

[Multiple-choice Question – Select One Answer Choice Only]
7. What is the area of the equilateral triangle if the base $BC = 6$?

(A) $9\sqrt{3}$
(B) $18\sqrt{3}$
(C) $26\sqrt{3}$
(D) $30\sqrt{3}$
(E) $36\sqrt{3}$

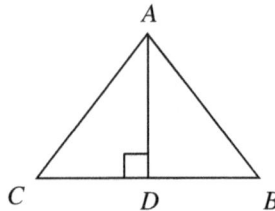

[Multiple-choice Question – Select One Answer Choice Only]
8. What is the perimeter of △ABC shown in the figure?

(A) $2 + 4\sqrt{2}$
(B) $4 + 2\sqrt{2}$
(C) 8
(D) $4 + 4\sqrt{2}$
(E) $4 + 4\sqrt{3}$

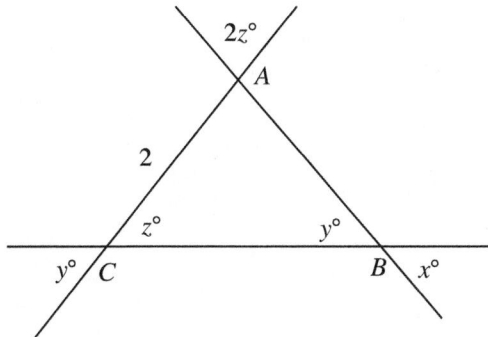

[Multiple-choice Question – Select One Answer Choice Only]
9. A closed rectangular tank contains a certain amount of water. When the tank is placed on its 3 ft. by 4 ft. side, the height of the water in the tank is 5 ft. When the tank is placed on another side of dimensions 4 ft. by 5 ft. what is the height, in feet, of the surface of the water above the ground?

(A) 2
(B) 3
(C) 4
(D) 5
(E) 6

[Multiple-choice Question – Select One Answer Choice Only]

10. In the figure, O is the center of the circle. If the area of the circle is 9π, then the perimeter of the sector PRQO is

    (A) $\pi/2 - 6$
    (B) $\pi/2 + 6$
    (C) $3\pi/4 + 6$
    (D) $\pi/2 + 18$
    (E) $3\pi/4 + 18$

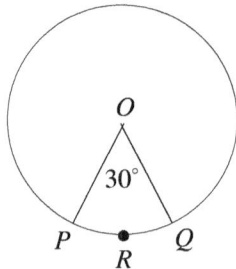

11.   Column A $x/15 > y/25$ Column B
      $6y + 5x$ $10x + 3y$

[Multiple-choice Question – Select One Answer Choice Only]

12. If it is true that $1/55 < x < 1/22$ and $1/33 < x < 1/11$, then which of the following numbers could x equal?

    I. $1/54$
    II. $1/23$
    III. $1/12$

    (A) I only
    (B) II only
    (C) III only
    (D) I and II only
    (E) II and III only

[Multiple-choice Question – Select One Answer Choice Only]
13. There are 87 balls in a jar. Each ball is painted with at least one of two colors, red or green. It is observed that 2/7 of the balls that have red color also have green color, while 3/7 of the balls that have green color also have red color. What fraction of the balls in the jar have both red and green colors?

   (A) 6/14
   (B) 2/7
   (C) 6/35
   (D) 6/29
   (E) 6/42

14. Column A $x + y = 1$ Column B
     x y

[Multiple-choice Question – Select One or More Answer Choices]
15. If $(x+5) \div \left(\dfrac{1}{x} + \dfrac{1}{5}\right) = 5$, then $x =$

   (A) −5
   (B) 1/2
   (C) 1
   (D) 5
   (E) 10

GRE Math Tests

[Multiple-choice Question – Select One Answer Choice Only]
16. A group of 30 employees of Cadre A has a mean age of 27. A different group of 70 employees of Cadre B has a mean age of 23. What is the mean age of the employees of the two groups together?

   (A) 23
   (B) 24.2
   (C) 25
   (D) 26.8
   (E) 27

[Multiple-choice Question – Select One or More Answer Choices]
17. In a set of three variables, the average of the first two variables is greater than 2, the average of the last two variables is greater than or equal to 3, and the average of the first and the last variables is 4. Which of the following could the average of the three numbers be?

   (A) 1
   (B) 2
   (C) 3
   (D) 4
   (E) 5

[Multiple-choice Question – Select One Answer Choice Only]
18. Joseph bought two varieties of rice, costing 5 cents per ounce and 6 cents per ounce each, and mixed them in some ratio. Then he sold the mixture at 7 cents per ounce, making a profit of 20 percent. What was the ratio of the mixture?

   (A) 1 : 10
   (B) 1 : 5
   (C) 2 : 7
   (D) 3 : 8
   (E) 5 : 7

[Multiple-choice Question – Select One or More Answer Choices]
19. If $(x-2)^2 = 1$, then $(x-2)(x-4) =$

    (A) −3
    (B) −1
    (C) 3
    (D) 8
    (E) 15

[Multiple-choice Question – Select One Answer Choice Only]
20. If x and y are both prime and greater than 2, then which one of the following CANNOT be a divisor of xy?

    (A) 2
    (B) 3
    (C) 11
    (D) 15
    (E) 17

21.	Column A	Mr. Chang sold 100 oranges to customers for $300 and earned a profit.	Column B
	The percentage of the profit that Chang got		The selling price of the oranges expressed as a percentage of the cost

[Multiple-choice Question – Select One Answer Choice Only]

22. Sarah cannot completely remember her four-digit ATM pin number. She does remember the first two digits, and she knows that each of the last two digits is greater than 5. The ATM will allow her three tries before it blocks further access. If she randomly guesses the last two digits, what is the probability that she will get access to her account?

    (A) 1/2
    (B) 1/4
    (C) 3/16
    (D) 3/18
    (E) 1/32

[Multiple-choice Question – Select One Answer Choice Only]

23. To halve the value of the expression $\dfrac{v+w}{x/yz}$ by doubling exactly one of the variables, one must double which one of the following variables?

    (A) v
    (B) w
    (C) x
    (D) y
    (E) z

[Multiple-choice Question – Select One or More Answer Choices]

24. The minimum temperatures from Monday through Sunday in the first week of July in southern Iceland are observed to be –2°C, 4°C, 4°C, 5°C, 7°C, 9°C, 10°C. What is the range of the temperatures?

    (A) –10°C
    (B) –8°C
    (C) 8°C
    (D) 10°C
    (E) 12°C

Answers and Solutions Test 6:

Question	Answer
1.	B, D
2.	B, C, E
3.	D
4.	B
5.	C
6.	18
7.	A
8.	B
9.	B
10.	A
11.	B
12.	B
13.	D
14.	D
15.	C
16.	B
17.	D, E
18.	B
19.	B, C
20.	A
21.	B
22.	C
23.	C
24.	E

1. If p is not divisible by 9 and q is not divisible by 10, then $p/9$ results in a non-terminating decimal and $q/10$ results in a terminating decimal and the sum of the two would not result in an integer. [Because (a terminating decimal) + (a non-terminating decimal) is always a non-terminating decimal, and a non-terminating decimal is not an integer.]

Since we are given that the expression is an integer, p must be divisible by 9.

For example, if $p = 1$ and $q = 10$, the expression equals $1/9 + 10/10 = 1.11...$, not an integer.

If $p = 9$ and $q = 5$, the expression equals $9/9 + 5/10 = 1.5$, not an integer.

If $p = 9$ and $q = 10$, the expression equals $9/9 + 10/10 = 2$, an integer.

In short, p must be a positive integer divisible by 9. Choices (B) and (D) are both divisible by 9. They must be chosen.

---

2. The prime numbers are 2, 3, 5, 7, 11, 13, 17, 19, 23, ...
From the list, both a and $a + 1$ are prime only when a is 2.
From the list, both b and $b + 2$ are prime when b is 3 or 5 or 11 or so on.
From the list, both c and $c + 3$ are prime when only c is 2.
Choice (A): $a = 2$, and Choice (B) is 3 or 5 or 11 ... Choice (C) 3, Choice (D) 4, or 6 or 12, ...Choice (E) $c + 1 = 3$.
From the analysis above, choices (B), (C), and (E) could be equal, the rest cannot. Select (B), (C), and (E).

3. Suppose a and b are the two digits of the number x, and let a represent the greater of the two. From the given information, we have

$$a - b = 6 \qquad (1)$$
$$a^2 - b^2 = 60$$

Applying the Difference of Squares formula, $a^2 - b^2 = (a - b)(a + b)$, to the second equation yields

$$(a - b)(a + b) = 60$$
$$6(a + b) = 60 \qquad \text{since } a - b = 6$$
$$a + b = 60/6 = 10 \qquad (2)$$

Adding equations (1) and (2) yields $2a = 16$. Dividing by 2 yields $a = 8$. Substituting this in equation (2) yields $8 + b = 10$. Solving for b yields $b = 2$. Hence, the two digits are 2 and 8. They can be arranged in any order. Hence, 28 and 82 are two feasible solutions. Now, 28 is less than Column B, and 82 is greater than Column B. Hence, we have a double case, and the answer is (D).

4. Let a, b, c, d, e, and f be the numbers in the set, and let f be the smallest number in the set.

When the smallest number (f) in the set is replaced by 0, the numbers in the set are a, b, c, d, e, and 0. Column A equals the average of these six numbers, which equals $\dfrac{a+b+c+d+e+0}{6} = \dfrac{a+b+c+d+e}{6}$.

Instead, if the smallest number in the set is removed, the remaining numbers in the set would be $a, b, c, d,$ and e. Now there are only 5 numbers in the set. Hence, Column B, which equals the average of the remaining numbers (five numbers) in the set, equals $\dfrac{a+b+c+d+e}{5}$.

Since all the numbers in the set are positive (given), the sum of the five numbers $a + b + c + d + e$ is also positive. Note that dividing a positive number by 5 yields a greater result than dividing it by 6. Hence, $\dfrac{a+b+c+d+e}{5}$ is greater than $\dfrac{a+b+c+d+e}{6}$. Thus, Column B is greater than Column A, and the answer is (B).

5. The eye-catcher is Column A since we are looking for the smallest factor and q is smaller than q^3. Let's use substitution to solve this problem. Since $q > 1$, we need to look at only 2, 3, and 4 (see Substitution Special Cases). If $q = 2$, then $<q> = <2> = 2$ and $<q^3> = <2^3> = <8> = 2$. In this case, the two columns are equal. If $q = 3$, then $<q> = 3$ and $<q^3> = 3$. In this case, the two columns are again equal. If $q = 4$, then $<q> = 2$ and $<q^3> = 2$. Once again, the two columns are equal. Hence, the answer is (C).

6. The angle made by a line is 180°. Hence, from the figure, we have

$$a + 2a + 3a + 4a = 180$$
$$10a = 180$$
$$a = 18$$

Enter in the grid.

Test 6—Solutions

7. Since the side BC of the equilateral triangle measures 6 (given), the other side AC also measures 6. Since the altitude AD bisects the base (this is true in all equilateral or isosceles triangles), $CD = BD = (1/2)BC = 1/2 \times 6 = 3$. Applying The Pythagorean Theorem to $\triangle ADC$ yields $AD^2 = AC^2 - CD^2 = 6^2 - 3^2 = 36 - 9 = 27$, or $AD = 3\sqrt{3}$. Hence, the area of the triangle is $1/2 \times base \times height = 1/2 \times BC \times AD = 1/2 \times 6 \times 3\sqrt{3} = 9\sqrt{3}$. The answer is (A).

8. Equating the vertical angles at points A and C in the figure yields $\angle A = 2z$ and $y = z$. Summing the angles of the triangle to 180° yields

$\angle A + \angle B + \angle C = 180$
$2z + y + z = 180$ we know that $\angle A = 2z$, $\angle B = y$, and $\angle C = z$
$2z + z + z = 180$ we know that $y = z$
$4z = 180$
$z = 180/4 = 45$

So, $\angle A = 2z = 2(45) = 90$, $\angle B = y = z = 45$ and $\angle C = z = 45$. Hence, $\triangle ABC$ is a right triangle. Also, since angles $\angle C$ and $\angle B$ are equal (equal to 45), the sides opposite these two angles, AB and AC, must be equal. Since AC equals 2 (from the figure), AB also equals 2. Now, applying The Pythagorean Theorem to the triangle yields

$BC^2 = AB^2 + AC^2$
$= 2^2 + 2^2$ given that $AB = AC = 2$
$= 4 + 4$
$= 8$
$BC = 2\sqrt{2}$ square rooting both sides

Now, the perimeter of $\triangle ABC = AB + BC + CA = 2 + 2\sqrt{2} + 2 = 4 + 2\sqrt{2}$. The answer is (B).

9. When based on the 3 ft · 4 ft side, the height of water inside the rectangular tank is 5 ft. Hence, the volume of the water inside tank is $length \times width \times height = 3 \cdot 4 \cdot 5$ cu. ft.

When based on 4 ft · 5 ft side, let the height of water inside the rectangular tank be h ft. Then the volume of the water inside tank would be $length \times width \times height = 4 \cdot 5 \cdot h$ cu. ft.

Equating the results for the volume of water, we have $3 \cdot 4 \cdot 5 = v = 4 \cdot 5 \cdot h$. Solving for h yields $h = (3 \cdot 4 \cdot 5)/(4 \cdot 5) = 3$ ft.

The answer is (B).

10. Since x is the radius of the larger circle, the area of the larger circle is πx^2. Since x is the diameter of the smaller circle, the radius of the smaller circle is $x/2$. Therefore, the area of the smaller circle is $\pi\left(\dfrac{x}{2}\right)^2 = \pi\dfrac{x^2}{4}$. Subtracting the area of the smaller circle from the area of the larger circle gives $\pi x^2 - \pi\dfrac{x^2}{4} = \dfrac{4}{4}\pi x^2 - \pi\dfrac{x^2}{4} = \dfrac{4\pi x^2 - \pi x^2}{4} = \dfrac{3\pi x^2}{4}$. The answer is (A).

11. Multiplying the given inequality $x/15 > y/25$ by 75 yields $5x > 3y$.

Now, subtracting $3y$ and $5x$ from both columns yields

Column A	$5x > 3y$	Column B
$3y$		$5x$

Since we know that $5x > 3y$, Column B is greater than Column A and the answer is (B).

---

12. Combining the two given inequalities $1/55 < x < 1/22$ and $1/33 < x < 1/11$ yields $1/33 < x < 1/22$. Since among the three positive denominators 12, 23, and 54, the number 23 is the only one in the positive range between 22 and 33, only the number $1/23$ lies between $1/33$ and $1/22$, and the numbers $1/54$ and $1/12$ do not. Hence, $x = 1/23$. The answer is (B), II only.

---

13. Let T be the total number of balls, R the number of balls having red color, G the number having green color, and B the number having both colors.

So, the number of balls having only red is $R - B$, the number having only green is $G - B$, and the number having both is B. Now, the total number of balls is $T = (R - B) + (G - B) + B = R + G - B$.

We are given that 2/7 of the balls having red color have green also. This implies that $B = 2R/7$. Also, we are given that 3/7 of the green balls have red color. This implies that $B = 3G/7$. Solving for R and G in these two equations yields $R = 7B/2$ and $G = 7B/3$. Substituting this into the equation $T = R + G - B$ yields $T = 7B/2 + 7B/3 - B$. Solving for B yields $B = 6T/29$. Hence, 6/29 of all the balls in the jar have both colors. The answer is (D). Note that we did not use the information: "There are 87 balls." Sometimes, not all information in a problem is needed.

---

14. If $x = y = 1/2$, then $x + y = 1/2 + 1/2 = 1$ and the columns are equal.

But if $x = 1$ and $y = 0$, then $x + y = 1 + 0 = 1$ and Column A is larger than Column B.

Hence, we have a double case, and the answer is (D).

---

15. We are given the equation

$$(x+5) \div \left(\frac{1}{x} + \frac{1}{5}\right) = 5$$

$$(x+5) \div \left(\frac{x+5}{5x}\right) = 5$$

$$(x+5) \cdot \left(\frac{5x}{x+5}\right) = 5$$

$$5x = 5$$

$$x = 1$$

The answer is (C). Note: If you solved the equation without getting a common denominator, you may have gotten -5 as a possible solution. But, -5 is not a solution. Why? *

---

* Because -5 is not in the domain of the original equation since it causes the denominator to be 0. When you solve an equation, you are only finding possible solutions. The "solutions" may not work when plugged back into the equation.

Test 6—Solutions

16. Cadre A has 30 employees whose mean age is 27. Hence, the sum of their ages is 30 × 27 = 810. Cadre B has 70 employees whose mean age is 23. Hence, the sum of their ages is 23 × 70 = 1610. Now, the total sum of the ages of the 100 (= 30 + 70) employees is 810 + 1610 = 2420. Hence, the average age is

The sum of the ages divided by the number of employees =

2420/100 =

24.2

The answer is (B).

17. Let a, b, and c be the variables. Given

$(a + b)/2 > 2$
$(b + c)/2 \geq 3$
$(a + c)/2 = 4$

Adding the top two inequalities and the last equation yields

$(a + b)/2 + (b + c)/2 + (a + c)/2 > 2 + 3 + 4$
$(a + b)/2 + (b + c)/2 + (a + c)/2 > 9$
$(a + b) + (b + c) + (a + c) > 18$
$a + b + b + c + a + c > 18$
$2a + 2b + 2c > 18$
$a + b + c > 9$
$(a + b + c)/3 > 9/3 = 3$
Average > 3

Choose (D) and (E).

18. Q28. Let $1 : k$ be the ratio in which Joseph mixed the two types of rice. Then a sample of $(1 + k)$ ounces of the mixture should equal 1 ounce of rice of the first type, and k ounces of rice of the second type. The rice of the first type costs 5 cents an ounce and that of the second type costs 6 cents an ounce. Hence, it cost him

(1 ounce × 5 cents per ounce) + (k ounces × 6 cents per ounce) = $5 + 6k$

Since he sold the mixture at 7 cents per ounce, he must have sold the net $1 + k$ ounces of the mixture at $7(1 + k)$.

Since he earned 20% profit doing this, $7(1 + k)$ must be 20% more than $5 + 6k$. Hence, we have the equation

$7(1 + k) = (1 + 20/100)(5 + 6k)$
$7 + 7k = (120/100)(5 + 6k)$
$7 + 7k = (6/5)(5 + 6k)$
$7 + 7k = 6/5 \cdot 5 + 6/5 \cdot 6k$
$7 + 7k = 6 + 36k/5$
$1 = k/5$
$k = 5$

Hence, the required ratio is $1 : k = 1 : 5$. The answer is (B).

19. Square rooting both sides of the equation $(x-2)^2 = 1$ yields $x - 2 = \pm 1$.

If $x - 2 = 1$, then $x = 2 + 1 = 3$.
If $x - 2 = -1$, then $x = 2 - 1 = 1$.

If $x = 3$, then $(x - 2)(x - 4) = (3 - 2)(3 - 4) = (1)(-1) = -1$. Choose (B).
If $x = 1$, then $(x - 2)(x - 4) = (1 - 2)(1 - 4) = (-1)(-3) = 3$. Choose (C).

The answer is (B) and (C).

---

20. Since x and y are prime and greater than 2, xy is the product of two odd numbers and is therefore odd. Hence, 2 cannot be a divisor of xy. The answer is (A).

---

21. We are given that $a/2$ is $b\%$ of 30. Now, $b\%$ of 30 is $\frac{30}{100}b$. Hence, $\frac{a}{2} = \frac{30}{100}b$. Solving for a yields $a = \frac{3}{5}b$. We are also given that a is $c\%$ of 50. Now, $c\%$ of 50 is $\frac{c}{100} \cdot 50 = \frac{c}{2}$. Hence, $a = c/2$. Plugging this into the equation $a = \frac{3}{5}b$ yields $\frac{c}{2} = \frac{3}{5}b$. Multiplying both sides by 2 yields $c = \frac{6}{5}b$. Since b is positive, c is also positive; and since $6/5 > 1$, $c > b$. Hence, the answer is (B).

---

22. Randomly guessing either of the last two digits does not affect the choice of the other, which means that these events are independent and we are dealing with consecutive probabilities. Since each of the last two digits is greater than 5, Sarah has four digits to choose from: 6, 7, 8, 9. Her chance of guessing correctly on the first choice is 1/4, and on the second choice also 1/4. Her chance of guessing correctly on both choices is

$$\frac{1}{4} \cdot \frac{1}{4} = \frac{1}{16}$$

Since she gets three tries, the total probability is $\frac{1}{16} + \frac{1}{16} + \frac{1}{16} = \frac{3}{16}$. The answer is (C).

---

23. Doubling the x in the expression yields $\frac{v+w}{2x/yz} = \frac{1}{2}\left(\frac{v+w}{x/yz}\right)$. Since we have written the expression as 1/2 times the original expression, doubling the x halved the original expression. The answer is (C).

---

24. The *range* is the greatest measurement minus the smallest measurement. The greatest of the seven temperature measurements is 10°C, and the smallest is –2°C. Hence, the required range is $10 - (-2) = 12$°C. The answer is (E).

Test 7

GRE Math Tests

Questions: 24
Time: 45 minutes

---

1. Column A $a < 0$ Column B
    $1/a$ a

---

2. Column A $a * b = \dfrac{a}{b} - \dfrac{b}{a},$ Column B
                           $m > n > 0$

    $\dfrac{1}{m} * \dfrac{1}{n}$ $\dfrac{1}{n} * \dfrac{1}{m}$

---

3. Column A   Column B
    Least common multiple of the mn
    two positive integers m and n

[Multiple-choice Question – Select One Answer Choice Only]
4. A set has exactly five consecutive positive integers starting with 1. What is the percentage decrease in the average of the numbers when the greatest one of the numbers is removed from the set?

    (A)   5
    (B)   8.5
    (C)   12.5
    (D)   15.2
    (E)   16.66

5. | Column A | $2x + 1 > 3x + 2$ and $5x + 2 > 4x$ | Column B |
    |----------|------------------------------------|----------|
    | x | | 1 |

6. | Column A | In $\triangle ABC$, $\angle B = 72°$ | Column B |
    |----------|--------------------------------------|----------|
    | $\angle A$ | | $72°$ |

[Multiple-choice Question – Select One Answer Choice Only]

7. $A, B,$ and C are three unequal faces of a rectangular tank. The tank contains a certain amount of water. When the tank is based on the face A, the height of the water is half the height of the tank. The dimensions of the side B are 3 ft × 4 ft and the dimensions of side C are 4 ft × 5 ft. What is the measure of the height of the water in the tank in feet?

   (A) 2
   (B) 2.5
   (C) 3
   (D) 4
   (E) 5

[Multiple-choice Question – Select One Answer Choice Only]
8. In the figure, what is the value of *x* ?

   (A) 10°
   (B) 30°
   (C) 45°
   (D) 60°
   (E) 75°

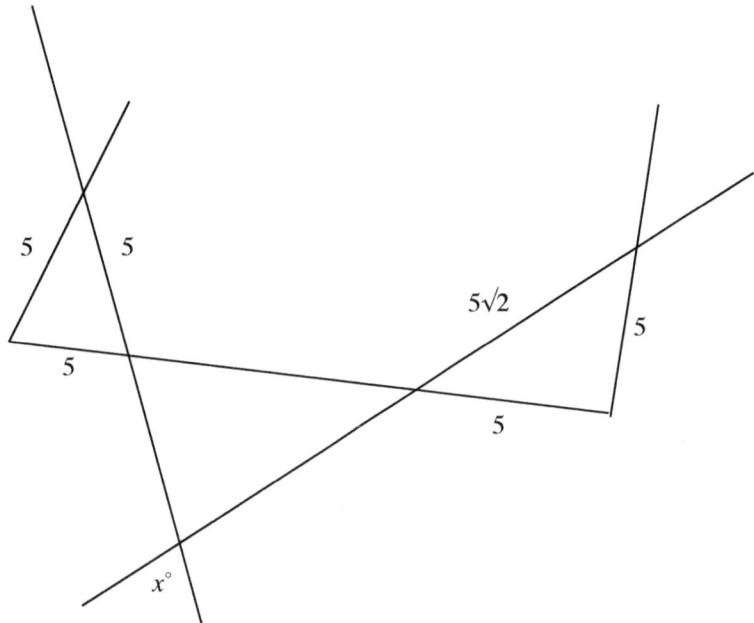

[Multiple-choice Question – Select One Answer Choice Only]
9. Which of the following indicates that $\triangle ABC$ is right angled?

   (I)   The angles of $\triangle ABC$ are in the ratio 1 : 2 : 3.
   (II) One of the angles of $\triangle ABC$ equals the sum of the other two angles.
   (III) $\triangle ABC$ is similar to the right triangle $\triangle DEF$.

   (A) I only
   (B) II only
   (C) III only
   (D) I and II only
   (E) I, II, and III

[Multiple-choice Question – Select One Answer Choice Only]
10. In the figure, the point $A(m, n)$ lies in Quadrant II as shown. In which region is the point $B(n, m)$?

    (A) Quadrant I
    (B) Quadrant II
    (C) Quadrant II
    (D) Quadrant IV
    (E) On the x-axis

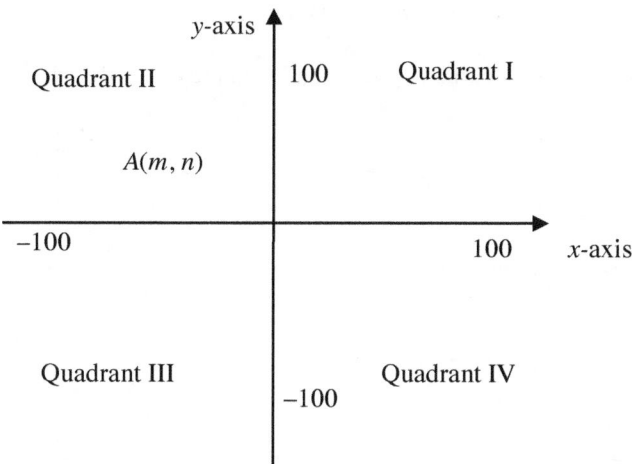

[Multiple-choice Question – Select One Answer Choice Only]
11. If $x^5 + x^2 < 0$, then which one of the following must be true?

    (A) $x < -1$
    (B) $x < 0$
    (C) $x > 0$
    (D) $x > 1$
    (E) $x^4 < x^2$

GRE Math Tests

[Multiple-choice Question – Select One Answer Choice Only]
12. Three workers $A, B,$ and C are hired for 4 days. The daily wages of the three workers are as follows:

A's first day wage is $4.
Each day, his wage increases by 2 dollars.

B's first day wage is $3.
Each day, his wage increases by 2 dollars.

C's first day wage is $1.
Each day, his wage increases by the prime numbers 2, 3, and 5 in that order.

Which one of the following is true about the wages earned by $A, B,$ and C in the first 4 days?

(A) $A > B > C$
(B) $C > B > A$
(C) $A > C > B$
(D) $B > A > C$
(E) $C > A > B$

[Multiple-choice Question – Select One Answer Choice Only]
13. 3/8 of a number is what fraction of 2 times the number?

(A) 3/16
(B) 3/8
(C) 1/2
(D) 4/6
(E) 3/4

[Multiple-choice Question – Select One Answer Choice Only]
14. $a, b,$ and c are three different numbers. None of the numbers equals the average of the other two. If $\dfrac{x}{a+b-2c} = \dfrac{y}{b+c-2a} = \dfrac{z}{c+a-2b}$, then $x + y + z =$

(A) 0
(B) 3
(C) 4
(D) 5
(E) 6

[Multiple-choice Question – Select One Answer Choice Only]
15. The difference between two angles of a triangle is 24°. The average of the same two angles is 54°. Which one of the following is the value of the greatest angle of the triangle?

   (A) 45°
   (B) 60°
   (C) 66°
   (D) 72°
   (E) 78°

[Multiple-choice Question – Select One or More Answer Choices]
16. If p and q are both positive integers such that $10p + 9q$ is a multiple of 90, then which of the following numbers could q equal?

   (A) 9
   (B) 10
   (C) 18
   (D) 20
   (E) 28
   (F) 30

[Multiple-choice Question – Select One or More Answer Choices]
17. If a is an integer, then which of the following are possible values for $\dfrac{a+3}{a+4}$?

   (A) 2/5
   (B) 1/2
   (C) 3/4
   (D) 11/14
   (F) 4/3

GRE Math Tests

[Multiple-choice Question – Select One Answer Choice Only]
18. If $b = a + c$ and $b = 3$, then $ab + bc =$

   (A) $\sqrt{3}$
   (B) 3
   (C) $3\sqrt{3}$
   (D) 9
   (E) 27

[Multiple-choice Question – Select One or More Answer Choices]
19. In January, the value of a stock increased by 25%; and in February, it changed by 20%. In March, it increased by 50%; and in April, it decreased by 40%. If Jack has invested $80 in the stock on January 1 and sold it at the end of April, which of the following could be the percentage change in the price of the stock?

   (A) 0%
   (B) 5%
   (C) 10%
   (D) 25%
   (E) 35%

20.   Column A   An off-season discount of 10% is   Column B
                                                being offered at a store for any
                                                purchase with list price above
                                                $500. No other discounts are
                                                offered at the store. John
                                                purchased a computer from the
                                                store for $459.

   The list price (in dollars) of the   500
   computer that John purchased

Test 7—Questions

[Multiple-choice Question – Select One Answer Choice Only]
21. Chelsea traveled from point *A* to point *B* and then from point *B* to point *C*. If she took 1 hour to complete the trip with an average speed of 50 mph, what is the total distance she traveled in miles?

   (A) 20
   (B) 30
   (C) 50
   (D) 70
   (E) 90

22.   Column A   Column B
   Fraction of numbers from 0   Fraction of numbers from 0
   through 1000 that are divisible   through 1000 that are divisible
   by both 7 and 10   by both 5 and 14

GRE Math Tests

[Multiple-choice Question – Select One or More Answer Choices]

23. A trainer on a Project Planning Module conducts batches of soft skill training for different companies. For each batch, the trainer chooses the batch size (which is the number of participants) in the batch such that he can always make teams of equal numbers leaving no participant. For a particular batch he decides to conduct 3 programs. The first program needs 3 participants per team, the second program needs 5 participants per team, and the third needs 6 participants per team. Which of the following better describe the batch size (number of participants) that he chooses for the batch?

(A) Exactly 14 participants.
(B) Exactly 30 participants.
(C) Batch size that is factor of 30.
(D) Batch size in multiples of 14 such as 14 or 28 or 42 ...
(E) Batch size in multiples of 30 such as 30 or 60 or 90 ...
(F) Batch size that divides by 30 evenly.

[Multiple-choice Question – Select One Answer Choice Only]

24. If the probability that Mike will miss at least one of the ten jobs assigned to him is 0.55, then what is the probability that he will do all ten jobs?

(A) 0.1
(B) 0.45
(C) 0.55
(D) 0.85
(E) 1

Answers and Solutions Test 7:

Question	Answer
1.	D
2.	B
3.	D
4.	E
5.	B
6.	D
7.	A
8.	E
9.	E
10.	D
11.	A
12.	A
13.	A
14.	A
15.	D
16.	B, D, F
17.	B, C, E
18.	D
19.	C, E
20.	D
21.	D
22.	C
23.	E, F
24.	B

---

1. If $a = -1$, both columns equal -1. If $a = -2$, the columns are unequal. The answer is (D).

---

2. The function $a * b$ is defined to be $a/b - b/a$ for any numbers a and b. Applying this definition to the columns gives

Column A	Column B
$\dfrac{1}{m} * \dfrac{1}{n} =$	$\dfrac{1}{n} * \dfrac{1}{m} =$
$\dfrac{\frac{1}{m}}{\frac{1}{n}} - \dfrac{\frac{1}{n}}{\frac{1}{m}} =$	$\dfrac{\frac{1}{n}}{\frac{1}{m}} - \dfrac{\frac{1}{m}}{\frac{1}{n}} =$
$\dfrac{n}{m} - \dfrac{m}{n} =$	$\dfrac{m}{n} - \dfrac{n}{m} =$
$\dfrac{nn - mm}{mn} =$	$\dfrac{mm - nn}{mn} =$
$\dfrac{n^2 - m^2}{mn}$	$\dfrac{m^2 - n^2}{mn}$

The given inequality $m > n > 0$ indicates that both m and n are positive and therefore their product mn is positive. Multiplying both columns by mn to clear fractions yields

$n^2 - m^2$ $\qquad\qquad\qquad\qquad\qquad\qquad\qquad\qquad$ $m^2 - n^2$

Adding n^2 and m^2 to both columns yields

$2n^2$ $\qquad\qquad\qquad\qquad\qquad\qquad\qquad\qquad$ $2m^2$

Finally, dividing both columns by 2 yields

n^2 $\qquad\qquad\qquad\qquad\qquad\qquad\qquad\qquad$ m^2

Since both m and n are positive and $m > n$, $m^2 > n^2$. Hence, Column B is greater than Column A, and the answer is (B).

---

3. Suppose the two positive integers m and n do not have a common factor, apart from 1. Then the LCM of m and n is mn. For example, when $m = 14 (= 2 \bullet 7)$ and $n = 15 (= 3 \bullet 5)$, the LCM is $14 \bullet 15$. In this case, the LCM equals mn, and the Column A equals Column B.

Now, suppose the two integers m and n have at least one common factor (other than 1). Then the LCM uses the common factors only once, unlike mn. Hence, the LCM is less than mn. For example, suppose $m = 10$ $(= 2 \bullet 5)$ and $n = 14 (= 2 \bullet 7)$. Here, m and n have 2 as a common factor. The LCM of m and n is $2 \bullet 5 \bullet 7 = 70$, and mn equals $2 \bullet 5 \bullet \underline{2} \bullet 7 = 140$. The LCM did not use the underlined 2 in the evaluation. Hence, here, Column A is less than Column B.

Since this is a double case, the answer is (D).

---

4. The average of the five consecutive positive integers 1, 2, 3, 4, and 5 is $(1 + 2 + 3 + 4 + 5)/5 = 15/5 = 3$. After dropping 5 (the greatest number), the new average becomes $(1 + 2 + 3 + 4)/4 = 10/4 = 2.5$. The percentage drop in the average is

$$\frac{\text{Old average} - \text{New average}}{\text{Old average}} \cdot 100 =$$

$$\frac{3 - 2.5}{3} \cdot 100 =$$

$$\frac{100}{6} =$$

$$16.66\%$$

The answer is (E).

---

5. We are given the two inequalities

$2x + 1 > 3x + 2$
$5x + 2 > 4x$

Subtracting $2x + 2$ from both sides of the top inequality and subtracting $4x + 2$ from both sides of the bottom inequality yields

$-1 > x$
$x > -2$

Combining these inequalities yields $-1 > x > -2$. Since any number between -1 and -2 is less than 1, Column B is greater than Column A. The answer is (B).

6.	Column A	In △ABC, ∠B = 72°	Column B
	∠A		72°

Summing the angles of △ABC to 180° yields ∠A + ∠B + ∠C = 180. Substituting the value of ∠B (= 72°) into this equation yields ∠A + 72 + ∠C = 180. Solving the equation for ∠A yields ∠A = 180 – 72 – ∠C = 108 – ∠C = Column A.

Now, Suppose ∠C = 1°. Then ∠A = 108° – 1° = 107°. In this case, ∠A is greater than 72°. Now, Suppose ∠C = 107°. Then ∠A = 108° – 107° = 1°. In this case, ∠A is less than 72°.

Hence, we have a double case, and the answer is (D).

---

7. Draw a rectangular tank as given (based on face A).

Mapping the corresponding given values to the sides of the faces yields

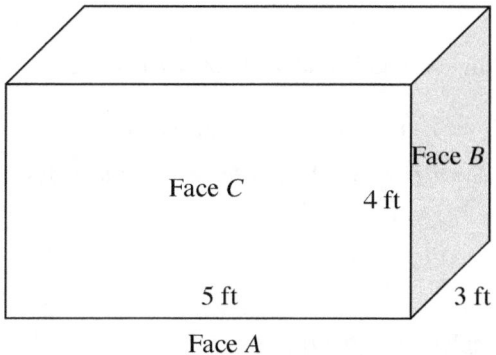

The height of the tank is 4 ft., and the height of the water is half the height of the tank, which is 4/2 = 2 ft. The answer is (A).

8. Let's name the vertices of the figure as shown

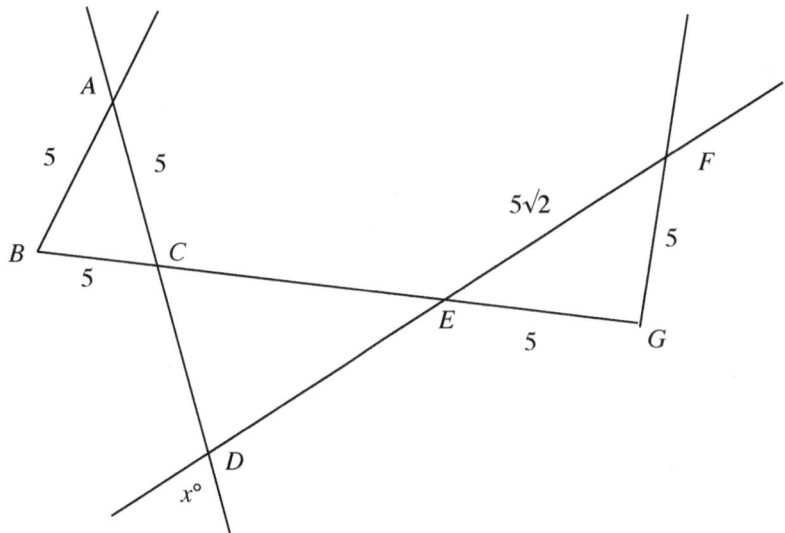

In △ABC, all the sides are equal (each equals 5). Hence, the triangle is equilateral, and ∠A = ∠B = ∠C = 60°.
Also, △EFG is a right-angled isosceles triangle (since EG = FG = 5), and The Pythagorean Theorem is satisfied ($EG^2 + FG^2 = 5^2 + 5^2 = 50 = EF^2 = (5\sqrt{2})^2$. Hence, ∠E = ∠F = 45° (Angles opposite equal sides of an isosceles right triangle measure 45° each).

Now, in △CED, we have:

∠D = x°	vertical angles, from the figure
∠C = ∠C in △ABC	vertical angles, from the figure
= 60°	we know
∠E in △CED = ∠E in △EFG	vertical angles
= 45°	we know

Now, summing these three angles of △CED to 180° yields 60 + 45 + x = 180. Solving this equation for x yields x = 75. The answer is (E).

---

9. From I, we have that the ratio of the three angles of the triangle is 1 : 2 : 3. Let k°, 2k°, and 3k° be the three angles. Summing the three angles to 180° yields k + 2k + 3k = 180; 6k = 180; k = 180/6 = 30. Now that we have the value of k, we can calculate the three angles and determine whether the triangle is a right triangle or not. Hence, I determines whether △ABC is a right triangle.

From II, we have that one angle of the triangle equals the sum of the other two angles. Let ∠A = ∠B + ∠C. Summing the angles of the triangle to 180° yields ∠A + ∠B + ∠C = 180; ∠A + ∠A = 180 (since ∠B + ∠C = ∠A); 2∠A = 180; ∠A = 180/2 = 90. Hence, the triangle is a right triangle. Hence, II determines whether △ABC is a right triangle.

From III, we have that △ABC is similar to the right triangle △DEF. Hence, △ABC is a triangle of same type as △DEF. So, III determines whether △ABC is a right triangle.

Hence, the answer is (E).

Test 7—Solutions

10. Since point $A(m, n)$ is in Quadrant II, the x-coordinate m is negative and the y-coordinate n is positive.

Hence, in point $B(n, m)$, the x-coordinate is positive, and the y-coordinate is negative. So, point B must be in Quadrant IV. The answer is (D).

11. Subtracting x^2 from both sides of the given inequality yields $x^5 < -x^2$. Dividing the inequality by the positive value x^2 yields $x^3 < -1$. Taking the cube root of both sides yields $x < -1$. The answer is (A).

12. The payments to Worker A for the 4 days are the four integers 4, 6, 8, and 10. The sum of the payments is $4 + 6 + 8 + 10 = 28$.

The payments to Worker B for the 4 days are the four integers 3, 5, 7, and 9. The sum of the payments is $3 + 5 + 7 + 9 = 24$.

The payments to Worker C for the 4 days are 1, $1 + 2 = 3$, $3 + 3 = 6$, and $6 + 5 = 11$. The sum of the payments is $1 + 3 + 6 + 11 = 21$.

From the calculations, $A > B > C$. The answer is (A).

13. Let the number be x. Now, 3/8 of the number is $3x/8$, and 2 times the number is $2x$. Forming the fraction yields $\dfrac{\frac{3}{8}x}{2x} = \dfrac{\frac{3}{8}}{2} = \dfrac{3}{8} \cdot \dfrac{1}{2} = \dfrac{3}{16}$. The answer is (A).

14. Let each part of the given equation $\dfrac{x}{a+b-2c} = \dfrac{y}{b+c-2a} = \dfrac{z}{c+a-2b}$ equal t. Then we have

$$\dfrac{x}{a+b-2c} = \dfrac{y}{b+c-2a} = \dfrac{z}{c+a-2b} = t$$

Simplifying, we get $x = t(a + b - 2c) = at + bt - 2ct$, $y = t(b + c - 2a) = bt + ct - 2at$, and $z = t(c + a - 2b) = ct + at - 2bt$.

Hence, $x + y + z = (at + bt - 2ct) + (bt + ct - 2at) + (ct + at - 2bt) = 0$. The answer is (A).

15. Let a and b be the two angles in the question, with $a > b$. We are given that the difference between the angles is $24°$, so $a - b = 24$. Since the average of the two angles is $54°$, we have $(a + b)/2 = 54$. Solving for b in the first equation yields $b = a - 24$, and substituting this into the second equation yields

$$\dfrac{a + (a - 24)}{2} = 54$$
$$\dfrac{2a - 24}{2} = 54$$
$$2a - 24 = 54 \times 2$$
$$2a - 24 = 108$$
$$2a = 108 + 24$$
$$2a = 132$$
$$a = 66$$

Also, $b = a - 24 = 66 - 24 = 42$.

Now, let c be the third angle of the triangle. Since the sum of the angles in the triangle is $180°$, $a + b + c = 180$. Plugging the previous results into the equation yields $66 + 42 + c = 180$. Solving for c yields $c = 72$. Hence, the greatest of the three angles a, b and c is c, which equals $72°$. The answer is (D).

16. We are given that "$10p + 9q$ is a multiple of 90." Hence, $(10p + 9q)/90$ must be an integer. Now, if p is not divisible by 9 and q is not divisible by 10, then $p/9$ results in a non-terminating decimal and $q/10$ results in a terminating decimal and the sum of the two would not result in an integer. [Because (a terminating decimal) + (a non-terminating decimal) is always a non-terminating decimal, and a non-terminating decimal is not an integer.]

Since we are given that the expression is an integer, q must be a 10 multiple.

For example, if $p = 1$ and $q = 10$, the expression equals $1/9 + 10/10 = 1.11...$, not an integer.

If $p = 9$ and $q = 5$, the expression equals $9/9 + 5/10 = 1.5$, not an integer.

If $p = 9$ and $q = 10$, the expression equals $9/9 + 10/10 = 2$, an integer.

The 10 multiples are (B), (D), and (F) must be chosen.

---

17. Choice (A): Suppose $\frac{a+3}{a+4} = \frac{2}{5}$; then $5(a+3) = 2(a+4); 3a = 8 - 5 \times 3 = -7; a = \frac{-7}{3}$, not integer. Reject.

Choice (B): Suppose $\frac{a+3}{a+4} = \frac{1}{2}$; then $\frac{a+3}{a+4} = \frac{1}{2}; 2(a+3) = 1(a+4); 2a+6 = a+4; a = 4-6 = -2$, is an integer. Accept.

Choice (C): Suppose $\frac{a+3}{a+4} = \frac{4}{5}$; then $5(a+3) = 4(a+4); 5a+15 = 4a+16; a = 16-15 = 1$, is an integer. Accept.

Choice (D): Suppose $\frac{a+3}{a+4} = \frac{11}{14}$;, then $14(a+3) = 11(a+4); 14a+42 = 11a+44; 3a = 2; a = \frac{2}{3}$, not an integer. Reject.

Choice (E): Suppose $\frac{a+3}{a+4} = \frac{4}{3}$; then $3(a+3) = 4(a+4); 3a+9 = 4a+16; a = 9-16 = -7$, is an integer. Accept.

The answers are (B), (C), and (E).

---

18. Factoring the common factor b from the expression $ab + bc$ yields $b(a + c) = b \cdot b$ [since $a + c = b$] = $b^2 = 3^2 = 9$. The answer is (D).

Test 7—Solutions

---

19. At the end of January, the value of the stock is $80 + 25\%(\$80) = \$80 + \$20 = \$100$.

In February, the value of the stock changed by 20%. This could mean either decrease by 20% or increase by 20%.

Suppose the change is the decrease by 20%:

At the end of February, the value of the stock is $\$100 - 20\%(\$100) = \$100 - \$20 = \$80$.

At the end of March, the value of the stock is $\$80 + 50\%(\$80) = \$80 + \$40 = \$120$.

At the end of April, the value of the stock is $\$120 - 40\%(\$120) = \$120 - \$48 = \$72$.

Now, the percentage change in price is

$$\frac{\text{change in price}}{\text{original price}} = \frac{80-72}{80} = \frac{8}{80} = \frac{1}{10} = 10\%$$

Select (C).

Now, suppose the change is the increase by 20%:

At the end of February, the value of the stock is $\$100 + 20\%(\$100) = \$100 + \$20 = \$120$.

At the end of March, the value of the stock is $\$120 + 50\%(\$120) = \$120 + \$60 = \$180$.

At the end of April, the value of the stock is $\$180 - 40\%(\$180) = \$180 - \$72 = \$108$.

Now, the percentage change in price is

$$\frac{\text{change in price}}{\text{original price}} = \frac{80-108}{80} = \frac{-28}{80} = -35\%$$

The answer is (E).

So, the correct answer is only (C) and (E).

---

20. We do not know whether the $459 price that John paid for the computer was with the discount offer or without the discount offer.

If he did not get the discount offer, the list price of the computer should be $459 and John paid the exact amount for the computer. In this case, Column A (= 459) is less than Column B.

If the price corresponds to the price after the discount offer, then $459 should equal a 10% discount on the list price. Hence, if l represents the list price, then we have $\$459 = l(1 - 10/100) = l(1 - 1/10) = (9/10)l$. Solving the equation for l yields $l = (10/9)459 = 510$ dollars (a case when discount was offered because the list price is greater than $500). Hence, it is also possible that John got the 10% discount on the computer originally list priced at $510. Here, list price (= Column A) is greater than 500 (= Column B).

Hence, we have a double case, and the answer is (D).

21. Let t be the entire time of the trip.

We have that the car traveled at 80 mph for $t/2$ hours and at 40 mph for the remaining $t/2$ hours. Remember that *Distance = Speed × Time*. Hence, the total distance traveled during the two periods equals $80 \times t/2 + 40 \times t/2 = 60t$. Now, remember that

$$\text{Average Speed} = \frac{\text{Total Distance}}{\text{Time Taken}} = \frac{60t}{t} = 60$$

The answer is (D).

---

22. Any number divisible by both 7 and 10 is a common multiple of 7 and 10. The least common multiple of 7 and 10 is 70. Hence, Column A reduces to the fraction of numbers from 0 through 1000 that are divisible by 70.

Similarly, any number divisible by both 5 and 14 is a common multiple of 5 and 14. The LCM of 5 and 14 is 70. Hence, Column B reduces to the fraction of numbers from 0 through 1000 that are divisible by 70.

Since the statement in each column is now the same, the columns are equal and the answer is (C).

---

23. The trainer wants to make teams of either 3 participants each or 5 participants each or 6 participants each successfully without leaving out any one of the participants in the batch. Hence, the batch size must be a multiple of all three numbers 3, 5, and 6. Therefore, the batch size must be a multiple of the least common multiple of 3, 5, and 6, which is 30. Select the choices (E) and (F).

---

24. There are only two cases:

    1)   Mike will miss at least one of the ten jobs.
    2)   Mike will not miss any of the ten jobs.

Hence,

(The probability Mike will miss at least one of the jobs) + (The probability he will not miss any job) = 1

Since the probability that Mike will miss at least one of the ten jobs is 0.55, this equation becomes

0.55 + (The probability that he will not miss any job) = 1

(The probability that he will not miss any job) = 1 − 0.55

(The probability that he will not miss any job) = 0.45

The answer is (B).

Test 8

GRE Math Tests

Questions: 24
Time: 45 minutes

1.   Column A   Column B
                 $x^2 + 2$ $x^3 - 2$

[Multiple-choice Question – Select One Answer Choice Only]
2. A two-digit even number is such that reversing its digits creates an odd number greater than the original number. Which one of the following cannot be the first digit of the original number?

    (A)   1
    (B)   3
    (C)   5
    (D)   7
    (E)   9

[Multiple-choice Question – Select One or More Answer Choices]
3. Which of the following choices does not equal any of the other choices?

    (A) $5.43 + 4.63 - 3.24 - 2.32$
    (B) $5.53 + 4.73 - 3.34 - 2.42$
    (C) $5.53 + 4.53 - 3.342 - 2.22$
    (D) $5.43 + 4.73 - 3.24 - 2.42$
    (E) $5.49 + 4.78 - 3.10 - 2.21$

[Multiple-choice Question – Select One Answer Choice Only]
4. What is the remainder when $7^2 \cdot 8^2$ is divided by 6?

    (A) 1
    (B) 2
    (C) 3
    (D) 4
    (E) 5

5.

Column A	$x = 3.635 \cdot 10^{16}$	Column B
$\dfrac{x+1}{x-1}$		$\dfrac{x-1}{x+1}$

6.

Column A	Line segments *AB* and *CD* are both parallel and congruent. The mid-point of AB is M.	Column B
The length of segment *CM*		The length of segment *DM*

7.   Column A   Column B
     The perimeter of quadrilateral   The circumference of the
     ABCD   circle

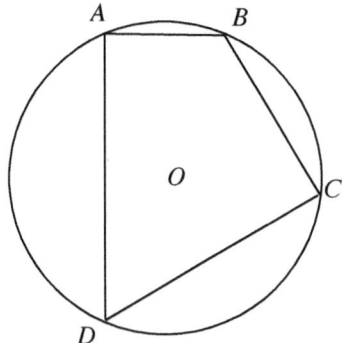

[Multiple-choice Question – Select One Answer Choice only]
8.   In the figure, the area of rectangle ABCD is 100. What is the area of the square EFGH?

    (A) 256
    (B) 275
    (C) 309
    (D) 399
    (E) 401

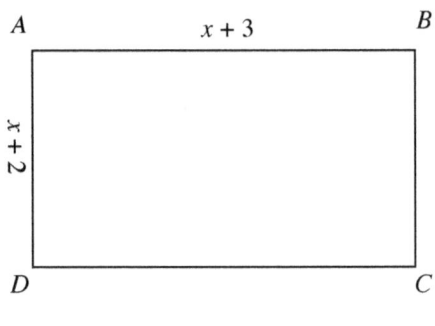
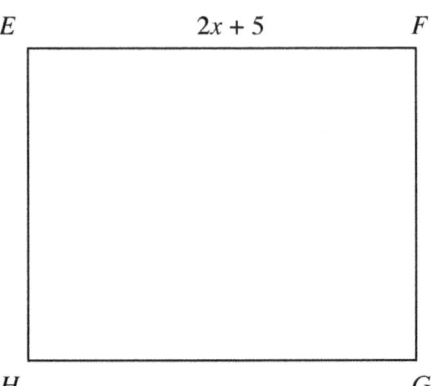

[Multiple-choice Question – Select One Answer Choice Only]
9. In the figure, ABCD and PQRS are two rectangles inscribed in the circle as shown and AB = 4, AD = 3, and QR = 4. What is the value of l ?

(A) 3/2
(B) 8/3
(C) 3
(D) 4
(E) 5

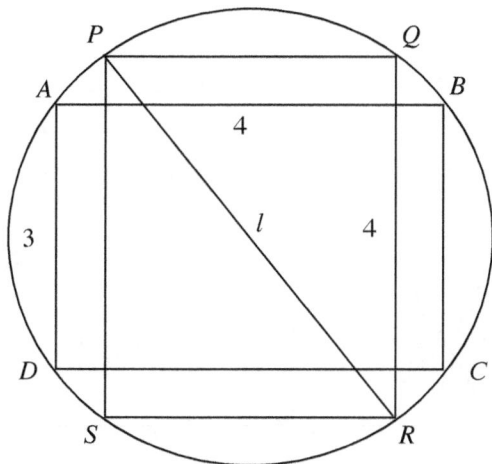

USE THIS SPACE FOR SCRATCHWORK.

[Multiple-choice Question – Select One Answer Choice Only]
10. In the figure, what is the area of $\triangle ABC$ if EC/CD = 3 ?

(A) 12
(B) 24
(C) 81
(D) 121.5
(E) 143

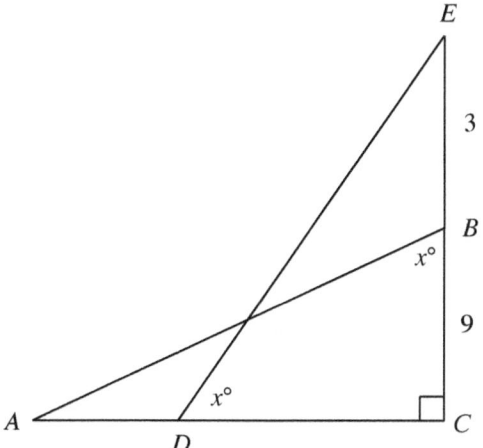

USE THIS SPACE FOR SCRATCHWORK.

[Numeric Entry Question]
11. In the figure, if $y = 60$, then what is the value of z?

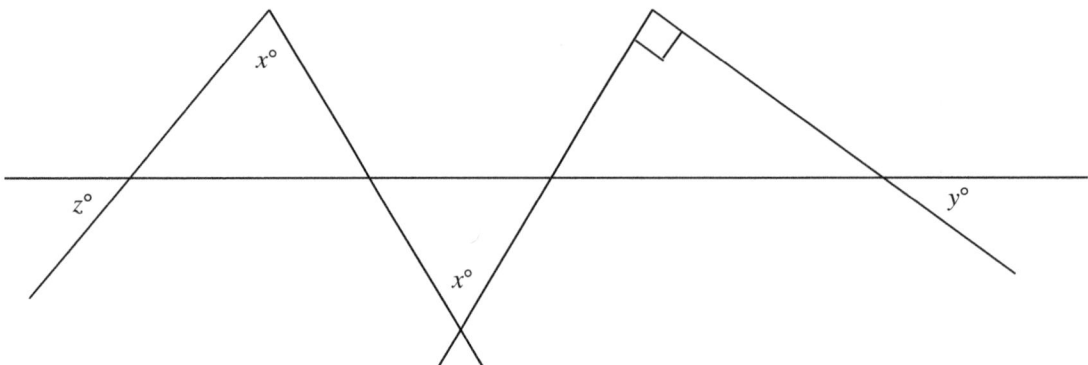

[Multiple-choice Question – Select One Answer Choice Only]
12. If $(x - y)^3 > (x - y)^2$, then which one of the following must be true?

  (A) $x^3 < y^2$
  (B) $x^5 < y^4$
  (C) $x^3 > y^2$
  (D) $x^5 > y^4$
  (E) $x^3 > y^3$

13. | Column A | $x < 1/y$, and x and y are positive | Column B |
    | $\dfrac{2+x-x^2}{x}$ | | $\dfrac{2y^2+y-1}{y}$ |

14. | Column A | $x+y=7$ | Column B |
    | $x+y^2$ | $x^2+y^2=25$ | $3+4^2$ |

[Multiple-choice Question – Select One Answer Choice Only]

15. In quadrilateral $ABCD$, $\angle A$ measures 20 degrees more than the average of the other three angles of the quadrilateral. Then $\angle A =$

(A) 70°
(B) 85°
(C) 95°
(D) 105°
(E) 110°

[Multiple-choice Question – Select One Answer Choice Only]

16. Bank X pays a simple interest of $80 on a principal of $1,000 annually. Bank Y pays a simple interest of $140 on a principal of $1,000 annually. What is the ratio of the interest rates of Bank X to Bank Y?

    (A) $5:8$
    (B) $8:5$
    (C) $14:8$
    (D) $4:7$
    (E) $5:7$

[Multiple-choice Question – Select One or More Answer Choices]

17. A *perfect square* is a number that becomes an integer when square rooting it. A, B, and C are three positive integers. The ratio of the three numbers is $1:2:3$, respectively. Which of the following expressions must be a perfect square?

    (A) $A + B + C$
    (B) $A^2 + B^2 + C^2$
    (C) $A^3 + B^3 + C^3$
    (D) $3A^2 + B^2 + C^2$
    (E) $3A^2 + 4B^2 + 4C^2$
    (F) $8A^2 + 5B^2 + 3C^2$
    (G) $A^2 + 4B^2 + C^2$

[Numeric Entry]
18. If $2(2^2) + 3^2 + 2^3 = 5^x$, then $x =$

[Multiple-choice Question – Select One Answer Choice Only]
19. In an acoustics class, 120 students are male and 100 students are female. 25% of the male students and 20% of the female students are engineering students. 20% of the male engineering students and 25% of the female engineering students passed the final exam. What percentage of engineering students passed the exam?

(A) 5%
(B) 10%
(C) 16%
(D) 22%
(E) 25%

[Multiple-choice Question – Select One Answer Choice Only]
20. Selling 12 candies at a price of $10 yields a loss of a%. Selling 12 candies at a price of $12 yields a profit of a%. What is the value of a?

(A) 11/1100
(B) 11/100
(C) 100/11
(D) 10
(E) 11

[Multiple-choice Question – Select One or More Answer Choices]

21. In *x* hours and *y* minutes a car traveled *z* miles. What is the car's speed in miles per hour?

   (A) $\dfrac{z}{60+y}$

   (B) $\dfrac{60z}{60x+y}$

   (C) $\dfrac{60}{60+y}$

   (D) $\dfrac{z}{x+y}$

   (E) $\dfrac{60+y}{60z}$

[Multiple-choice Question – Select One or More Answer Choices]

22. Each year, funds A and B grow by a particular percentage based on the following policy of the investment company:

   (1) The allowed percentages of growths on the two funds are 20% or 30%.
   (2) The growth percentages of the two funds are not the same in any year.
   (3) No fund will grow by same percentage growth in any two consecutive years.

   In the first year, Fund B was offered a growth of 30%.

   Bob doesn't know in which fund he invested 3 years back. What is the possible percentage by which the fund of $1000 that he invested might have grown in the last 3 years?

   (A) 70%
   (B) 80%
   (C) 87.2%
   (D) 92.2%
   (E) 108.2%

	Column A		Column B
23.	The number of distinct prime factors of 12		The number of distinct prime factors of 36

	Column A	A bowl contains 500 marbles. There are x red marbles and y blue marbles in the bowl.	Column B
24.	The number marbles in the bowl that are neither red nor blue		$500 - x - y$

Answers and Solutions Test 8:

Question	Answer
1.	D
2.	E
3.	C, E
4.	D
5.	A
6.	D
7.	B
8.	E
9.	E
10.	D
11.	30
12.	E
13.	A
14.	D
15.	D
16.	D
17.	D, F, G
18.	2
19.	D
20.	C
21.	B
22.	C, E
23.	C
24.	C

If you answered 18 out of 24 questions in this test, you are likely to score 750+ in your GRE.

---

1. Since $x > 0$, we need only look at $x = 1, 2$, and $1/2$. If $x = 1$, then $x^2 + 2 = 3$ and $x^3 - 2 = -1$. In this case, Column A is larger. Next, if $x = 2$, then $x^2 + 2 = 6$ and $x^3 - 2 = 6$. In this case, the two columns are equal. This is a double case and therefore the answer is (D).

---

2. Let the original number be represented by xy. (Note: here xy does not denote multiplication, but merely the position of the digits: x first, then y.). Reversing the digits of xy gives yx. We are told that $yx > xy$. This implies that $y > x$. (For example, $73 > 69$ because $7 > 6$.) If $x = 9$, then the condition $y > x$ cannot be satisfied. Hence, x cannot equal 9. The answer is (E).

Method II:
Let the original number be represented by xy. In expanded form, xy can be written as $10x + y$. For example, $53 = 5(10) + 3$. Similarly, $yx = 10y + x$. Since $yx > xy$, we get $10y + x > 10x + y$. Subtracting x and y from both sides of this equation yields $9y > 9x$. Dividing this equation by 9 yields $y > x$. Now, if $x = 9$, then the inequality $y > x$ cannot be satisfied. The answer is (E).

Test 8—Solutions

---

3. Choice (A) = 5.43 + 4.63 − 3.24 − 2.32 = 4.5. Short-list choice (A).

   Choice (B) = 5.53 + 4.73 − 3.34 − 2.42 = 4.5 = Choice (A). Reject choices (A) and (B).

   Choice (C) = 5.53 + 4.53 − 3.34$\underline{2}$ − 2.22. There is a third-digit in one of the terms in this expression (4.53$\underline{2}$) unlike any other terms. Hence, the result must be in three digits after the decimal. Clearly, the choice is unequal to the others. Accept the choice. Though not necessary now, the result is 4.498.

   Choice (D) = 5.43 + 4.73 − 3.24 − 2.42 = 4.5 = Choice (A). Reject choice (D).

   Choice (E) = 5.4$\underline{9}$ + 4.7$\underline{8}$ − 3.1$\underline{0}$ − 2.2$\underline{1}$. The result of the expression in the second place (0.0$\underline{9}$ + 0.0$\underline{8}$ − 0.0$\underline{0}$ − 0.0$\underline{1}$ = 0.16) after the decimal does not end with 0 or a multiple of 10 (Here, it ends with 6). Hence, the result contains up to a second digit after the decimal. No second digit exists in 4.5. Hence, accept this choice. Though not necessary now, the result is 4.96.

The answers are (C) and (E).

---

4. $7^2 \cdot 8^2 = (7 \cdot 8)^2 = 56^2$.

The number immediately before 56 that is divisible by 6 is 54. Now, writing 56^2 as $(54 + 2)^2$, we have

$$\begin{aligned}
56^2 &= (54 + 2)^2 \\
&= 54^2 + 2(2)(54) + 2^2 \quad \text{by the formula } (a + b)^2 = a^2 + 2ab + b^2 \\
&= 54[54 + 2(2)] + 2^2 \\
&= 6 \times 9[54 + 2(2)] + 4 \quad \text{here, the remainder is 4}
\end{aligned}$$

Since the remainder is 4, the answer is (D).

---

5. Since x is a large positive number, both $x + 1$ and $x − 1$ are positive. Hence, we can clear fractions by multiplying both columns by $(x + 1)(x − 1)$, which yields

$(x + 1)^2$ $\hspace{8cm}$ $(x − 1)^2$

Performing the multiplication yields

$x^2 + 2x + 1$ $\hspace{7cm}$ $x^2 − 2x + 1$

Subtracting x^2 and 1 from both columns yields

$2x$ $\hspace{9cm}$ $−2x$

Since x is a positive number, Column A is positive and Column B is negative.

Since all positive numbers are greater than all negative numbers, Column A is greater than Column B and the answer is (A).

6. Most people will draw the figure as follows:

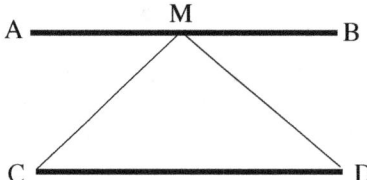

In this drawing, CM equals DM. But that is too ordinary. There must be a way to draw the lines so that the lengths are not equal. One such drawing is as follows:

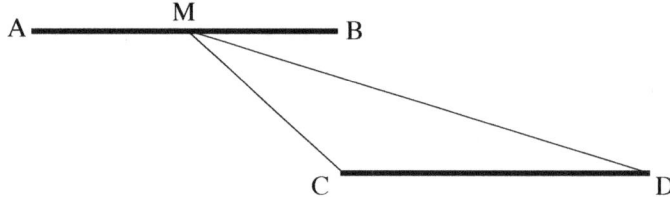

This is a double case, and therefore the answer is (D). (Note: When drawing a geometric figure, be careful not to assume more than what is given. In this problem, we are told only that the two lines are parallel and congruent; we cannot assume that they are aligned.)

---

7. Since the shortest distance between two points is a straight line, the length of a chord is always shorter than the length of the arc that it makes on the circle. Hence, from the figure, we have

$$AB < \text{arc } AB$$
$$BC < \text{arc } BC$$
$$CD < \text{arc } CD$$
$$DA < \text{arc } DA$$

Adding the four inequalities yields $AB + BC + CD + DA < \text{arc } AB + \text{arc } BC + \text{arc } CD + \text{arc } DA$. The left side of the inequality is the perimeter of quadrilateral $ABCD$ (which Column A equals), and the right side is the circumference of the circle (which Column B equals). Hence, Column A is less than Column B, and the answer is (B).

---

8. The area of the rectangle $ABCD$, $length \times width$, is $(x + 3)(x + 2) = x^2 + 5x + 6 = 100$ (given the area of the rectangle $ABCD$ equals 100). Now, subtracting 6 from both sides yields $x^2 + 5x = 94$.

The area of square $EFGH$ is

$$(2x + 5)^2 = 4x^2 + 20x + 25 = 4(x^2 + 5x) + 25 = 4(94) + 25 = 401$$

The answer is (E).

You could also solve for x in the top equation and substitute in the bottom equation—but this would take too long. The GRE wants to see whether you can find the shorter solution. The premise being that if you spend a lot of time doing long calculations you will not have as much time to solve the problems and therefore will not score as high as someone who has the insight to find the shorter solutions.

9. *PQRS* is a rectangle inscribed in the circle. Hence, diagonal *PR* must pass through the center of the circle. So, *PR* is a diameter of the circle.

Similarly, *BD* is a diagonal of rectangle *ABCD*, which is also inscribed in the same circle. Hence, the two diagonals must be diameters and equal. So, we have *PR* = *BD*.

Now, in the figure, let's join the opposite vertices *B* and *D* of the rectangle *ABCD*:

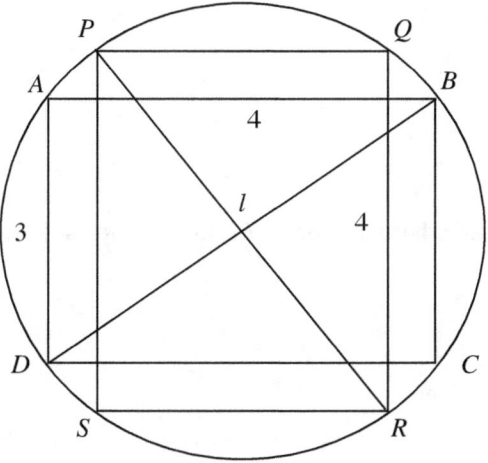

Applying The Pythagorean Theorem to the right triangle *ABD* yields $BD^2 = AB^2 + AD^2 = 4^2 + 3^2 = 16 + 9 = 25$. By square rooting, we get $BD = \sqrt{25} = 5$. Hence, *PR* also equals 5. Since, from the figure, *l* equals *PR*, *l* equals 5. The answer is (E).

---

10. In right triangle *EDC*, ∠*C* is right angled, ∠*D* measures $x°$, and angle ∠*E* measures $180° - (90° + x°) = 90° - x°$.

In right triangle *ABC*, ∠*C* is right-angled, ∠*B* measures $x°$, and angle ∠*A* equals $180° - (90° + x°) = 90° - x°$.

Since corresponding angles in the two triangles are equal, both are similar triangles and the corresponding angle sides must be same. Hence, we have

$AC/BC = EC/CD$

$AC/9 = 3$

$AC = 3 \cdot 9 = 27$

Now, the area of △*ABC* = 1/2 ·*base* · *height* = 1/2 · *AC* · *BC* = 1/2 · 27 · 9 = 121.5.

The answer is (D).

11. Let's name the vertices in the figure as shown below.

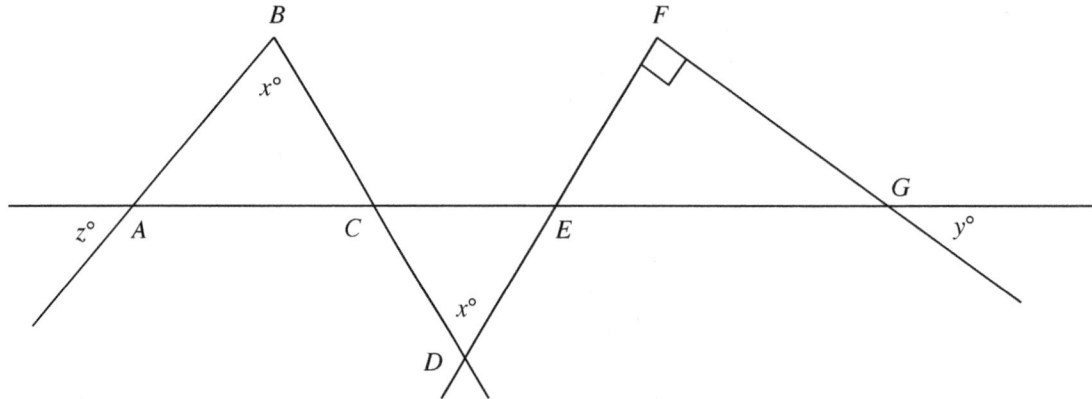

In the figure, ∠FGE equals y° (vertical angles). Summing the angles of △GEF to 180° yields

∠GEF + ∠EFG + ∠FGE = 180
∠GEF + 90 + y = 180
∠GEF = 180 − (90 + y)
 = 90 − y = 90 − 60   (since y = 60, given)
 = 30°

Now, since ∠ABD = ∠BDF (both equal x, from the figure), lines AB and DF are parallel (alternate interior angles). Hence, ∠CAB = ∠GEF (corresponding angles) = 30° (we know ∠GEF = 30°). Now, z = ∠CAB (vertical angles), and therefore z = 30. Enter in the grid.

12. If $x - y$ equaled 0, the inequality $(x - y)^3 > (x - y)^2$ would not be valid. Hence, $x - y$ is nonzero. Since the square of a nonzero number is positive, $(x - y)^2$ is positive. Hence, $(x - y)^3 > 0$. Taking the square root of both sides of this equation preserves the direction of the inequality:

$$\sqrt[3]{(x-y)^3} > \sqrt[3]{0}$$

$$x - y > 0$$

So, $x > y$. Cubing both sides of this inequality will also preserve the direction of the inequality $x^3 > y^3$. The answer is (E). The remaining choices need *not* be true.

13. Breaking up the fractions in both columns yields

Column A: $\dfrac{2 + x - x^2}{x} = \dfrac{2}{x} + \dfrac{x}{x} - \dfrac{x^2}{x} = \dfrac{2}{x} + 1 - x$

Column B: $\dfrac{2y^2 + y - 1}{y} = \dfrac{2y^2}{y} + \dfrac{y}{y} - \dfrac{1}{y} = 2y + 1 - \dfrac{1}{y}$

Now, multiplying both sides of the given inequality $x < 1/y$ by -1 and flipping the direction of the inequality yields

$-x > -1/y$ ... (1)

Since x and $1/y$ are both positive (since y is positive, so is its reciprocal $1/y$), we can safely invert both sides of the inequality $x < 1/y$ and flip the direction of the inequality to yield $1/x > y$. Multiplying both sides of this inequality by 2 yields

$2/x > 2y$ ... (2)

Now, adding inequalities (1) and (2) yields

$\dfrac{2}{x} - x > 2y - \dfrac{1}{y}$

Adding 1 to both sides of this inequality yields

$\dfrac{2}{x} + 1 - x > 2y + 1 - \dfrac{1}{y}$

Column A > Column B   from the known results

Hence, the answer is (A).

---

14. We have the system of equations

$x + y = 7$
$x^2 + y^2 = 25$

Solving the top equation for y yields $y = 7 - x$. Substituting this into the bottom equation yields

$x^2 + (7 - x)^2 = 25$
$x^2 + 49 - 14x + x^2 = 25$
$2x^2 - 14x + 24 = 0$
$x^2 - 7x + 12 = 0$
$(x - 3)(x - 4) = 0$
$x - 3 = 0$ or $x - 4 = 0$
$x = 3$ or $x = 4$

Now, if $x = 3$, then $y = 7 - 3 = 4$ and Column A equals $x + y^2 = 3 + 4^2$ (= Column B); and if $x = 4$, then $y = 7 - 4 = 3$ and Column A $= x + y^2 = 4 + 3^2$ ($\neq$ Column B). Hence, we have a double case, and the answer is (D).

GRE Math Tests

15. Setting the angle sum of the quadrilateral to 360° yields $\angle A + \angle B + \angle C + \angle D = 360$. Subtracting $\angle A$ from both sides yields $\angle B + \angle C + \angle D = 360 - \angle A$. Forming the average of the three angles $\angle B$, $\angle C$, and $\angle D$ yields $(\angle B + \angle C + \angle D)/3$ and this equals $(360 - \angle A)/3$, since we know that $\angle B + \angle C + \angle D = 360 - \angle A$. Now, we are given that $\angle A$ measures 20 degrees more than the average of the other three angles. Hence, $\angle A = (360 - \angle A)/3 + 20$. Solving the equation for $\angle A$ yields $\angle A = 105$. The answer is (D).

16. The formula for simple interest rate (in percentage) equals $\dfrac{\text{Interest}}{\text{Principal}} \times 100$. From this formula, it is clear that the principal being the same (= \$1000 in the two given cases), the interest rates are directly proportional to the interests earned on them (\$80 in case of Bank A and \$140 in case of Bank B). Hence, the ratio of the interest rates is $80 : 140 = 4 : 7$. The answer is (D).

17. Forming the given ratio yields

$$A/1 = B/2 = C/3 = k, \text{ for some integer}$$
$$A = k, B = 2k, \text{ and } C = 3k$$

Choice (A): $A + B + C = k + 2k + 3k = 6k$. This is a perfect square only when k is a product of 6 and a perfect square number. For example, when k is $6 \cdot 9^2$, $6k = 6^2 \cdot 9^2$, a perfect square. In all other cases (suppose $k = 2$, then $6k = 12$), it is not a perfect square. Hence, reject.

Choice (B): $A^2 + B^2 + C^2 = k^2 + (2k)^2 + (3k)^2 = k^2 + 4k^2 + 9k^2 = 14k^2$. This is surely not a perfect square. For example, suppose k equals 2. Then $14k^2 = 56$, which is not a perfect square. Hence, reject.

Choice (C): $A^3 + B^3 + C^3 = k^3 + (2k)^3 + (3k)^3 = k^3 + 8k^3 + 27k^3 = 36k^3$. This is a perfect square only when k^3 is perfect square. For example, suppose $k = 2$. Then $36k^3 = 288$, which is not a perfect square. Hence, reject.

Choice (D): $3A^2 + B^2 + C^2 = 3k^2 + (2k)^2 + (3k)^2 = 3k^2 + 4k^2 + 9k^2 = 16k^2 = 4^2 k^2 = (4k)^2$. The square root of $(4k)^2$ is $4k$ and is an integer for any integer value of k. Hence, this expression must always result in a perfect square. Choose (D).

Choice (E): $3A^2 + 4B^2 + 4C^2 = 3k^2 + 4(2k)^2 + 4(3k)^2 = 3k^2 + 16k^2 + 36k^2 = 55k^2$. This is surely not a perfect square. Hence, reject.

Choice (F): $8A^2 + 5B^2 + 3C^2 = 8k^2 + 5(2k)^2 + 4(3k)^2 = 8k^2 + 20k^2 + 36k^2 = 64k^2 = (8k)^2$. The square root of $(8k)^2$ is $8k$ and is an integer for any integer value of k. Hence, this expression must always result in a perfect square. Choose (F).

Choice (G): $4A^2 + 3B^2 + C^2 = 4k^2 + 3(2k)^2 + (3k)^2 = 4k^2 + 12k^2 + 9k^2 = 25k^2 = (5k)^2$. The square root of $(5k)^2$ is $5k$ and is an integer for any integer value of k. Hence, this expression must always result in a perfect square. Choose (G).

The answer is (D), (F), and (G).

Test 8—Solutions

18. $2(2^2) + 3^2 + 2^3 = 5^x$
    $8 + 9 + 8 = 5^x$
    $25 = 5^x$
    $5^2 = 5^x$

Since the bases are the same, 5, the exponents must equal each other: $x = 2$. Enter 2 in the grid.

---

19. There are 100 female students in the class, and 20% of them are engineering students. Now, 20% of 100 equals $20/100 \times 100 = 20$. Hence, the number of female engineering students in the class is 20.

Now, 25% of the female engineering students passed the final exam: 25% of $20 = 25/100 \times 20 = 5$. Hence, the number of female engineering students who passed is 5.

There are 120 male students in the class. And 25% of them are engineering students. Now, 25% of 120 equals $25/100 \times 120 = 1/4 \times 120 = 30$. Hence, the number of male engineering students is 30.

Now, 20% of the male engineering students passed the final exam: 20% of $30 = 20/100 \times 30 = 6$. Hence, the number of male engineering students who passed is 6.

Hence, the total number of Engineering students who passed is

(Female Engineering students who passed) + (Male Engineering students who passed) =

$5 + 6 =$

11

The total number of Engineering students in the class is

(Number of female engineering students) + (Number of male engineering students) =
$30 + 20 =$
50

Hence, the percentage of engineering students who passed is

$$\frac{\text{Total number of engineering students who passed}}{\text{Total number of engineering students}} \times 100 =$$

$11/50 \times 100 =$

22%

The answer is (D).

161

GRE Math Tests

20. Let c be the cost of each candy. Then the cost of 12 candies is $12c$. We are given that selling 12 candies at \$10 yields a loss of $a\%$. The formula for the loss percentage is $\dfrac{\text{cost} - \text{selling price}}{\text{cost}} \cdot 100$. Hence, $a = \dfrac{12c - 10}{12c} \cdot 100 = -a\%$ (Loss). Let this be equation (1).

We are also given that selling 12 candies at \$12 yields a profit of $a\%$. The formula for profit percent is $\dfrac{\text{selling price} - \text{cost}}{\text{cost}} \cdot 100$. Hence, we have $\dfrac{12 - 12c}{12c} \cdot 100 = a\%$. Let this be equation (2).

Equating equations (1) and (2), we have

$$\dfrac{12 - 12c}{12c} \cdot 100 = \dfrac{12c - 10}{12c} \cdot 100$$
$$12 - 12c = 12c - 10 \quad \text{by canceling } 12c \text{ and } 100 \text{ from both sides}$$
$$24c = 22$$
$$c = 22/24$$

From equation (1), we have

$$a = \dfrac{12c - 10}{12c} \cdot 100$$
$$= \dfrac{12 \cdot \frac{22}{24} - 10}{12 \cdot \frac{22}{24}} \cdot 100$$
$$= \dfrac{11 - 10}{11} \cdot 100$$
$$= \dfrac{100}{11}$$

The answer is (C).

Method II: Since both loss percentage and the gain percentage are based on cost price and since both are equal, the cost price must be in the middle at

$$(\$10 + \$12)/2 = \$22/2 = \$11$$

Hence, a, the gain percentage, must be

$$a = \dfrac{11 - 10}{11} \cdot 100 = \dfrac{1}{11} \cdot 100 = \dfrac{100}{11}$$

Test 8—Solutions

21. Since the time is given in mixed units, we need to change the minutes into hours. Since there are 60 minutes in an hour, y minutes is equivalent to $y/60$ hours. Hence, the car's travel time, "x hours and y minutes," is $x + y/60$ hours. Plugging this along with the distance traveled, z, into the formula $d = rt$ yields

$$z = r\left(x + \frac{y}{60}\right)$$

$$z = r\left(\frac{60}{60}x + \frac{y}{60}\right)$$

$$z = r\left(\frac{60x + y}{60}\right)$$

$$\frac{60z}{60x + y} = r$$

The answer is (B).

22. We have two cases here:

    I.   Bob invested in Fund A.
    II.   Bob invested in Fund B.

In the first year, fund B was given a growth of 30%. Hence, according to clauses (1) and (2), fund A must have grown by 20% (the other allowed growth percentage clause (1)).

In the second year, according to the clauses (1) and (3), the growth percentages of the two funds will swap between the only allowed values 30% and 20% (clause (1)). Hence, fund A grows by 30% and fund B grows by 20%.

In the third year, according to clauses (1) and (3), the growth percentages will again swap between the only two allowed values 20% and 30% (clause (1)). Hence, fund A grows by 20% and fund B grows by 30%.

The growth in Fund A is

$1000(100 + 20% of 100)(100 + 30% of 100)(100 + 20% of 100) = $1000 (1.2)(1.3)(1.2) = 1.872 times = 87.2%.

The growth in Fund B is

$1000(100 + 30% of 100)(100 + 20% of 100)(100 + 30% of 100) = $1000 (1.3)(1.2)(1.3) = 2.082 times = 108.2%.

Select the choices (C) and (E).

23. Prime factoring 12 and 36 gives

$$12 = 2 \cdot 2 \cdot 3$$
$$36 = 2 \cdot 2 \cdot 3 \cdot 3$$

Thus, each number has two distinct prime factors, namely 2 and 3. The answer is (C).

---

24. There are $x + y$ red and blue marbles in the bowl. Subtracting this from the total of 500 marbles gives the number of marbles that are neither red nor blue: $500 - (x + y) = 500 - x - y$. Hence, the columns are equal, and the answer is (C).

Test 9

GRE Math Tests

Questions: 24
Time: 45 minutes

---

1. Column A $x = y \neq 0$ Column B
    0 x/y

[Numeric Entry Question]
2. Each of the two positive integers a and b ends with the digit 2. Enter the last two digits of $(a - b)^2$ in the grid below.

[Numeric Entry Question]
3. In the figure, lines *l* and *m* are parallel. If $y - z = 60$, then what is the value of *x* ?

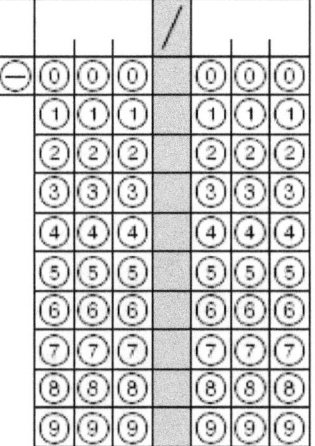

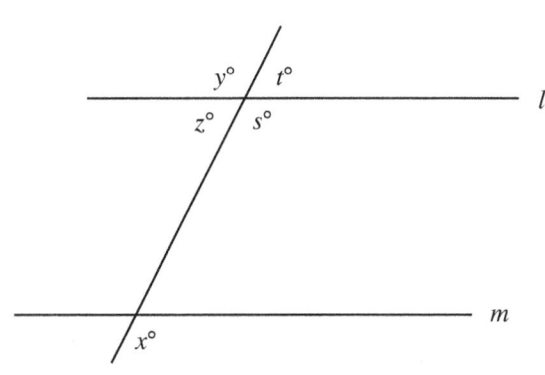

[Multiple-choice Question – Select One or More Answer Choices]
4. If $\triangle ABD$ shown must be a right triangle, then which of the following line-segments cannot be ruled out as being the longest?

(A) *AB*
(B) *AC*
(C) *AD*
(D) *CD*
(E) *BD*

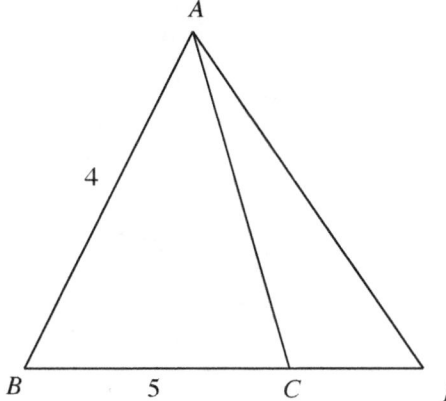

The figure is not drawn to scale

[Multiple-choice Question – Select One Answer Choice Only]
5.   In the figure, ABCD is a rectangle, and the area of ΔACE is 10. What is the area of the rectangle?

   (A)   18
   (B)   22.5
   (C)   36
   (D)   44
   (E)   45

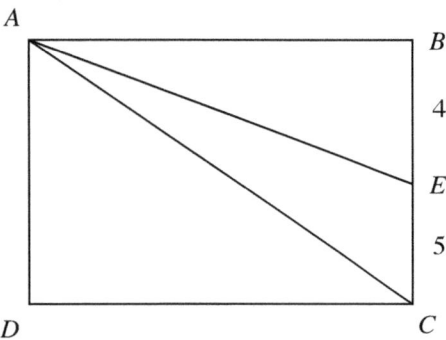

[Numeric Entry Question]
6.   In the figure A, B and C are points on the circle. What is the value of x ?

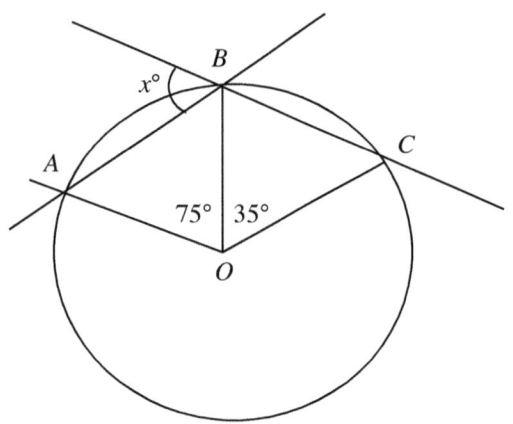

[Multiple-choice Question – Select One Answer Choice Only]
7. In the figure shown, *ABCDEF* is a regular hexagon and *AOF* is an equilateral triangle. The perimeter of △*AOF* is 2*a* feet. What is the perimeter of the hexagon in feet?

   (A) 2*a*
   (B) 3*a*
   (C) 4*a*
   (D) 6*a*
   (E) 12*a*

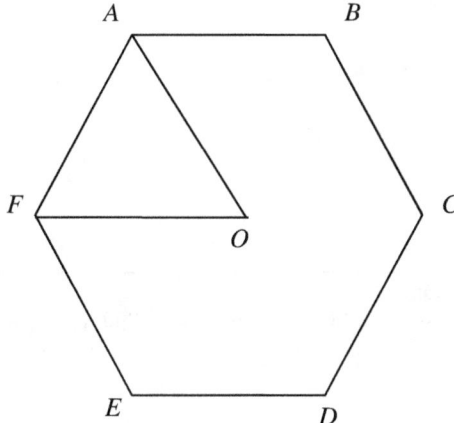

[Multiple-choice Question – Select One Answer Choice Only]
8. In triangle *ABC*, *AB* = 5 and *AC* = 3. Which one of the following is the measure of the length of side *BC* ?

   (A) *BC* < 7
   (B) *BC* = 7
   (C) *BC* > 7
   (D) *BC* ≤ 7
   (E) It cannot be determined from the information given

[Multiple-choice Question – Select One Answer Choice Only]
9. Let C and K be constants. If $x^2 + Kx + 5$ factors into $(x + 1)(x + C)$, the value of K is

    (A) 0
    (B) 5
    (C) 6
    (D) 8
    (E) not enough information

[Multiple-choice Question – Select One Answer Choice Only]
10. The product of two numbers x and y is twice the sum of the numbers. What is the sum of the reciprocals of x and y?

    (A) 1/8
    (B) 1/4
    (C) 1/2
    (D) 2
    (E) 4

[Multiple-choice Question – Select One Answer Choice Only]
11. If $l + t = 4$ and $l + 3t = 9$, then which one of the following equals $l + 2t$?

    (A) 13/2
    (B) 19/2
    (C) 15/2
    (D) 17/3
    (E) 21/4

[Multiple-choice Question – Select One Answer Choice Only]

12. A system of equations is as shown below

$x + l = 6$
$x - m = 5$
$x + p = 4$
$x - q = 3$

What is the value of $l + m + p + q$?

(A) 2
(B) 3
(C) 4
(D) 5
(E) 6

USE THIS SPACE FOR SCRATCHWORK.

[Multiple-choice Question – Select One or More Answer Choices]

13. A precious stone if dropped breaks into pieces of equal size and weight. However, the stone is of a rare kind and the price of the stone is always evaluated as a proportion of the square of its weight. A stone can break in any number of pieces resulting in a new price per piece. Which of the following ratios of the original stone price and the net price of pieces can the stone break into?

(A) 1 : 1
(B) 2 : 1
(C) 1 : 2
(D) 4 : 1
(E) 3 : 1
(F) 3 : 2
(G) 5 : 3

USE THIS SPACE FOR SCRATCHWORK.

14.

Column A		Column B
	a and b are positive. $(a + 6) : (b + 6) = 5 : 6$	
b		1

USE THIS SPACE FOR SCRATCHWORK.

15.

Column A	$x \neq 3$ and $x \neq 6$	Column B
$\dfrac{2x^2 - 72}{x - 6}$		$\dfrac{2x^2 - 18}{x - 3}$

16.

Column A	12 students from section A and 15 students from section B failed an Anthropology exam. Thus, equal percentage of attendees failed the exam from the sections.	Column B
Number of attendees for the exam from section A		Number of attendees for the exam from section B

17.

Column A	The annual exports of the company NeuStar increased by 25% last year. This year, it increased by 20%.	Column B
Increase in exports last year		Increase in exports in the current year

[Multiple-choice Question – Select One Answer Choice Only]
18. The costs of equities of type A and type B (in dollars) are positive integers. If 4 equities of type A and 5 equities of type B together costs 27 dollars, what is the total cost of 2 equities of type A and 3 equities of type B in dollars?

   (A) 15
   (B) 24
   (C) 35
   (D) 42
   (E) 55

[Multiple-choice Question – Select One or More Answer Choices]
19. In a sequence of positive integers, a_n, the nth term is defined as $(a_{n-1} - 1)^2$. If 9 is one of the terms of the sequence, then what are the two terms immediately next to 9?

   (A) 4
   (B) 9
   (C) 63
   (D) 64
   (E) 63^2

[Multiple-choice Question – Select One Answer Choice Only]
20. In the town of Windsor, 250 families have at least one car while 60 families have at least two cars. How many families have exactly one car?

   (A) 30
   (B) 190
   (C) 280
   (D) 310
   (E) 420

GRE Math Tests

[Multiple-choice Question – Select One Answer Choice Only]

21. The probability that Tom will win the Booker prize is 0.5, and the probability that John will win the Booker prize is 0.4. There is only one Booker prize to win. What is the probability that at least one of them wins the prize?

   (A) 0.2
   (B) 0.4
   (C) 0.7
   (D) 0.8
   (E) 0.9

[Multiple-choice Question – Select One Answer Choice Only]

22. A certain brand of computer can be bought with or without a hard drive. The computer with the hard drive costs 2,900 dollars. The computer without the hard drive costs 1,950 more than the hard drive alone. What is the cost of the hard drive?

   (A) 400
   (B) 450
   (C) 475
   (D) 500
   (E) 525

23.
Column A	Column B
The square root of 7/8	The square of 7/8

[Multiple-choice Question – Select One Answer Choice Only]

24. For all $p \neq 2$ define $p*$ by the equation $p* = \dfrac{p+5}{p-2}$. If $p = 3$, then $p* =$

   (A) 8/5
   (B) 8/3
   (C) 4
   (D) 5
   (E) 8

Answers and Solutions Test 9:

Question	Answer
1.	B
2.	00
3.	120
4.	C, E
5.	C
6.	B
7.	C
8.	E
9.	C
10.	C
11.	A
12.	A
13.	B, D, E
14.	A
15.	A
16.	B
17.	C
18.	A
19.	A, D
20.	B
21.	D
22.	C
23.	A
24.	E

If you answered 18 out of 24 questions in this test, you are likely to score 750+ in your GRE.

1. If x and y are positive, then Column B is positive and therefore larger than zero. If x and y are negative, then Column B is still positive since a negative divided by a negative yields a positive. This covers all possible signs for x and y. The answer is (B).

2. Since each of the two integers a and b ends with the same digit, the difference of the two numbers ends with 0. For example $642 - 182 = 460$, and 460 ends with 0. The square of a number ending with 0 also ends with 0. For example, $20^2 = 400$. Fill the grid with 00.

3. Since the angle made by a line is 180°, $z + y = 180$. Also, we are given that $y - z = 60$. Adding the equations yields

$$z + y + y - z = 180 + 60$$
$$2y = 240$$
$$y = 120$$

Since the lines l and m are parallel, the alternate exterior angles x and y are equal. Hence, x equals 120. Enter in the grid.

4. In a right triangle, the angle opposite the longest side is the right angle. Since from figure $AB = 4 < BC = 5 < BD$, AB is not the longest side. Hence, $\angle D$ is not the right angle. Hence, one of the other angles $\angle A$ or $\angle B$ is right angled. Hence, BD or AD could be the longest. The answer is (C) and (E).

5. The formula for the area of a triangle is 1/2 × *base* × *height*. Hence, the area of △*ACE* (which is given to equal 10) is 1/2 × *CE* × *AB*. Hence, we have

$$1/2 \times CE \times AB = 10$$
$$1/2 \times 5 \times AB = 10 \quad \text{(from the figure, } CE = 5\text{)}$$
$$AB = 4$$

Now, the formula for the area of a rectangle is *length* × *width*. Hence, the area of the rectangle *ABCD* = *BC* × *AB*

$$= (BE + EC) \times (AB) \quad \text{from the figure, } BC = BE + EC$$
$$= (4 + 5) \times 4 \quad \text{from the figure, } BE = 4 \text{ and } EC = 5$$
$$= 9 \times 4$$
$$= 36$$

The answer is (C).

---

6. *OA* and *OB* are radii of the circle. Hence, angles opposite them in △*AOB* are equal: ∠*OAB* = ∠*ABO*. Summing the angles of △*AOB* to 180° yields ∠*OAB* + ∠*ABO* + ∠*AOB* = 180 or 2∠*ABO* + 75° = 180 [since ∠*OAB* = ∠*OBA*); ∠*OBA* = (180 − 75)/2 = 105/2].

Similarly, *OB* equals *OC* (radii of a circle are equal) and angles opposite them in △*BOC* are equal: ∠*OBC* = ∠*BCO*. Summing angles of the triangle to 180° yields ∠*OBC* + ∠*BCO* + 35 = 180 or 2∠*OBC* + 35 = 180 [since ∠*OBC* = ∠*BCO*; ∠*OBC* = (180 − 35)/2 = 145/2].

Now, since an angle made by a line is 180°, we have

$$x + \angle ABO + \angle OBC = 180$$

$$x + 105/2 + 145/2 = 180$$

$$x + 250/2 = 180$$

$$x + 125 = 180$$

$$x = 180 - 125 = 55$$

The answer is (B).

---

7. We are given that *AOF* is an equilateral triangle. In an equilateral triangle, all three sides are equal and therefore the perimeter of the triangle equals (*number of sides*) × (*side length*) = 3*AF* (where *AF* is one side of the equilateral triangle). Now, we are given that the perimeter of △*AOF* is 2*a*. Hence, 3*AF* = 2*a*, or *AF* = 2*a*/3.

We are given that *ABCDEF* is a regular hexagon. In a regular hexagon, all six sides are equal and therefore the perimeter of the hexagon equals (*number of sides*) × (*side length*) = 6*AF* (where *AF* is also one side of the hexagon). Substituting *AF* = 2*a*/3 into this formula yields

$$6AF = 6(2a/3) = 4a$$

The answer is (C).

Test 9—Solutions

8. The most natural drawing is the following:

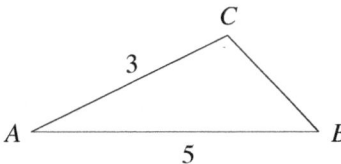

In this case, the length of side BC is less than 7. However, there is another drawing possible, as follows:

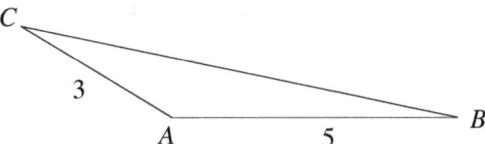

In this case, the length of side BC is greater than 7. Hence, there is not enough information to decide, and the answer is (E).

---

9. Since the number 5 is merely repeated from the problem, we eliminate (B). Further, since this is a hard problem, we eliminate (E), "not enough information."

Now, since 5 is prime, its only factors are 1 and 5. So, the constant C in the expression $(x + 1)(x + C)$ must be 5:

$$(x + 1)(x + 5)$$

Multiplying out this expression yields

$$(x + 1)(x + 5) = x^2 + 5x + x + 5$$

Combining like terms yields

$$(x + 1)(x + 5) = x^2 + 6x + 5$$

Hence, $K = 6$, and the answer is (C).

---

10. We are given that the product of x and y is twice the sum of x and y. Hence, we have $xy = 2(x + y)$.

Now, the sum of the reciprocals of x and y is

$$\frac{1}{x} + \frac{1}{y} =$$

$$\frac{y + x}{xy} =$$

$$\frac{x + y}{2(x + y)} =$$

$$\frac{1}{2}$$

The answer is (C).

177

11. Adding the two given equations $l + t = 4$ and $l + 3t = 9$ yields

$$(l + t) + (l + 3t) = 4 + 9$$
$$2l + 4t = 13$$
$$l + 2t = 13/2 \quad \text{by dividing both sides by 2}$$

The answer is (A).

---

12. The given system of equations is

$$x + l = 6$$
$$x - m = 5$$
$$x + p = 4$$
$$x - q = 3$$

Subtracting the second equation from the first one yields

$$(x + l) - (x - m) = 6 - 5$$
$$l + m = 1 \quad \ldots (1)$$

Subtracting the fourth equation from the third one yields

$$(x + p) - (x - q) = 4 - 3$$
$$p + q = 1 \quad \ldots (2)$$

Adding equations (1) and (2) yields

$$(l + m) + (p + q) = 1 + 1 = 2$$
$$l + m + p + q = 2$$

The answer is (A).

---

13. Suppose $p = kw^2$, where p is price of a single piece of weight w, and k is the constant of proportionality.

Now, suppose the stone breaks into n pieces of equal sizes. Then, the weight of each of the n pieces must be w/n. The price of each piece must be $k(w/n)^2 = kw^2/n^2$, and the price of n of those pieces will be

$$n \times kw^2/n^2 = kw^2/n$$

The ratio of the price of the bigger piece to the net price of the n pieces must be $kw^2 : kw^2/n = 1 : 1/n = n : 1 =$ Positive integer more than $1 : 1$ (n is a positive integer greater than 1, we know the stone has broken).

Choice (A) $1 : 1$ should be eliminated because n is not 1.
Choice (B) $2 : 1$ should be acceptable assuming n is 2.
Choice (C) $1 : 2 = 1/2 : 1$, n is integer, not fraction. Reject.
Choice (D) $4 : 1$—Certainly in $n : 1$ format assuming n is 4. Accept.
Choice (E) $3 : 1$—Certainly in $n : 1$ format assuming n is 3. Accept.
Choice (F) $3 : 2 = 3/2 : 1$—n is not fraction. Eliminate.
Choice (G) $5 : 3 = 5/3 : 1$—n is not fraction. Eliminate.

The answers are (B), (D), and (E).

Test 9—Solutions

---

14. Forming the ratio yields $\dfrac{a+6}{b+6} = \dfrac{5}{6}$. Multiplying both sides of the equation by $6(b+6)$ yields

$$6(a+6) = 5(b+6)$$
$$6a + 36 = 5b + 30$$
$$6a = 5b - 6$$
$$a = 5b/6 - 1$$
$$0 < 5b/6 - 1 \qquad \text{since } a \text{ is positive}$$
$$1 < 5b/6$$
$$6/5 < b$$
$$1.2 < b$$
$$1 < 1.2 < b \qquad \text{since } 1 < 1.2$$
$$\text{Column B} < 1.2 < \text{Column A}$$

Hence, the answer is (A).

---

15. Start by factoring 2 from the numerator of each fraction:

$$\dfrac{2(x^2 - 36)}{x - 6} \qquad\qquad\qquad \dfrac{2(x^2 - 9)}{x - 3}$$

Next, apply the Difference of Squares Formula $a^2 - b^2 = (a+b)(a-b)$ to the expressions in both columns:

$$\dfrac{2(x+6)(x-6)}{x-6} \qquad\qquad\qquad \dfrac{2(x+3)(x-3)}{x-3}$$

Next, cancel the term $x - 6$ in Column A and the term $x - 3$ in Column B:

$$2(x + 6) \qquad\qquad\qquad 2(x + 3)$$

Next, distribute the 2 in each expression:

$$2x + 12 \qquad\qquad\qquad 2x + 6$$

Finally, cancel $2x$ from both columns:

$$12 \qquad\qquad\qquad 6$$

Hence, Column A is greater than Column B, and answer is (A).

---

16. Given that an equal percent of attendees failed the exam in sections A and B. Let x be the percent. If a students took the exam from section A and b students took the exam from section B, then number of students who failed from the sections would be $a(x/100)$ and $b(x/100)$, respectively. Given that the two equal 12 and 15, respectively, we have $a(x/100) = 12$ and $b(x/100) = 15$. Since $12 < 15$, $a(x/100) < b(x/100)$. Canceling $x/100$ from both sides yields $a < b$. Hence, Column B > Column A, and the answer is (B).

---

17. Let x be the annual exports of the company before last year. It is given that the exports increased by 25% last year. The increase (Column A) equals $(25/100)x = x/4$, and the net exports equals $x + x/4 = 5x/4$. Now, exports increased by 20% this year. So, the increase (Column B) equals $(20/100)(5x/4) = x/4$. Hence, both columns equal $x/4$, and the answer is (C).

18. Let m and n be the costs of the equities of type A and type B, respectively. Since the costs are integers (given), m and n must be positive integers.

We have that 4 equities of type A and 5 equities of type B together cost 27 dollars. Hence, we have the equation $4m + 5n = 27$. Since m is a positive integer, $4m$ is a positive integer; and since n is a positive integer, $5n$ is a positive integer. Let $p = 4m$ and $q = 5n$. So, p is a multiple of 4 and q is a multiple of 5 and $p + q = 27$. Subtracting q from both sides yields $p = 27 - q$ [(a positive multiple of 4) equals $27 -$ (a positive multiple of 5)]. Let's seek such a solution for p and q:

If $q = 5, p = 27 - 5 = 22$, not a multiple of 4. Reject.

If $q = 10, p = 27 - 10 = 17$, not a multiple of 4. Reject.

If $q = 15, p = 27 - 15 = 12$, a multiple of 4. Acceptable. So, $n = p/4 = 3$ and $m = q/5 = 3$.

The following checks are not actually required since we already have an acceptable solution.

If $q = 20, p = 27 - 20 = 7$, not a multiple of 4. Reject.

If $q = 25, p = 27 - 25 = 2$, not a multiple of 4. Reject.

If $q \geq 30, p \leq 27 - 30 = -3$, not positive. Reject.

Hence, the cost of 2 equities of type A and 3 equities of type B is $2m + 3n = 2 \cdot 3 + 3 \cdot 3 = 15$. The answer is (A).

---

19. In the above solution, you seem to assume that 9 is the a_{n-1} term. Why? It seems more natural to assume that 9 is the a_n term. Consider the following rewrite:

The sequence is defined as $a_n = (a_{n-1} - 1)^2$. Suppose 9 is the a_n term. Then the term immediately after it is a_{n+1}. To create the a_{n+1} term, replace n with $n + 1$ in the formula $a_n = (a_{n-1} - 1)^2$:

$a_{n+1} = (a_n - 1)^2 = (9 - 1)^2 = 8^2 = 64$
Select (D).

Since 9 is the a_n term, the term immediately before it is a_{n-1}. Replacing a_n with 9 in the formula $a_n = (a_{n-1} - 1)^2$ yields

$(a_{n-1} - 1)^2 = 9$
$a_{n-1} - 1 = \pm 3$ by taking the square root of both sides of the equation
$a_{n-1} = 1 \pm 3 = -2$ or 4 we reject -2 because the sequence is given to be positive
Select (A).

So, the term immediately before 9 is 4 and the one immediately after 9 is 64. The select choices (A) and (D).

Test 9—Solutions

20. Let A be the set of families having exactly one car. Then the question is how many families are there in set A.

Next, let B be the set of families having exactly two cars, and let C be the set of families having more than two cars.

Then the set of families having at least one car is the collection of the three sets A, B, and C.

The number of families in the three sets A, B, and C together is 250 (given) and the number of families in the two sets B and C together is 60 (given).

Now, since set A is the difference between a set containing the three families of A, B, and C and a set of families of B and C only, the number of families in set A equals

(the number of families in sets A, B, and C together) – (the number of families in sets B and C) =

250 – 60 =

190

The answer is (B).

---

21. The probability that Tom passes is 0.3. Hence, the probability that Tom does not pass is $1 - 0.3 = 0.7$.

The probability that John passes is 0.4. Hence, the probability that John does not pass is $1 - 0.4 = 0.6$.

At least one of them gets a degree in three cases:

    1) Tom passes and John does not
    2) John passes and Tom does not
    3) Both Tom and John pass

Hence, the probability of at least one of them passing equals

    (The probability of Tom passing and John not) +
    (The probability of John passing and Tom not) +
    (The probability of both passing)

    (The probability of Tom passing and John not) =
    (The probability of Tom passing) × (The probability of John not) =
    $0.3 \times 0.6 =$
    0.18

    (The probability of John passing and Tom not) =
    (The probability of John passing) × (The probability of Tom not) =
    $0.4 \times 0.7 =$
    0.28

    (The probability of both passing) =
    (The probability of Tom passing) × (The probability of John passing) =
    $0.3 \times 0.4 =$
    0.12

Hence, the probability of at least one passing is $0.18 + 0.28 + 0.12 = 0.58$. The answer is (D).

Method II:

The probability of Tom passing is 0.3. Hence, the probability of Tom not passing is $1 - 0.3 = 0.7$.

The probability of John passing is 0.4. Hence, the probability of John not passing is $1 - 0.4 = 0.6$.

At least one of Tom and John passes in all the cases except when both do not pass.

Hence,

The probability of at least one passing =

$1 -$ (the probability of neither passing) =

$1 -$ (The probability of Tom not passing) $\times$ (The probability of John not passing) =

$1 - 0.7 \times 0.6 =$

$1 - 0.42 =$

0.58

The answer is (D).

---

22. Let C be the cost of the computer without the hard drive, and let H be the cost of the hard drive. Then translating *"The computer with the hard drive costs 2,900 dollars"* into an equation yields $C + H = 2,900$.

Next, translating *"The computer without the hard drive costs 1,950 dollars more than the hard drive alone"* into an equation yields $C = H + 1,950$.

Combining these equations, we get the system:

$$C + H = 2,900$$
$$C = H + 1,950$$

Solving this system for H, yields $H = 475$. The answer is (C).

---

23. Squaring a fraction between 0 and 1 makes it smaller, and taking the square root of it makes it larger. Therefore, Column A is greater. The answer is (A).

---

24. Substituting $p = 3$ into the equation $p^* = \dfrac{p+5}{p-2}$ gives $3^* = \dfrac{3+5}{3-2} = \dfrac{8}{1} = 8$. The answer is (E).

Test 10

GRE Math Tests

Questions: 24
Time: 45 minutes

[Multiple-choice Question – Select One or More Answer Choices]
1. If $(x - 3)(x + 2) = (x - 2)(x + 3)$, then x could be

   (A) -3
   (B) -2
   (C) 0
   (D) 2
   (E) 3

USE THIS SPACE FOR SCRATCHWORK.

[Quantitative Comparison Question]

2. | Column A | p and q are two positive integers and $p/q = 7.5$ | Column B |
    |---|---|---|
    | q | | 15 |

USE THIS SPACE FOR SCRATCHWORK.

[Numeric Entry Question]
3. The number 3072 is divisible by both 6 and 8. What is the first integer larger than 3072 that is also divisible by both 6 and 8?

USE THIS SPACE FOR SCRATCHWORK.

[Quantitative Comparison Question]

4.   Column A

A *palindrome number* is a number that reads the same forward or backward. For example, 787 is a palindrome number.

Column B

Smallest palindrome number greater than 233

Smallest palindrome greater than 239

[Quantitative Comparison Question]

5.   Column A

A set has exactly five consecutive positive integers.

Column B

The percentage decrease in the average of the numbers when one of the numbers is dropped from the set

20%

[Quantitative Comparison Question]

6.

Column A

Column B

1

A 10-foot ladder is leaning against a vertical wall. The top of the ladder touches the wall at a point 8 feet above the ground. The base of the ladder slips 1 foot away from the wall.

The distance the top of the ladder slides down the wall

[Multiple-choice Question – Select One or More Answer Choices]
7. In the figure, lines *l* and *m* are parallel. Which of the following, if true, makes lines *p* and *q* parallel?

   (A) $a = b$
   (B) $a = c$
   (C) $a = d$
   (D) $d = b$
   (E) $b = c$

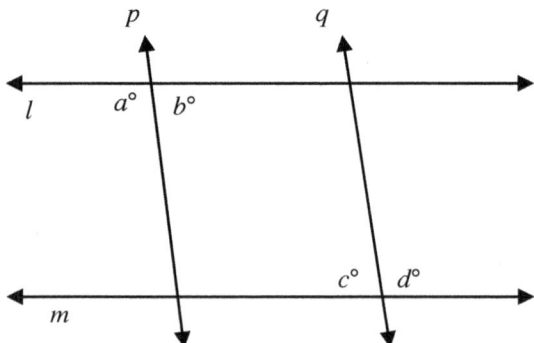

[Numeric Entry Question]
8. In the figure, the area of rectangle ABCD is 45. What is the area of the square EFGH in terms of the same units?

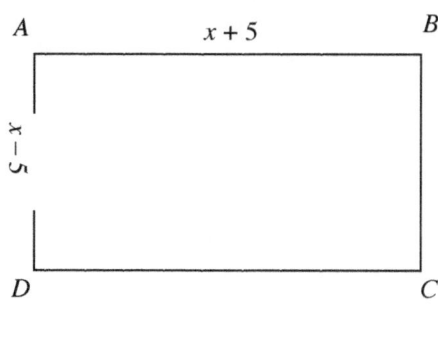

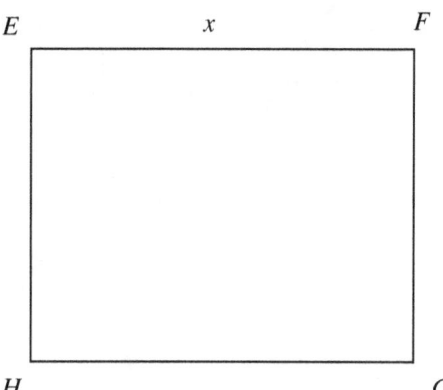

USE THIS SPACE FOR SCRATCHWORK.

[Multiple-choice Question – Select One Answer Choice Only]

9. If *ABCD* is a square and the area of △*AFG* is 10, then what is the area of △*AEC*?

(A) 5
(B) $\dfrac{10}{\sqrt{2}}$
(C) $\dfrac{10}{\sqrt{3}}$
(D) 10
(E) 20

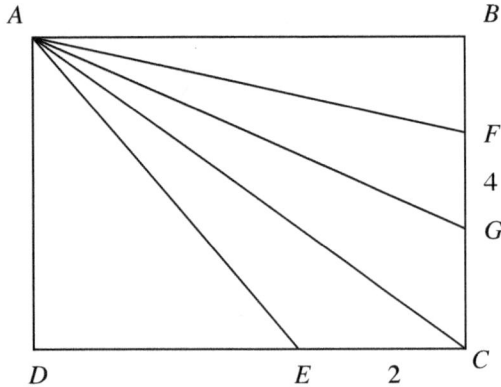

[Multiple-choice Question – Select One or More Answer Choices]
10. Which of the following relations is true regarding the angles of the quadrilateral shown in the figure?

   (A) $\angle A = \angle C$
   (B) $\angle B > \angle D$
   (C) $\angle A < \angle C$
   (D) $\angle B = \angle D$
   (E) $\angle A = \angle B$

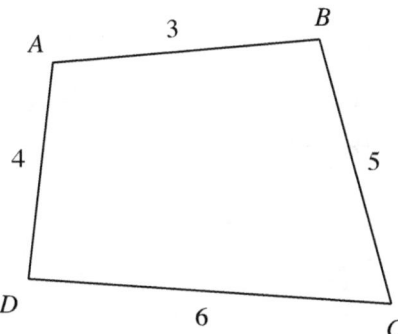

The figure is not drawn to scale.

[Multiple-choice Question – Select One Answer Choice Only]
11. If $x > y$ and $x < 0$, then which of the following must be true?

   (I) $\dfrac{1}{x} < \dfrac{1}{y}$
   (II) $\dfrac{1}{x-1} < \dfrac{1}{y-1}$
   (III) $\dfrac{1}{x+1} < \dfrac{1}{y+1}$

   (A)   I only
   (B)   II only
   (C)   III only
   (D)   I and II only
   (E)   I and III only

[Quantitative Comparison Question]

12. Column A
1/2 + 1/4 + 1/8 + 1/16

Column B
1

[Numeric Entry Question]

13. What is the numerical value of the expression $\dfrac{2^x + 2^{x-1}}{2^{x+1} - 2^x}$?

[Numeric Entry Question]
14. The sum of two numbers is 13, and their product is 30. What is the sum of the squares of the two numbers?

[Numeric Entry Question]
15. At Stephen Stores, 3 pounds of cashews cost $8. What is the cost in cents of a bag weighing 9 ounces?

[Multiple-choice Question – Select One or More Answer Choices]
16. The Savings of an employee equals Income minus Expenditure. If their Incomes ratio is 1 : 2 : 3 and their Expenses ratio is 3 : 2 : 1, then what is the order of the employees A, B, and C in the increasing order of the size of their savings?

   (A) $A > B > C$
   (B) $A > C > B$
   (C) $B > A > C$
   (D) $B > C > A$
   (E) $C > B > A$

[Numeric Entry Question]
17. If $x \neq 3$ and $x \neq 6$, then $\dfrac{2x^2 - 72}{x - 6} - \dfrac{2x^2 - 18}{x - 3} =$

[Multiple-choice Question – Select One or More Answer Choices]
18. 8 is 4% of a, and 4 is 8% of b. c equals b/a. What is the value of c?

   (A)   1/32
   (B)   1/4
   (C)   1
   (D)   4
   (E)   32

[Multiple-choice Question – Select One Answer Choice Only]
19. The total income of Mr. Teng in the years 2003, 2004, and 2005 was $36,400. His income increased by 20% each year. What was his income in 2005?

    (A) 5,600
    (B) 8,800
    (C) 10,000
    (D) 12,000
    (E) 14,400

[Multiple-choice Question – Select One Answer Choice Only]
20. Waugh jogged to a restaurant at x miles per hour, and jogged back home along the same route at y miles per hour. He took 30 minutes for the whole trip. If the restaurant is 2 miles from home along the route he took, what is the average speed in miles per hour at which he jogged for the whole trip?

    (A) 0.13
    (B) 0.5
    (C) 2
    (D) 4
    (E) 8

[Multiple-choice Question – Select One Answer Choice Only]
21. When the price of oranges is lowered by 40%, 4 more oranges can be purchased for $12 than can be purchased for the original price. How many oranges can be purchased for 24 dollars at the original price?

    (A) 8
    (B) 12
    (C) 16
    (D) 20
    (E) 24

[Multiple-choice Question – Select One Answer Choice Only]
22. In the sequence a_n, the nth term is defined as $(a_{n-1} - 1)^2$. If $a_3 = 64$, then what is the value of a_2?

   (A) 2
   (B) 3
   (C) 4
   (D) 5
   (E) 9

[Multiple-choice Question – Select One Answer Choice Only]
23. Ana is a girl and has the same number of brothers as sisters. Andrew is a boy and has twice as many sisters as brothers. Ana and Andrew are the children of Emma. How many children does Emma have?

   (A) 2
   (B) 3
   (C) 5
   (D) 7
   (E) 8

[Multiple-choice Question – Select One Answer Choice Only]
24. If $x + y = 5$, then what is the probability that x is positive?

   (A) 1/5
   (B) 4/5
   (C) 1/2
   (D) 1/4
   (E) 1/3

Answers and Solutions Test 10:

Question	Answer
1.	C
2.	D
3.	3096
4.	C
5.	B
6.	A
7.	C, E
8.	70
9.	A
10.	B
11.	D
12.	B
13.	3/2 OR 1.5
14.	109
15.	150
16.	E
17.	6
18.	B
19.	E
20.	E
21.	B
22.	E
23.	D
24.	C

If you got 18/24 correct on this test, you are likely to get 750+ on the actual GRE by the time you complete all the tests in the book.

---

1. If $x = 0$, then the equation $(x - 3)(x + 2) = (x - 2)(x + 3)$ becomes

$$(0 - 3)(0 + 2) = (0 - 2)(0 + 3)$$

$$(-3)(2) = (-2)(3)$$

$$-6 = -6$$

The answer is (C). The other choices yield zero on one-side of the equation and a non-zero number on the other side.

Method II:
Expanding the equation $(x - 3)(x + 2) = (x - 2)(x + 3)$ yields

$$x^2 - 3x + 2x - 6 = x^2 - 2x + 3x - 6$$
$$x^2 - x - 6 = x^2 + x - 6$$
$$-x - 6 = x - 6$$
$$-2x = 0$$
$$x = 0$$

The equation $(x - 3)(x + 2) = (x - 2)(x + 3)$ appears to be of degree 2, so two solutions may be possible. However, the x^2 cancels from both sides of the equation, so it is actually of degree 1. Hence, only one solution is possible. Choose just Choice (C).

GRE Math Tests

2. Let's solve the equation $p/q = 7.5$ for q. Multiplying both sides by q yields $p = 7.5q$. Now, dividing both sides by 7.5 yields $q = p/7.5$. Since q is given to be an integer, 7.5 must divide into p evenly. That is, p is a multiple of 7.5. The smallest such integer multiple is 15 (= 7.5 x 2). In this case, $q = 15/7.5 = 2$. Here, q (Column A) is smaller than 15 (Column B). But there are much larger multiples of 7.5. For example, 120 (= 7.5 x 16). In this case, $q = 120/7.5 = 16$. Here, Column A is larger than Column B. So, we have a double case, and the answer is (D).

3. Any number divisible by both 6 and 8 must be a multiple of the least common multiple of the two numbers, which is 24. Hence, any such number can be represented as $24n$. If 3072 is one such number and is represented as $24n$, then the next such number should be $24(n + 1) = 24n + 24 = 3072 + 24 = 3096$. Hence, enter 3096 in the grid.

4. A palindrome number reads the same forward or backward. There is no palindrome number between 233 through 239 since none of the numbers read the same both forward and backward. Hence, the palindrome number immediately after 233 is the same as the palindrome number immediate after 239. Hence, Column A and Column B refer to the same number, and the answer is (C).

5. The average of the five consecutive positive integers, say, $a, a + 1, a + 2, a + 3$, and $a + 4$ is

$$\frac{a + (a+1) + (a+2) + (a+3) + (a+4)}{5} =$$
$$\frac{5a + 10}{5} =$$
$$a + 2$$

The average decrease is a maximum when the greatest number in the set is dropped. Hence, after dropping $a + 4$, the average of the remaining numbers $a, a + 1, a + 2$, and $a + 3$ is

$$\frac{a + (a+1) + (a+2) + (a+3)}{4} =$$
$$\frac{4a + 6}{4} =$$
$$a + \frac{3}{2}$$

The percentage decrease in the average is

$$\frac{\text{Old Average} - \text{New Average}}{\text{Old Average}} \times 100 =$$
$$\frac{(a+2) - \left(a + \frac{3}{2}\right)}{a + 2} \times 100 =$$
$$\frac{\frac{1}{2}}{a + 2} \times 100$$

The percentage is a maximum when a takes the minimum possible value. Since a is a positive integer, the minimum value of a is 1. Hence, the maximum possible percentage equals

$$\frac{\frac{1}{2}}{1+2} \times 100 =$$

$$\frac{\frac{1}{2}}{3} \times 100 =$$

$$\frac{1}{2 \cdot 3} \times 100 =$$

$$\frac{100}{6} =$$

$$16.66\%$$

Hence, the maximum possible value of Column A is 16.66%, which is less than Column B. Hence, the answer is (B).

Method II:
Since we are not told what the five consecutive positive integers are, we can chose any five consecutive positive integers. Let the five consecutive positive integers be 1, 2, 3, 4, 5. Then the average is

$$\frac{1+2+3+4+5}{5} = \frac{15}{5} = 3$$

The percentage decrease in the average of these numbers will be greatest when the largest number is deleted. To this end, we delete 5 from the set and form the new average:

$$\frac{1+2+3+4}{4} = \frac{10}{4} = \frac{5}{2}$$

The percentage decrease in the average is

$$\frac{\text{Old Average} - \text{New Average}}{\text{Old Average}} \times 100 =$$

$$\frac{3-\frac{5}{2}}{3} \times 100 =$$

$$\frac{\frac{1}{2}}{3} \times 100 =$$

$$\frac{1}{6} \times 100 =$$

$$\frac{100}{6} <$$

$$16.67\%$$

Hence, the maximum possible value of Column A is less than 16.67%, which is less than Column B. Hence, the answer is (B).

6. We can immediately eliminate (C) because that would be too easy. Let y be the distance the top of the ladder slides down the wall, let h be the height of the new resting point of the top of the ladder, and x be the original distance of the bottom of the ladder from the wall:

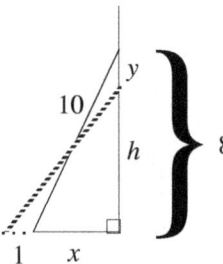

Applying The Pythagorean Theorem to the original triangle yields	$x^2 + 8^2 = 10^2$
Solving this equation for x yields	$x = 6$
Hence, the base of the final triangle is	$1 + 6 = 7$
Applying The Pythagorean Theorem to the final triangle yields	$h^2 + 7^2 = 10^2$
Solving this equation for h yields	$h = \sqrt{51}$

Adding this information to the drawing yields

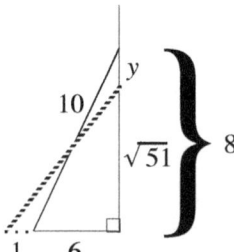

From the drawing, $y = 8 - \sqrt{51} < 8 - 7 = 1$, since $\sqrt{51} \approx 7.1$. Hence, Column A is larger, and the answer is (A).

7.

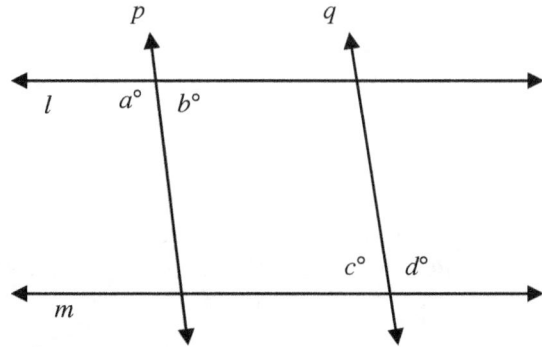

Superimposing parallel line m on line l yields a figure like this:

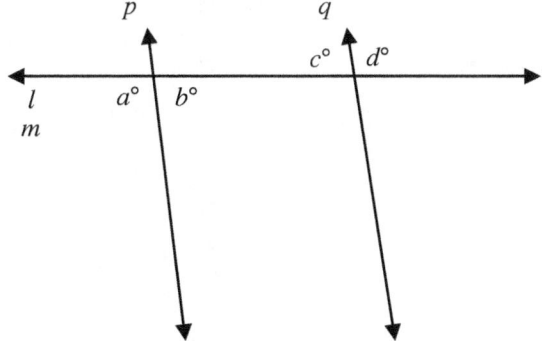

Now, when two lines (here p and q) cut by a transversal (here l) are parallel, we have

    (I)   Corresponding angles are equal: No corresponding angles are listed in the figure.
    (II)   Alternate interior angles are equal: $b = c$. In choice (E).
    (III) Alternate exterior angles are equal: $a = d$. In choice (C).
    (IV) Interior angles are supplementary.
    (V)   Exterior angles are supplementary.

The answer is (C) and (E).

---

8. The formula for the area of the rectangle is *length × width*. Hence, the area of rectangle *ABCD* is

$$AB \times AD = (x + 5)(x - 5) = x^2 - 5^2$$

We are given that the area is 45, so $x^2 - 5^2 = 45$. Solving the equation for x^2 yields

$$x^2 = 45 + 25 = 70$$

Now, the formula for the area of a square is $side^2$. Hence, the area of square *EFGH* is $EF^2 = x^2$. Now, as shown earlier, x^2 equals 70. Enter in the grid.

9. The formula for the area of a triangle is $1/2 \times base \times height$. By the formula, the area of $\triangle AFG$ (which is given to be 10) is $1/2 \times FG \times AB$. Hence, we have

$1/2 \times 4 \times AB = 10$ given that the area of $\triangle AFG = 10$
$AB = 5$

Also, by the same formula, the area of $\triangle AEC$ is

$1/2 \times EC \times DA$
$= 1/2 \times 2 \times DA$ from the figure, $EC = 2$ units
$= 1/2 \times 2 \times AB$ $ABCD$ is a square. Hence, side DA = side AB
$= 1/2 \times 2 \times 5$
$= 5$

The answer is (A).

---

10. Joining the opposite vertices B and D on the quadrilateral yields the following figure:

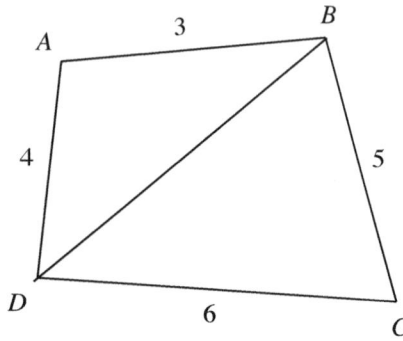

Since the angle opposite the longer side in a triangle is greater, we have

$AD (= 4) > AB (= 3)$ (from the figure). Hence, $\angle ABD > \angle BDA$ and

$CD (= 6) > BC (= 5)$ (from the figure). Hence, $\angle DBC > \angle CDB$.

Adding the two known inequalities $\angle ABD > \angle BDA$ and $\angle DBC > \angle CDB$ yields

$\angle ABD + \angle DBC > \angle BDA + \angle CDB$

$\angle B > \angle D$ Since from the figure, $\angle ABD + \angle DBC$ equals $\angle ABC (= \angle B)$ and $\angle BDA + \angle CDB$ equals $\angle CDA (= \angle D)$

Hence, the answer is (B).

Test 10—Solutions

11. We are given the inequality $x > y$ and that x is negative. Since $x > y$, y must also be negative. Hence, xy, the product of two negative numbers, must be positive. Dividing the inequality by the positive expression xy yields $\frac{x}{xy} > \frac{y}{xy}$, or $\frac{1}{y} > \frac{1}{x}$. Rearranging yields $\frac{1}{x} < \frac{1}{y}$. Hence, I is true.

Since x is negative, $x - 1$ is also negative. Similarly, since y is negative, $y - 1$ is also negative. Hence, the product of the two, $(x - 1)(y - 1)$, must be positive. Subtracting -1 from both sides of the given inequality $x > y$ yields $x - 1 > y - 1$. Dividing the inequality by the positive value $(x - 1)(y - 1)$ yields $\frac{1}{y-1} > \frac{1}{x-1}$. Rearranging the inequality yields $\frac{1}{x-1} < \frac{1}{y-1}$. Hence, II must be true.

Though x is negative, it is possible that $x + 1$ is positive while $y + 1$ is still negative. Here, $\frac{1}{x+1} < \frac{1}{y+1}$ is false because the left-hand side is positive while the right-hand side is negative. Hence, III need not be true.

Hence, the answer is (D), I and II must be true.

---

12. Let's multiply both columns by 16 to clear the fractions. (Remember, this can only be done if the number you are multiplying by is positive.)

$8 + 4 + 2 + 1$ 16

15   16

Hence, Column A is less than Column B, and the answer is (B).

---

13. The term 2^{x-1} equals $\frac{2^x}{2}$, and the term 2^{x+1} equals $2^x \cdot 2$. Hence, the given expression $\frac{2^x + 2^{x-1}}{2^{x+1} - 2^x}$ becomes

$$\frac{2^x + \frac{2^x}{2}}{2^x \cdot 2 - 2^x} =$$

$$\frac{2^x\left(1 + \frac{1}{2}\right)}{2^x(2-1)} = \quad \text{by factoring out } 2^x \text{ from both numerator and denominator}$$

$$\frac{\left(1 + \frac{1}{2}\right)}{2-1} = \quad \text{by canceling } 2^x \text{ from both numerator and denominator}$$

$$\frac{3/2}{1} =$$

$$\frac{3}{2}$$

Grid-in the value.

14. Let the two numbers be x and y. Since their sum is 13, $x + y = 13$. Since their product is 30, $xy = 30$. Solving the equation $xy = 30$ for y yields $y = 30/x$. Plugging this into the equation $x + y = 13$ yields

$x + 30/x = 13$
$x^2 + 30 = 13x$ by multiplying both sides of the equation by x
$x^2 - 13x + 30 = 0$ by subtracting $13x$ from both sides of the equation
$(x - 3)(x - 10) = 0$
$x = 3$ or $x = 10$

Now, if $x = 3$, then $y = 13 - x = 13 - 3 = 10$. Hence, $x^2 + y^2 = 3^2 + 10^2 = 9 + 100 = 109$. Grid in the value.

Method II:
$(x + y)^2 = x^2 + y^2 + 2xy$. Hence, $x^2 + y^2 = (x + y)^2 - 2xy = 13^2 - 2(30) = 169 - 60 = 109$.

---

15. This problem can be solved by setting up a proportion. Note that 1 pound has 16 ounces, so 3 pounds has 48 (= 3 × 16) ounces. Now, the proportion, in cents to ounces, is

$$\frac{800}{48} = \frac{\text{cents}}{9}$$

or

$$\text{cents} = 9 \cdot \frac{800}{48} = 150$$

Grid-in the value.

---

16. We have that the incomes of A, B, and C are in the ratio $1 : 2 : 3$. Let their incomes be i, $2i$, and $3i$, respectively. Also, their expenses ratio is $3 : 2 : 1$. Hence, let their expenses be $3e$, $2e$, and e. Since the Saving = Income − Expenditure, the savings of the three employees A, B, and C is $i - 3e$, $2i - 2e$, and $3i - e$, respectively.

Now, the saving of C is greater the saving of B when $3i - e > 2i - 2e$, or $i + e > 0$ which surely is correct, since the income and expenditure, i and e, are both money and therefore positive.

Now, the saving of B is greater the saving of A when $2i - 2e > i - 3e$, or $i + e > 0$ which is surely correct, since the income and the expenditure, i and e, are both money and therefore positive.

Hence, the employees A, B, and C in the order of their savings is $C > B > A$. The answer is (E).

17. Start by factoring 2 from the numerators of each fraction:

$$\frac{2(x^2-36)}{x-6} - \frac{2(x^2-9)}{x-3}$$

Next, apply the Difference of Squares Formula $a^2 - b^2 = (a + b)(a - b)$ to both fractions in the expression:

$$\frac{2(x+6)(x-6)}{x-6} - \frac{2(x+3)(x-3)}{x-3}$$

Next, cancel the term $x - 6$ from the first fraction and $x - 3$ from the second fraction:

$$2(x + 6) - 2(x + 3) = 2x + 12 - 2x - 6 = 6$$

Grid-in the value.

---

18. 4% of a is $4a/100$. Since this equals 8, we have $4a/100 = 8$. Solving for a yields $a = 8 \cdot \frac{100}{4} = 200$.

Also, 8% of b equals $8b/100$, and this equals 4. Hence, we have $\frac{8}{100} \cdot b = 4$. Solving for b yields $b = 50$. Now, $c = b/a = 50/200 = 1/4$. The answer is (B).

---

19. Let p be the income of Mr. Teng in the year 2003.

We are given that his income increased by 20% each year. So, the income in the second year, 2004, must be $p(1 + 20/100) = p(1 + 0.2) = 1.2p$. The income in the third year, 2005, must be

$$1.2p(1 + 20/100) =$$

$$1.2p(1 + 0.2) =$$

$$1.2p(1.2) =$$

$$1.44p$$

Hence, the total income in the three years equals $p + 1.2p + 1.44p$. Since the total income is 36,400, we have the equation $p + 1.2p + 1.44p = 36,400$, or $3.64p = 36,400$, or $p = 36,400/3.64 = 10,000$. Hence, the income in the third year equals $1.44p = 1.44 \times 10,000 = 14,400$. The answer is (E).

---

20. Remember that

Average Speed = Net Distance ÷ Time Taken

We are given that the time taken for the full trip is 30 minutes. Hence, we only need the distance traveled. We are given that the restaurant is 2 miles from home. Since Waugh jogs back along the same route, the net distance he traveled equals $2 + 2 = 4$ miles. Hence, the Average Speed equals 4 miles ÷ 30 minutes = 4 miles ÷ 1/2 hour = 8 miles per hour. The answer is (E).

21. Let the original price of each orange be x dollars. Remember that *Quantity = Amount ÷ Rate*. Hence, we can purchase $12/x$ oranges for 12 dollars. After a 40% drop in price, the new price is $x(1 - 40/100) = 0.6x$ dollars per orange. Hence, we should be able to purchase $12/(0.6x) = 20/x$ oranges for the same 12 dollars. The excess number of oranges we get (for \$12) from the lower price is

$$20/x - 12/x = (1/x)(20 - 12) = (1/x)(8) = 8/x = 4 \text{ (given)}$$

Solving the equation $8/x = 4$ for x yields $x = 2$. Hence, the number of oranges that can be purchased for 24 dollars at original price x is $24/2 = 12$. The answer is (B).

---

22. Replacing n with 3 in the formula $a_n = (a_{n-1} - 1)^2$ yields $a_3 = (a_{3-1} - 1)^2 = (a_2 - 1)^2$. We are given that $a_3 = 64$. Putting this in the formula $a_3 = (a_2 - 1)^2$ yields

$$64 = (a_2 - 1)^2$$

$$a_2 - 1 = \pm 8$$

$$a_2 = -7 \text{ or } 9$$

Since we know that a_2 is the result of the square of number $[a_2 = (a_1 - 1)^2]$, it cannot be negative. Hence, pick the positive value 9 for a_2. The answer is (E).

---

23. Let the number of female children Emma has be n. Since Anna herself is one of them, she has $n - 1$ sisters. Hence, as given, she must have the same number ($= n - 1$) of brothers. Hence, the number of male children Emma has is $n - 1$. Since Andrew is one of them, Andrew has $(n - 1) - 1 = n - 2$ brothers. Now, the number of sisters Andrew has (includes Anna) is n (= the number of female children). Since Andrew has twice as many sisters as brothers, we have the equation $n = 2(n - 2)$. Solving the equation for n yields $n = 4$. Hence, Emma has 4 female children, and the number of male children she has is $n - 1 = 4 - 1 = 3$. Hence, the total number of children Emma has is $4 + 3 = 7$. The answer is (D).

---

24. Clearly, there is no constrain on x. The variable x is just as likely to be negative as positive. Hence, the probability is 1/2. The answer is (C).

Test 11

GRE Math Tests

Questions: 25
Time: 45 minutes

---

1. Column A $x < 0$ Column B
    $x^2 - x^5$ 0

---

2. Column A a and b are the digits of a two-digit number ab, and $a = b + 3$.   Column B

    The two-digit number ab 40

[Numeric Entry Question]

3. What is the minimum value of the sum of two integers whose product is 36?

[Numeric Entry Question]
4. What is the maximum value of m such that 7^m divides into 14! evenly? ($n!$ means $1 \cdot 2 \cdot 3 \cdot \ldots \cdot n$)

5. Column A — In the rectangular coordinate system shown, points A and E lie on the x-axis, and points B and D lie on the y-axis. Point C is the midpoint of the line AB, and point F is the midpoint of the line DE. — Column B

The slope of the line AB — The slope of the line DE

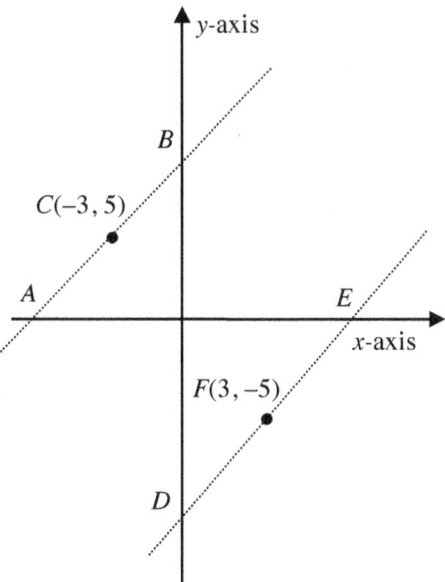

GRE Math Tests

6.   Column A   Column B

   B ―― C
   /   /
  / 65° / 5
 A 3 D

   15   The area of parallelogram ABCD

USE THIS SPACE FOR SCRATCHWORK.

[Multiple-choice Question – Select One Answer Choice only]
7. In the figure, what is the average of the five angles shown inside the circle?
   (A)   36
   (B)   45
   (C)   60
   (D)   72
   (E)   90

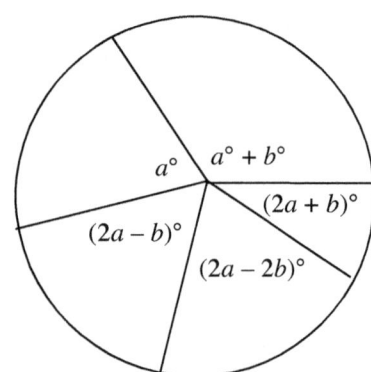

USE THIS SPACE FOR SCRATCHWORK.

[Multiple-choice Question – Select One or More Answer Choices]
8. In the figure, A, B, C, and D are points on a line in that order. If $AC = 5$, $BD = 10$, and $AD = 13$, then what is the length of BC ?

    (A) 2
    (B) 8
    (C) 15
    (D) 18
    (E) 28

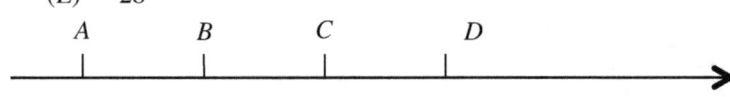

[Numeric Entry Question]
9. In the figure, $ABCD$ is a rectangle, and the area of quadrilateral $AFCE$ is equal to the area of $\triangle ABC$. What is the value of x ?

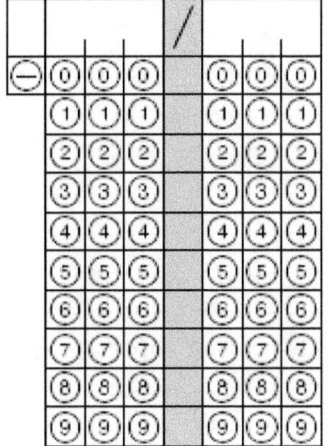

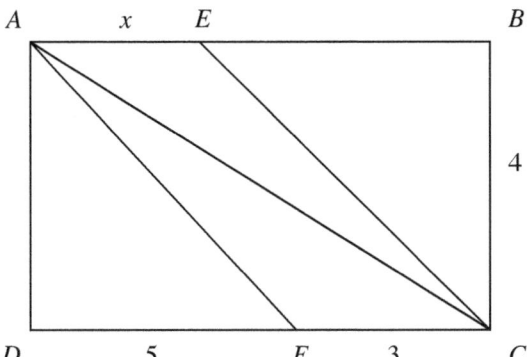

[Multiple-choice Question – Select One Answer Choice Only]

10. In the figure, *ABCD* is a square, and *BC* is tangent to the circle with radius 3. *P* is the point of intersection of the line *OC* and the circle. If *PC* = 2, then what is the area of square *ABCD* ?

(A) 9
(B) 13
(C) 16
(D) 18
(E) 25

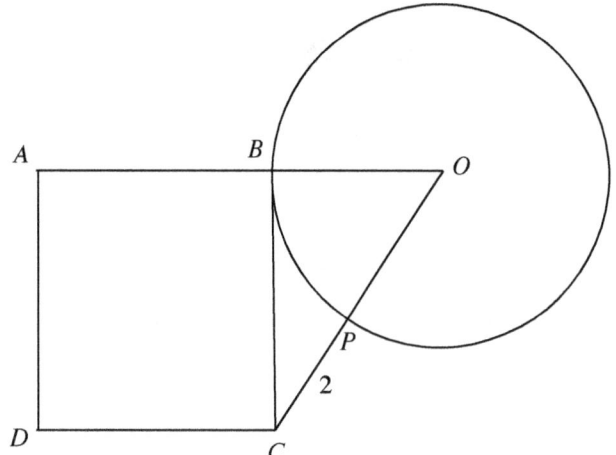

[Multiple-choice Question – Select One Answer Choice Only]
11. In the figure, h denotes the height and b the base of the triangle. If $2b + h = 6$, what is the area of the triangle?

   (A)   1
   (B)   2
   (C)   3
   (D)   4
   (E)   Not enough information

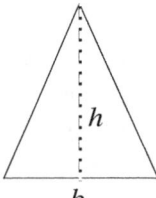

---

12.   Column A   C is the midpoint of points   Column B
                                          $A(-3, -4)$ and $B(-5, 6)$
             Slope of AC   Slope of BC

[Multiple-choice Question – Select One Answer Choice Only]
13. If $ab > 0$, then which one of the following must be true?

   (A) $a/b > 0$
   (B) $a - b > 0$
   (C) $a + b > 0$
   (D) $b - a > 0$
   (E) $a + b < 0$

[Multiple-choice Question – Select One Answer Choice Only]

14. If $p + q = 7$ and $pq = 12$, then what is the value of $\dfrac{1}{p^2} + \dfrac{1}{q^2}$?

    (A) 1/6
    (B) 25/144
    (C) 49/144
    (D) 7/12
    (E) 73/144

[Multiple-choice Question – Select One Answer Choice Only]

15. In a set of three numbers, the average of first two numbers is 2, the average of the last two numbers is 3, and the average of the first and the last numbers is 4. What is the average of three numbers?

    (A) 2
    (B) 2.5
    (C) 3
    (D) 3.5
    (E) 4

[Multiple-choice Question – Select One Answer Choice Only]
16. In the figure, what is the value of y if $x : y = 2 : 3$?

    (A) 16
    (B) 32
    (C) 48
    (D) 54
    (E) 72

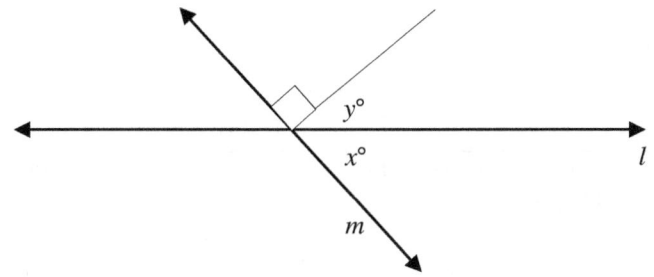

USE THIS SPACE FOR SCRATCHWORK.

[Multiple-choice Question – Select One or More Answer Choices]
17. The ratio of the numbers a, b, c, d, and e is $1 : 2 : 3 : 4 : 5$. Which of the following equal $2c$?

    (A) $a + b$
    (B) $b + d$
    (C) $a + e$
    (D) $a + b + c$
    (E) $2a + b + c$

USE THIS SPACE FOR SCRATCHWORK.

[Multiple-choice Question – Select One Answer Choice Only]
18. If x, y, and z are consecutive integers in that order, which of the following must be true?

    I. xy is even.
    II. $x - z$ is even.
    III. x^z is even.

    (A) I only
    (B) II only
    (C) III only
    (D) I and II only
    (E) I and III only

USE THIS SPACE FOR SCRATCHWORK.

19. | Column A | | Column B |
|---|---|---|
| x | $x = \dfrac{1}{1+\dfrac{1}{1+\dfrac{1}{2}}}$ | 1 |

20. | Column A | 500% of *a* equals 500*b*. *a* and *b* are positive. | Column B |
|---|---|---|
| *a* | | *b* |

[Multiple-choice Question – Select One Answer Choice Only]

21. A wheat bag weighs 5 pounds and 12 ounces. How much does the bag weigh in pounds? [1 pound = 16 ounces]

  (A) 5 1/4
  (B) 5 1/2
  (C) 5 3/4
  (D) 6 ¼
  (E) 6 3/4

[Multiple-choice Question – Select One Answer Choice Only]

22. Cyclist M leaves point P at 12 noon and travels in a straight path at a constant velocity of 20 miles per hour. Cyclist N leaves point P at 2 PM, travels the same path at a constant velocity, and overtakes M at 4 PM. What was the average speed of N?

  (A) 15
  (B) 24
  (C) 30
  (D) 35
  (E) 40

[Multiple-choice Question – Select One Answer Choice Only]
23. The sum of the first *n* terms of a series is 31, and the sum of the first *n* – 1 terms of the series is 20. What is the value of *n*th term in the series?

   (A) 9
   (B) 11
   (C) 20
   (D) 31
   (E) 51

[Multiple-choice Question – Select One Answer Choice Only]
24. What is the probability that the sum of two different numbers randomly picked (without replacement) from the set $S = \{1, 2, 3, 4\}$ is 5?

   (A) 1/5
   (B) 3/16
   (C) 1/4
   (D) 1/3
   (E) 1/2

[Multiple-choice Question – Select One Answer Choice Only]
25. In how many of ways can 5 balls be placed in 4 tins if any number of balls can be placed in any tin?

   (A) 4^3
   (B) 4^4
   (C) 5^4
   (D) 4^5
   (E) 5^5

GRE Math Tests

Answers and Solutions Test 11:

Question	Answer
1.	A
2.	D
3.	12
4.	2
5.	C
6.	A
7.	D
8.	A
9.	5
10.	C
11.	E
12.	C
13.	A
14.	B
15.	C
16.	D
17.	B, C, E
18.	D
19.	B
20.	A
21.	C
22.	E
23.	B
24.	D
25.	D

***If you got 16/25 correct on this test, you are likely to get 750+ on the actual GRE** by the time you complete all the tests in the book.*

1. If $x = -1$, then $x^2 - x^5 = 2$ and Column A is larger. If $x = -2$, then $x^2 - x^5 = (-2)^2 - (-2)^5 = 4 + 32 = 36$ and Column A is again larger. Finally, if $x = -1/2$, then $x^2 - x^5 = \frac{1}{4} + \frac{1}{32} = \frac{9}{32}$ and Column A is still larger. This covers the three types of negative numbers, so we can confidently conclude the answer is (A).

2. Suppose b equals 0. Then $a = b + 3 = 0 + 3 = 3$. Hence, Column A equals $ab = 30^*$. Here, Column A is less than Column B, which equals 40.

Now, suppose b equals 1. Then $a = b + 3 = 1 + 3 = 4$. Hence, Column A equals $ab = 41$. Here, Column A is greater than Column B, which equals 40.

Hence, we have a double case, and the answer is (D).

3. List all possible factors x and y whose product is 36, and calculate the corresponding sum $x + y$:

x	y	xy	x + y
1	36	36	37
2	18	36	20
3	12	36	15
4	9	36	13
6	6	36	12

From the table, the minimum sum is 12. Enter in the grid.

---

* Note: Here ab does not represent the product of a and b; rather it denotes the positions of a and b in the two-digit number: a is the tens digit, and b is the units digit.

4. The term 14! equals the product of the numbers 1, 2, 3, 4, 5, 6, 7, 8, 9, 10, 11, 12, 13, and 14. Only two of these numbers are divisible by 7. The numbers are 7 and 14. Hence, 14! can be expressed as the product of $k \cdot 7 \cdot 14$, where k is not divisible by 7. Now, since there are two 7s in 14!, the numbers 7 and 7^2 divide 14! evenly. 7^3 and further powers of 7 leave a remainder when divided into 14!. Hence, the maximum value of m is 2. Enter the same in the grid.

5. Let the coordinate representations of points A and E (which are on the x-axis) be $(a, 0)$, and $(e, 0)$, respectively. Also, let the coordinates of the points B and D (which are on the y-axis) be $(0, b)$ and $(0, d)$, respectively.

The formula for the midpoint of two points (x_1, y_1) and (x_2, y_2) in a coordinate system is $\left(\dfrac{x_1 + x_2}{2}, \dfrac{y_1 + y_2}{2}\right)$.

Hence, the midpoint of AB is $\left(\dfrac{a+0}{2}, \dfrac{0+b}{2}\right) = \left(\dfrac{a}{2}, \dfrac{b}{2}\right) = C(-3, 5)$. Equating the x- and y-coordinates on both sides yields $a/2 = -3$ and $b/2 = 5$. Solving for a and b yields $a = -6$ and $b = 10$.

Similarly, the midpoint of DE is $\left(\dfrac{e+0}{2}, \dfrac{0+d}{2}\right) = \left(\dfrac{e}{2}, \dfrac{d}{2}\right) = F(3, -5)$. Equating the x- and y-coordinates on both sides yields $e/2 = 3$ and $d/2 = -5$. Solving for e and d yields $e = 6$ and $d = -10$.

The slope of a line through two points (x_1, y_1) and (x_2, y_2) in a coordinate system is $\dfrac{y_2 - y_1}{x_2 - x_1}$.

Hence, Column A, which is the slope of AB, equals $\dfrac{b-0}{0-a} = -\dfrac{b}{a} = -\dfrac{10}{-6} = \dfrac{5}{3}$.

Column B, which is the slope of DE, equals $\dfrac{d-0}{0-e} = -\dfrac{d}{e} = -\dfrac{-10}{6} = \dfrac{5}{3}$.

Since both columns equal 5/3, the answer is (C).

6. If the parallelogram were a rectangle, then its area would be 15 and the columns would be equal. But as the rectangle is tilted to the right, its area decreases:

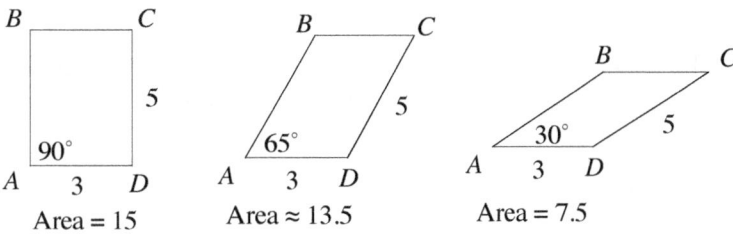

The answer is (A).

7. The average of the five angles is

$$\frac{\text{Sum of the five angles}}{5} = \frac{360}{5} = 72$$

The answer is (D).

8. From the number line, we have the equations

$AC = AB + BC$
$BD = BC + CD$
$AD = AB + BC + CD$

Adding the first two equations yields

$AC + BD = (AB + BC) + (BC + CD) = AB + 2BC + CD$

Subtracting the third equation from this equation yields

$AC + BD - AD = AB + 2BC + CD - (AB + BC + CD)$
$= BC$

Hence, $BC = AC + BD - AD = 5 + 10 - 13 = 2$ (given $AC = 5$, $BD = 10$, and $AD = 13$).

The answer is (A).

9. The formula for the area of a triangle is $1/2 \times base \times height$. Hence, the areas of triangles AEC, ACF, and ABC are

The area of $\triangle AEC = (1/2)(x)(BC) = (1/2)(x)(4) = 2x$

The area of $\triangle ACF = (1/2)(3)(AD)$

$\quad = (1/2)(3)(BC)$ $\quad AD = BC$, since opposite sides a rectangle are equal
$\quad = (3/2)BC = (3/2)(4) = 6$

The area of $\triangle ABC = (1/2)(AB)(AD)$

$\quad = (1/2)(CD)(BC)$ $\quad AB = CD$ and $AD = BC$ since opposite sides in a rectangle are equal
$\quad = (1/2)(DF + FC)(BC)$
$\quad = (1/2)(5 + 3)(4)$
$\quad = (1/2)(8)(4)$
$\quad = 16$

Now, the area of the quadrilateral AECF equals (the area of $\triangle AEC$) + (the area of $\triangle ACF$) = $2x + 6$, and the area of $\triangle ABC = 16$. Since we are given that the area of the triangle ABC equals the area of the quadrilateral AFCE, $2x + 6 = 16$. Solving for x in this equation yields $x = 5$. Enter in the grid.

Test 11—Solutions

10. In the figure, since OB and OP are radii of the circle, both equal 3 units. Also, the length of line segment OC is $OP + PC = 3 + 2 = 5$. Now, since BC is tangent to the circle, $\angle OBC = 90$. Hence, triangle OBC is a right triangle. Applying The Pythagorean Theorem yields

$$BC^2 + BO^2 = OC^2$$

$$BC^2 + 3^2 = 5^2$$

$$BC^2 = 5^2 - 3^2$$

$$BC^2 = 25 - 9$$

$$BC^2 = 16$$

$$BC = 4$$

By the formula for the area of a square, the area of square $ABCD$ is $side^2 = BC^2 = 4^2 = 16$. The answer is (C).

11. The area of a triangle is $\frac{1}{2} base \times height$. For the given triangle, this becomes

$$Area = \frac{1}{2} b \times h$$

Solving the equation $2b + h = 6$ for h gives $h = 6 - 2b$. Plugging this into the area formula gives

$$Area = \frac{1}{2} b(6 - 2b)$$

Since the value of b is not given, we cannot determine the area. Hence, there is not enough information, and the answer is (E).

12. Any two points and their midpoint always lie on the same line. So, the slope of AC equals the slope of BC. Hence, the columns are equal, and the answer is (C).

13. The product ab is positive when both a and b are positive or when both a and b are negative; in either case, a/b is positive. Hence, choice (A) is always positive, and the answer is (A).

14. Solving the equation $p + q = 7$ for q yields $q = 7 - p$. Plugging this into the equation $pq = 12$ yields

$$p(7 - p) = 12$$
$$7p - p^2 = 12$$
$$p^2 - 7p + 12 = 0$$
$$(p - 3)(p - 4) = 0$$
$$p - 3 = 0 \text{ or } p - 4 = 0$$
$$p = 3 \text{ or } p = 4$$

If $p = 3$, then $q = 7 - p = 7 - 3 = 4$. Plugging these values into the expression $\dfrac{1}{p^2} + \dfrac{1}{q^2}$ yields

$$\frac{1}{3^2} + \frac{1}{4^2} =$$
$$\frac{1}{9} + \frac{1}{16} =$$
$$\frac{25}{144}$$

The result is the same for the other solution $p = 4$ (and then $q = 7 - p = 7 - 4 = 3$). The answer is (B).

Method II:

$$\frac{1}{p^2} + \frac{1}{q^2} = \frac{(q^2 + p^2)}{p^2 q^2} = \frac{(p+q)^2 - 2pq}{(pq)^2} = \frac{(7)^2 - 2(12)}{12^2} = \frac{49 - 24}{144} = \frac{25}{144}.$$ The answer is (B).

15. Let the three numbers be x, y, and z. We are given that

$$\frac{x + y}{2} = 2$$
$$\frac{y + z}{2} = 3$$
$$\frac{x + z}{2} = 4$$

Summing the three equations yields

$$\frac{x + y}{2} + \frac{y + z}{2} + \frac{x + z}{2} = 2 + 3 + 4$$
$$\frac{x}{2} + \frac{y}{2} + \frac{y}{2} + \frac{z}{2} + \frac{x}{2} + \frac{z}{2} = 9$$
$$x + y + z = 9$$

The average of the three numbers is $(x + y + z)/3 = 9/3 = 3$. The answer is (C).

Test 11—Solutions

16. We know that the angle made by a line is 180°. Applying this to line m yields $x + y + 90 = 180$. Subtracting 90 from both sides of this equation yields $x + y = 90$. We are also given that $x : y = 2 : 3$. Hence, $x/y = 2/3$. Multiplying this equation by y yields $x = 2y/3$. Plugging this into the equation $x + y = 90$ yields

$$2y/3 + y = 90$$
$$5y/3 = 90$$
$$5y = 270$$
$$y = 54$$

The answer is (D).

17. Given that $a : b : c : d : e = 1 : 2 : 3 : 4 : 5$. Let $a : b : c : d : e = k : 2k : 3k : 4k : 5k$.

Now, $2c = 2(3k) = 6k$. Substitute into the choices:

(A) $a + b = k + 2k = 3k \neq 6k$. Reject.
(B) $b + d = 2k + 4k = 6k = 6k$. Accept.
(C) $a + e = k + 5k = 6k$. Accept.
(D) $a + b + c = k + 2k + 3k = 6k$. Accept.
(E) $2a + b + c = 2k + 2k + 3k = 7k$. Reject.

The answer is (B), (C), and (E).

18. Since x and y are consecutive integers, one of them must be even. Hence, the product xy is even and Statement I is true. As to Statement II, suppose z is odd, then x must be odd as well. Now, the difference of two odd numbers is an even number. Next, suppose z is even, then x must be even as well. Now, the difference of two even numbers is again an even number. Hence, Statement II is true. As to Statement III, let $x = 1$, then $z = 3$ and $x^z = 1^3 = 1$, which is odd. Thus, Statement III is not necessarily true. The answer is (D).

19.
$$x = \cfrac{1}{1+\cfrac{1}{1+\cfrac{1}{2}}}$$
$$= \cfrac{1}{1+\cfrac{1}{\cfrac{3}{2}}}$$
$$= \cfrac{1}{1+\cfrac{2}{3}}$$
$$= \cfrac{1}{\cfrac{5}{3}}$$
$$= \cfrac{3}{5} < 1$$

Hence, Column A is less than Column B, and the answer is (B).

GRE Math Tests

20. We are given that 500% of a equals $500b$. Now, 500% of a is $\frac{500}{100}a = 5a$. Setting this equal to $500b$ yields $5a = 500b$. Dividing both sides of this equation by 5 yields $a = 100b$. Since both a and b are positive (given) and a is 100 times b, a is greater than b. Hence, Column A is greater than Column B, and the answer is (A).

21. There are 16 ounces in a pound. Hence, each ounce equals 1/16 pounds. Now, 12 ounces equals $12 \times 1/16 = 3/4$ pounds. Hence, 5 pounds + 12 ounces equals 5 3/4 pounds. The answer is (C).

22. Recall the formula *Distance = Rate × Time*, or $D = R \cdot T$. From the second sentence, we get for Cyclist N:

$$D = R \cdot 2$$

Now, Cyclist M traveled at 20 miles per hour and took 4 hours. Hence, Cyclist M traveled a total distance of

$$D = R \cdot T = 20 \cdot 4 = 80 \text{ miles}$$

Since the cyclists covered the same distance at the moment they met, we can plug this value for D into the equation $D = R \cdot 2$:

$$80 = R \cdot 2$$
$$40 = R$$

The answer is (E).

23. (The sum of the first n terms of a series) = (The sum of the first $n - 1$ terms) + (The nth term).

Substituting the given values in the equation yields $31 = 20 + n$th term. Hence, the nth term is $31 - 20 = 11$. The answer is (B).

24. The first selection can be done in 4 ways (by selecting any one of the numbers 1, 2, 3, and 4 of the set S). Hence, there are 3 elements remaining in the set. The second number can be selected in 3 ways (by selecting any one of the remaining 3 numbers in the set S). Hence, the total number of ways the selection can be made is $4 \times 3 = 12$.

The selections that result in the sum 5 are 1 and 4, 4 and 1, 2 and 3, 3 and 2, a total of 4 selections. So, 4 of the 12 possible selections have a sum of 5. Hence, the probability is the fraction $4/12 = 1/3$. The answer is (D).

25. The first ball can be placed in any one of the four tins.

Similarly, the second, third, fourth, and fifth balls can be placed in any one of the 4 tins.

In other words, each of the four tins can hold any number of balls from 1 through 5.

Hence, the number of ways of placing the balls is $5 \cdot 5 \cdot 5 \cdot 5 = 5^4$. The answer is (C).

Test 12

GRE Math Tests

Questions: 24
Time: 45 minutes

[Quantitative Comparison Question]
1.   Column A $m > 0$ Column B
      m^{10} m^{100}

USE THIS SPACE FOR SCRATCHWORK.

[Multiple-choice Question – Select One Answer Choice Only]
2. If $A*B$ is the greatest common factor of A and B, $A\$B$ is defined as the least common multiple of A and B, and $A \cap B$ is defined as equal to $(A*B) \$ (A\$B)$, then what is the value of $12 \cap 15$?
    (A) 42
    (B) 45
    (C) 48
    (D) 52
    (E) 60

USE THIS SPACE FOR SCRATCHWORK.

[Quantitative Comparison Question]
3.   Column A   Column B
      The first number larger than   324
      300 that is a multiple of both 6
      and 8

USE THIS SPACE FOR SCRATCHWORK.

[Quantitative Comparison Question]
4.   Column A $\dfrac{l}{m+n} = \dfrac{m}{n+l} = \dfrac{n}{l+m} = k$ Column B
                          where k is a real number, and
                          $l + m + n \neq 0$
            k $1/3$

USE THIS SPACE FOR SCRATCHWORK.

[Numeric Entry Question]
5. A rectangular field is 3.2 yards long. A fence marking the boundary is 11.2 yards in length. What is the area of the field in square yards?

[Quantitative Comparison Question]
6.   Column A   A circle is inscribed in a triangle.   Column B
    The perimeter of the triangle   The circumference of the circle

GRE Math Tests

[Multiple-choice Question – Select One Answer Choice Only]
7. In the figure, what is the area of $\triangle ABC$?

(A) 2
(B) $\sqrt{2}$
(C) 1
(D) $\dfrac{1}{\sqrt{2}}$
(E) 1/2

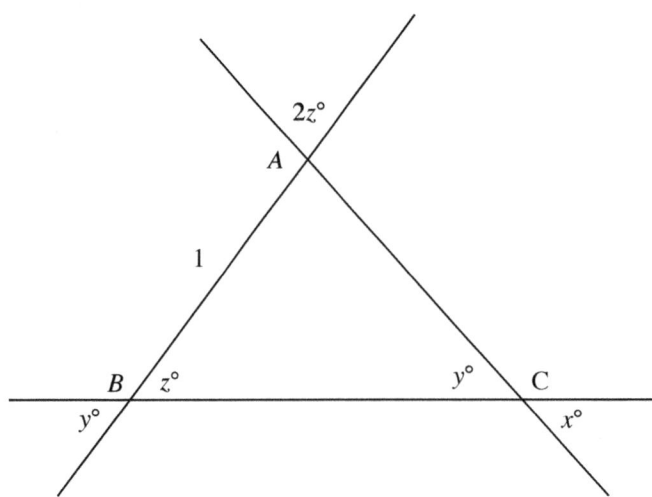

USE THIS SPACE FOR SCRATCHWORK.

[Multiple-choice Question – Select One Answer Choice Only]

8. In the figure, *ABCD* is a rectangle. Points *P*, *Q*, *R*, *S*, and *T* cut side *AB* of the rectangle such that *AP* = 3, *PQ* = *QR* = *RS* = *ST* = 1. *E* is a point on *AD* such that *AE* = 3. Which one of the following line segments is parallel to the diagonal *BD* of the rectangle?

(A) *EP*
(B) *EQ*
(C) *ER*
(D) *ES*
(E) *ET*

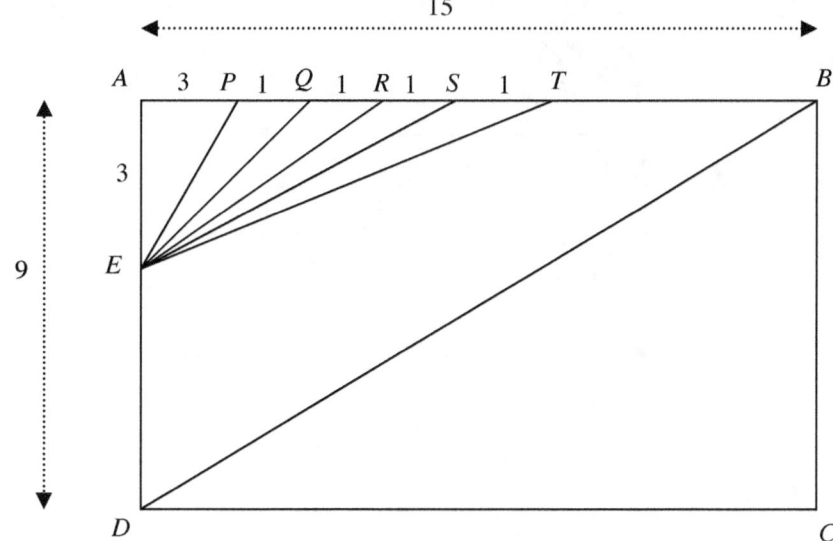

The figure is not drawn to scale.

GRE Math Tests

[Multiple-choice Question – Select One Answer Choice Only]
9. All the lines in the rectangular coordinate system shown in the figure are either horizontal or vertical with respect to the x-axis. What is the area of the figure ABCDEFGH ?

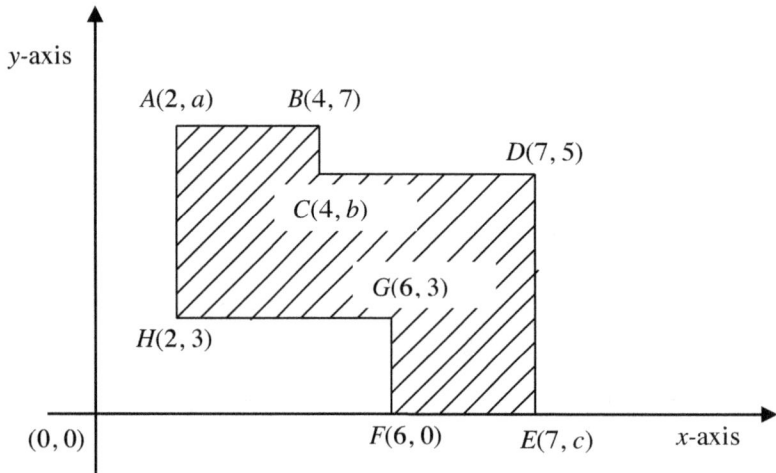

(A) 15
(B) 15.5
(C) 16
(D) 16.5
(E) 17

[Multiple-choice Question – Select One Answer Choice Only]
10. If $x < y < -1$, then which one of the following expressions is positive?

(A) $-x^2$
(B) y
(C) $x^2 y$
(D) $\dfrac{x^2}{y^2}$
(E) $y - x^2$

228

[Multiple-choice Question – Select One Answer Choice Only]

11. If $p + q = 12$ and $pq = 35$, then $\dfrac{1}{p} + \dfrac{1}{q} =$

    (A) 1/5
    (B) 1/7
    (C) 1/35
    (D) 12/35
    (E) 23/35

[Numeric Entry Question]

12. If $x = y$ and $x + y = 10$, then $2x + y =$

[Multiple-choice Question – Select One Answer Choice Only]

13. If $2x + 3y = 11$ and $3x + 2y = 9$, then $x + y =$

    (A) 4
    (B) 7
    (C) 8
    (D) 9
    (E) 11

[Multiple-choice Question – Select One Answer Choice Only]

14. Which one of the following numbers can be removed from the set $S = \{0, 2, 4, 5, 9\}$ without changing the average of set S?

    (A) 0
    (B) 2
    (C) 4
    (D) 5
    (E) 9

[Multiple-choice Question – Select One Answer Choice Only]

15. In the figure, the ratio of the area of parallelogram *ABCD* to the area of rectangle *AECF* is 5 : 3. What is the area of the rectangle *AECF* ?

    (A) 18
    (B) 24
    (C) 25
    (D) 50
    (E) 54

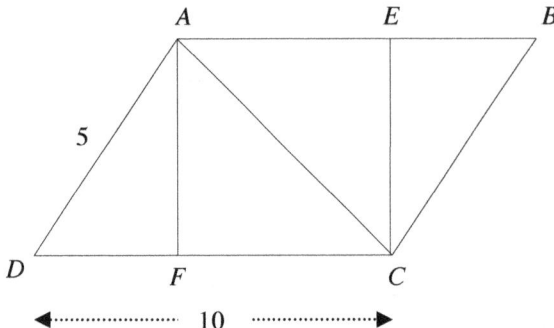

Test 12—Questions

[Multiple-choice Question – Select One Answer Choice Only]
16. Two alloys A and B are composed of two basic elements. The ratios of the compositions of the two basic elements in the two alloys are 5 : 3 and 1 : 2, respectively. A new alloy X is formed by mixing the two alloys A and B in the ratio 4 : 3. What is the ratio of the composition of the two basic elements in alloy X?

(A) 1 : 1
(B) 2 : 3
(C) 5 : 2
(D) 4 : 3
(E) 7 : 9

[Numeric Entry Question]
17. If $x^2 - 4x + 3 = 0$, then what is the value of $(x - 2)^2$?

231

[Multiple-choice Question – Select One or More Answer Choices]

18. If $x = \dfrac{1}{1+\dfrac{1}{1+\dfrac{1}{x}}}$, then x could equal

   (A) -2
   (B) $-\dfrac{1}{2}$
   (C) $\dfrac{1}{\sqrt{2}}$
   (D) $\sqrt{2}$
   (E) $\sqrt{\dfrac{1}{2}}$

[Multiple-choice Question – Select One Answer Choice Only]

19. Train X leaves New York at 10:00AM and travels East at a constant speed of x miles per hour. If another Train Y leaves New York at 11:30AM and travels East along the same tracks at speed $4x/3$ mph, then at what time will Train Y catch Train X?

   (A) 2 PM of the same day
   (B) 3 PM of the same day
   (C) 3:30 PM of the same day
   (D) 4 PM of the same day
   (E) 8 PM of the same day

[Multiple-choice Question – Select One Answer Choice Only]

20. How many ounces of a solution that is 30 percent salt must be added to a 50-ounce solution that is 10 percent salt so that the resulting solution is 20 percent salt?

   (A) 20
   (B) 30
   (C) 40
   (D) 50
   (E) 60

[Multiple-choice Question – Select One Answer Choice Only]
21. A car traveled 65% of the way from Town A to Town B at an average speed of 65 mph. The car traveled at an average speed of v mph for the remaining part of the trip. The average speed for the entire trip was 50 mph. What is v in mph?

   (A) 65
   (B) 50
   (C) 45
   (D) 40
   (E) 35

The next two questions are based on the following frequency distribution.

x	f
0	1
1	5
2	4
3	3

[Multiple-choice Question – Select One Answer Choice Only]
22. What is the range of f? [Note, The *range* of a set is the greatest value of the set minus the smallest value of the set.]

   (A) 0
   (B) 1
   (C) 3
   (D) 4
   (E) 5

[Multiple-choice Question – Select One or More Answer Choices]
23. Removing which of the following rows from the frequency distribution table affects the range of f?

   (A) $x = 0, f(0) = 1$
   (B) $x = 1, f(1) = 5$
   (C) $x = 2, f(2) = 4$
   (D) $x = 3, f(3) = 3$

GRE Math Tests

[Multiple-choice Question – Select One Answer Choice Only]
24. A hat contains 15 marbles, and each marble is numbered with one and only one of the numbers 1, 2, 3. From a group of 15 people, each person selects exactly 1 marble from the hat.

Numbered Marble	Number of People Who Selected The Marble
1	4
2	5
3	6

Test 12—Solutions

Answers and Solutions Test 12:

Question	Answer
1.	D
2.	E
3.	B
4.	A
5.	7.68
6.	A
7.	E
8.	C
9.	E
10.	D
11.	D
12.	B
13.	A
14.	C
15.	A
16.	A
17.	C
18.	C, D
19.	D
20.	D
21.	E
22.	D
23.	A, B
24.	D

If you got 18/24 correct on this test, you are likely to get 750+ on the actual GRE by the time you complete all the tests in the book.

---

1. If $m = 1$, then $m^{10} = 1$ and $m^{100} = 1$. In this case, the two columns are equal. Next, if $m = 2$, then clearly m^{100} is greater than m^{10}. This is a double case, and the answer is (D).

---

2. According to the definitions given,

$$12 \cap 15 = (12*15) \$ (12\$15) =$$

$$(\text{GCF of 12 and 15}) \$ (\text{LCM of 12 and 15}) =$$

$$3\$60 =$$

$$\text{LCM of 3 and 60} = 60$$

The answer is (E).

---

3. The least common multiple of 6 and 8 is 24. Hence, a multiple of both 6 and 8 must also be a multiple of 24.

We know that a multiple of 24 exists once in every 24 consecutive numbers. For example, between the numbers 34 (not itself a multiple of 24) and 34 + 24, there must be exactly one multiple of 24. Similarly, exactly one multiple of 24 must exist between 300 (not itself a multiple of 24) and 300 + 24 (= 324, Column B), and this number is the first multiple of 24 larger than 300, Column A. So, Column A lies between 300 and 324 (Column B) and therefore is less than Column B. The answer is (B).

4. The given equation is $\dfrac{l}{m+n} = \dfrac{m}{n+l} = \dfrac{n}{l+m} = k$. Forming the three equations yields $l = (m+n)k$, $m = (n+l)k$, $n = (l+m)k$. Summing these three equations yields

$$l + m + n = (m+n)k + (n+l)k + (l+m)k$$
$$= k[(m+n) + (n+l) + (l+m)]$$
$$= k(m+n+n+l+l+m)$$
$$= k(2m+2n+2l)$$
$$= 2k(m+n+l)$$
$$1 = 2k \quad \text{by canceling } m+n+l \text{ from each side}$$
$$1/2 = k$$

Hence, Column A equals 1/2. Since 1/2 is greater than 1/3, Column A is greater than Column B, and the answer is (A).

---

5. Let l be the length of the rectangle. Then $l = 3.2$ yards. We are given that the length of the fence required (perimeter) for the field is 11.2 yards. The formula for the perimeter of a rectangle is $2(length + width)$. Hence, the perimeter of the field is $2(l + w) = 11.2$, or $2(3.2 + w) = 11.2$. Solving for w yields $w = 2.4$. The formula for the area of a rectangle is $length \times width$. Hence, the area of the rectangle is $lw = 3.2 \times 2.4 = 7.68$.

---

6. Let ABC be the triangle, and let P, Q, and R be the points at which the inscribed circle touches the sides of the triangle.

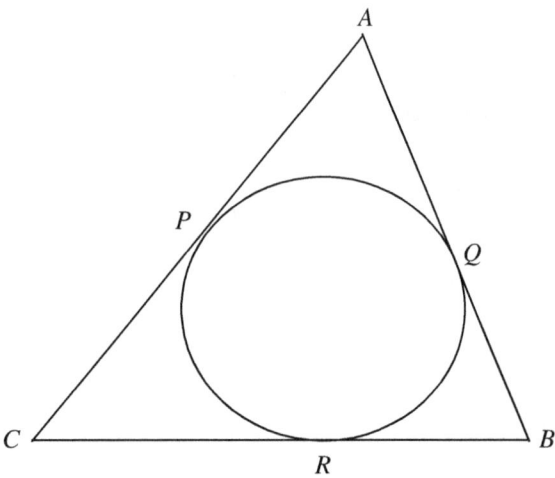

From the figure, it is clear that to go from one point on the circle, say, point P to another point, say, point Q, the shortest available path is the arc PQ. Hence, arc $PQ < PA + AQ$. Similarly, arc $QR < QB + BR$, and arc $RP < RC + CP$. Summing the three inequalities yields arc

$$PQ + \text{arc } QR + \text{arc } RP < (PA + AQ) + (QB + BR) + (RC + CP)$$

The right side of the inequality is the perimeter of the triangle ABC (which Column A equals), and the left side is the circumference of the circle (which Column B equals). Hence, Column A is greater than Column B, and the answer is (A).

Test 12—Solutions

---

7. Equating vertical angles at point B in the figure yields $y = z$ and $\angle A = 2z$. So, the triangle is isosceles ($\angle B = \angle C = y$). Now, the angle sum of a triangle is 180°, so

$$2z + z + z = 180$$

Solving for z yields $z = 180/4 = 45$. Hence, we have $2z = 2 \times 45 = 90$. So, $\triangle ABC$ is a right-angled isosceles triangle with right angle at vertex A and equal angles at $\angle B$ and $\angle C$ (both equaling 45°). Since sides opposite equal angles in a triangle are equal, sides AB and AC are equal and each is 1 unit (given that AB is 1 unit). Now, the area of the right triangle ABC is

$$1/2 \times AB \times AC =$$
$$1/2 \times 1 \times 1 =$$
$$1/2$$

The answer is (E).

---

8. Let X represent the point on AB that makes EX parallel to the diagonal BD. Then $\triangle AXE$ and $\triangle ABD$ must be similar because

$\angle EAX = \angle DAB$ common angles
$\angle AEX = \angle ADB$ corresponding angles are equal

Hence, the sets of corresponding angles are equal and the triangles are similar.

Hence, the corresponding sides of the triangles must be in the same ratio. This yields the following equations:

$AE/AD = AX/AB$
$3/9 = AX/15$
$AX = 15 \times 3/9 = 15 \times 1/3 = 5$

Hence, the point X is 5 units away from the point A on the side AB. Now,

$AP = 3 \neq AX$, which equals 5.
$AQ = AP + PQ = 3 + 1 = 4 \neq AX$, which equals 5.
$AR = AP + PQ + QR = 3 + 1 + 1 = 5 = AX$, which equals 5.

Hence, point R coincides with point X and therefore just like EX, ER must be parallel to BD. Hence, the answer is (C).

GRE Math Tests

9. Drop a vertical line from *C* on to the side *GH* (to meet at, say, *P*), and draw a horizontal line from *G* on to the side *DE* (to meet at, say, *Q*). The resultant figure is as follows:

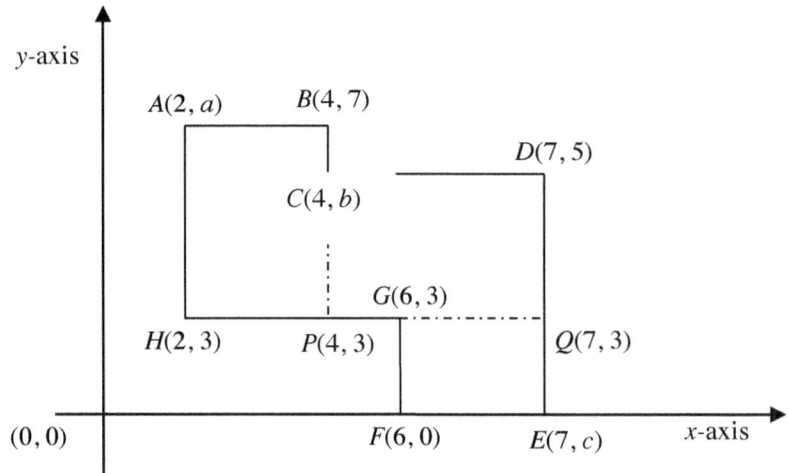

Since the point *P* is horizontal to the point *H*, its *y*-coordinate equals the *y*-coordinate of *H*, which is 3. Also, since *P* is vertical to the point *C*, its *x*-coordinate equals the *x*-coordinate of the point *C*, which is 4. Hence, the point *P* is (4, 3).

Similarly, the point *Q* is vertical to the point *D* and therefore its *x*-coordinate equals the *x*-coordinate of *D*, which is 7. Also, the point *Q* is horizontal to the point *H* and therefore takes its *y*-coordinate. Hence, the point *Q* is (7, 3).

Now, the shaded region in the given figure is the sum of the three rectangles *ABPH*, *CDQP*, and *GQEF*.

The area of the rectangle *ABPH* is *length·width* =

$BP \cdot AB$ =
(Difference in *y*-coordinates of *B* and *P*) · (Difference in *x*-coordinates of *A* and *B*) =
(7 − 3)(4 − 2) =
4 · 2 =
8

The area of the rectangle *CDQP* is *length·width* =

$CD \cdot DQ$ =
(Difference in *x*-coordinates of *C* and *D*) · (Difference in *y*-coordinates of *D* and *Q*) =
(7 − 4)(5 − 3) =
3 · 2 =
6

The area of the rectangle *GQEF* is *length · width* =

$GQ \cdot QE$ =
(Difference in *x*-coordinates of *G* and *Q*) · (Difference in *y*-coordinates of *Q* and *E*) =
(7 − 6)(3 − 0) =
1 · 3 =
3

Hence, the total area is 8 + 6 + 3 = 17 sq. units. The answer is (E).

10. From the inequality $x < y < -1$, we have that x and y do not equal zero. Since the square of a nonzero number is positive, x^2 is positive.

Hence, $-x^2$ must be negative, and therefore choice (A) is negative.

Since we are given that $y < -1$, y is negative. Hence, choice (B) is negative.

Multiplying the inequality $y < -1$ by the positive value x^2 yields $x^2 y < -x^2$, a negative value. Hence, choice (C) is also negative.

Dividing x^2, a positive value, by y^2 (also a positive value) yields a positive value. Hence, $\dfrac{x^2}{y^2}$ is positive, and choice (D) is positive.

Since y and $-x^2$ are negative, their sum $y + (-x^2) = y - x^2$ must also be negative. Hence, choice (E) is negative.

The only choice that is positive is (D), so the answer is (D).

---

11. Adding the fractions yields

$$\frac{1}{p} + \frac{1}{q} = \frac{p+q}{pq} = \frac{12}{35}$$

we are given that $p + q = 12$ and $pq = 35$

The answer is (D).

Method II:
Solving the given equation $pq = 35$ for q yields $q = 35/p$. Plugging this into the equation $p + q = 12$ yields

$p + 35/p = 12$
$p^2 + 35 = 12p$ by multiplying both sides by p
$p^2 - 12p + 35 = 0$ by subtracting $12p$ from both sides
$(p - 5)(p - 7) = 0$
$p - 5 = 0$ or $p - 7 = 0$
$p = 5$ or $p = 7$

When $p = 5$, the equation $p + q = 12$ shows that $q = 7$. Similarly, when $p = 7$, q equals 5. In either case, $\dfrac{1}{p} + \dfrac{1}{q} = \dfrac{1}{5} + \dfrac{1}{7} = \dfrac{7+5}{35} = \dfrac{12}{35}$. The answer is (D).

Notice how much shorter the first method of solution is than the second method. The strategy in the first method is to manipulate the expression we want to calculate, while hoping that the given information ($p + q = 12$ and $pq = 35$) will appear. In this case, all the given information did appear, making for a quick solution.

Whether the ability to perform these types of manipulations (solutions) is indicative of intellectual intelligence is debatable, but the writers of the test tacitly assume it is. The premise being that students who don't notice the shortcuts to the solutions will be more pressed for time and therefore will not perform as well as students who notice the shortcuts.

12. Substituting $x = y$ into the equation $x + y = 10$ yields $x + x = 10$. Combining like terms yields $2x = 10$. Finally, dividing by 2 yields $x = 5$. Hence, $2x + y = 2(5) + 5 = 15$. The answer is (B).

13. Adding the two equations $2x + 3y = 11$ and $3x + 2y = 9$ yields $5x + 5y = 20$, or $x + y = 20/5 = 4$. The answer is (A).

14. The average of the elements in the original set S is $(0 + 2 + 4 + 5 + 9)/5 = 20/5 = 4$. If we remove an element that equals the average, then the average of the new set will remain unchanged. The new set after removing 4 is $\{0, 2, 5, 9\}$. The average of the elements is $(0 + 2 + 5 + 9)/4 = 16/4 = 4$. The answer is (C).

15. The area of the parallelogram $ABCD$ equals $base \times height = DC \times AF = 10AF$ (from the figure, $DC = 10$).

The area of rectangle $AECF$ equals $length \times width = FC \times AF$.

We are given that the ratio of the two areas is 5 : 3. Forming the ratio gives

$$\frac{10AF}{FC \cdot AF} = \frac{5}{3}$$

$$\frac{10}{FC} = \frac{5}{3}$$

$$FC = 10 \cdot \frac{3}{5} = 6$$

From the figure, we have $DC = DF + FC$. Now, $DC = 10$ (from the figure) and $FC = 6$ (as we just calculated). So, $10 = DF + 6$. Solving this equation for DF yields $DF = 4$. Since F is one of the angles in the rectangle $AECF$, it is a right angle. Since $\angle AFC$ is right-angled, $\angle AFD$ must also measure $90°$. So, $\triangle AFD$ is right-angled, and The Pythagorean Theorem yields

$$AF^2 + DF^2 = AD^2$$

$$AF^2 + 4^2 = 5^2$$

$$AF^2 = 5^2 - 4^2 = 25 - 16 = 9$$

$$AF = 3$$

Now, the area of the rectangle $AECF$ equals $FC \times AF = 6 \times 3 = 18$. The answer is (A).

Test 12—Solutions

16. The new alloy X is formed from the two alloys A and B in the ratio 4 : 3. Hence, 7 parts of the alloy contains 4 parts of alloy A and 3 parts of alloy B. Let 7x ounces of alloy X contain 4x ounces of alloy A and 3x ounces of alloy B.

Now, alloy A is formed of the two basic elements mentioned in the ratio 5 : 3. Hence, 4x ounces of alloy A contains $\frac{5}{5+3} \cdot 4x = \frac{5x}{2}$ ounces of first basic element and $\frac{3}{5+3} \cdot 4x = \frac{3x}{2}$ ounces of the second basic element.

Also, alloy B is formed of the two basic elements mentioned in the ratio 1 : 2. Hence, let the 3x ounces of alloy A contain $\frac{1}{1+2} \cdot 3x = x$ ounces of the first basic element and $\frac{2}{1+2} \cdot 3x = 2x$ ounces of the second basic element.

Then the total compositions of the two basic elements in the 7x ounces of alloy X would contain 5x/2 ounces (from A) + x ounces (from B) = 7x/2 ounces of first basic element, and 3/2 x (from A) + 2x (from B) = 7x/2 ounces of the second basic element. Hence, the composition of the two basic elements in alloy X is 7x/2 : 7x/2 = 1 : 1. The answer is (A).

$$AF^2 + DF^2 = AD^2$$
$$AF^2 + 4^2 = 5^2$$
$$AF^2 = 5^2 - 4^2 = 25 - 16 = 9$$
$$AF = 3$$

Now, the area of the rectangle AECF equals FC×AF = 6 × 3 = 18. The answer is (A).

---

17. Adding 1 to both sides of the given equation $x^2 - 4x + 3 = 0$ yields $x^2 - 4x + 4 = 1$. Expanding $(x - 2)^2$ by the Perfect Square Trinomial formula $(a - b)^2 = a^2 - 2ab + b^2$ yields $x^2 - 4x + 2^2 = x^2 - 4x + 4 = 1$. Hence, $(x - 2)^2 = 1$, and the answer is (C).

Method II:
Factoring the equation $x^2 - 4x + 3 = 0$ yields

$$(x - 3)(x - 1) = 0$$

$$x - 3 = 0 \quad \text{or} \quad x - 1 = 0$$

$$x = 3 \quad \text{or} \quad x = 1$$

Plugging either 3 or 1 into the expression $(x - 2)^2$ yields the value 1, so the answer is (C).

18.
$$x = \cfrac{1}{1 + \cfrac{1}{1 + \cfrac{1}{x}}}$$

$$= \cfrac{1}{1 + \cfrac{1}{\cfrac{x+1}{x}}}$$

$$= \cfrac{1}{1 + \cfrac{x}{x+1}}$$

$$= \cfrac{1}{\cfrac{(x+1) + x}{x+1}}$$

$$= \cfrac{1}{\cfrac{2x+1}{x+1}}$$

$$= \cfrac{x+1}{2x+1}$$

Cross-multiplying both sides yields

$$x(2x + 1) = x + 1$$
$$2x^2 + x = x + 1$$
$$2x^2 = 1$$
$$x^2 = \frac{1}{2}$$
$$x = \sqrt{\frac{1}{2}} = \frac{1}{\sqrt{2}}$$

The answer is (C) and (E).

Test 12—Solutions

---

19. Train X started at 10:00AM. Let the time it has been traveling be t. Since Train Y started at 11:30AM, it has been traveling an hour and a half less. So, represent its time as $t - 1\ 1/2 = t - 3/2$.

Train X travels at speed x miles per hour, and Train Y travels at speed $4x/3$ miles per hour. By the formula *Distance = Speed · Time*, the respective distances they travel before meeting equal xt and $(4x/3)(t - 3/2)$. Since the trains started from the same point and traveled in the same direction, they will have traveled the same distance when they meet. Hence, we have

$$xt = (4x/3)(t - 3/2)$$

$$t = (4/3)(t - 3/2) \qquad \text{by canceling } x \text{ from both sides}$$

$$t = 4t/3 - 2 \qquad \text{by distributing 4/3 on the right side}$$

$$t - 4t/3 = -2$$

$$-t/3 = -2 \qquad \text{by subtracting the expressions on the left side}$$

$$t = 6 \text{ hours}$$

Hence, Train Y will catch Train X at 4PM (10AM plus 6 hours is 4PM). The answer is (D).

---

20. Let x be the ounces of the 30 percent solution. Then $30\%x$ is the amount of salt in that solution. The final solution will be $50 + x$ ounces, and its concentration of salt will be $20\%(50 + x)$. The original amount of salt in the solution is $10\% \cdot 50$. Now, the amount of salt in the original solution plus the amount of salt in the added solution must equal the amount of salt in the resulting solution:

$$10\% \cdot 50 + 30\%x = 20\%(50 + x)$$

Multiply this equation by 100 to clear the percent symbol yields

$$10 \cdot 50 + 30x = 20(50 + x)$$

$$500 + 30x = 1000 + 20x$$

$$10x = 500$$

$$x = 50$$

The answer is (D).

GRE Math Tests

21. Let d be the distance between the towns A and B. 65% of this distance (= 65% of d = 65/100 × d = 0.65d) was traveled at 65 mph and the remaining 100 − 65 = 35% of the distance was traveled at v mph. Now, Remember that *Time = Distance ÷ Rate*. Hence, the time taken by the car for the first 65% distance is 0.65d/65 = d/100, and the time taken by the car for the last 35% distance is 0.35d/v. Hence, the total time taken is d/100 + 0.35d/v = d(1/100 + 0.35/v).

Now, remember that *Average Speed* = $\dfrac{\text{Total Distance Traveled}}{\text{Total Time Taken}}$.

Hence, the average speed of the journey is $\dfrac{d}{d\left(\dfrac{1}{100} + \dfrac{0.35}{v}\right)}$. Equating this to the given value for the average speed yields

$$\dfrac{d}{d\left(\dfrac{1}{100} + \dfrac{0.35}{v}\right)} = 50$$

$$\dfrac{1}{\dfrac{1}{100} + \dfrac{0.35}{v}} = 50$$

$$1 = 50 \times \dfrac{1}{100} + 50 \times \dfrac{0.35}{v}$$

$$1 = \dfrac{1}{2} + 50 \times \dfrac{0.35}{v}$$

$$\dfrac{1}{2} = 50 \times \dfrac{0.35}{v}$$

$$v = 2 \times 50 \times 0.35 = 100 \times 0.35 = 35$$

The answer is (E).

---

22. The *range* of f is the greatest value of f minus the smallest value of f:

$$5 - 1 = 4$$

The answer is (D).

---

23. The range of f is

$$\text{Maximum } f - \text{Minimum } f =$$
$$5 - 1 =$$
$$4$$

Intermediate points, such as $f = 3$ and $f = 4$, do not affect this range.

Eliminate (C) and (D), and choose (A) and (B).

---

24. There are 11 (= 5 + 6) people who selected a number 2 or number 3 marble, and there are 15 total people. Hence, the probability of selecting a number 2 or number 3 marble is 11/15, and the answer is (D).

Test 13

GRE Math Tests

Questions: 25
Time: 45 minutes

[Quantitative Comparison Question]
1. For all numbers x, $<x>$ denotes the value of x^3 rounded to the nearest multiple of ten.

Column A	Column B
$<x + 1>$	$<x> + 1$

[Multiple-choice Question – Select One or More Answer Choices]
2. If x is divisible by both 3 and 4, then the number x must be a multiple of which of the following?

(A) 8
(B) 12
(C) 15
(D) 18
(E) 24

[Quantitative Comparison Question]
3.

Column A		Column B
	$a, b,$ and c are consecutive integers in increasing order of size.	
$a/5 - b/6$		$b/5 - c/6$

[Quantitative Comparison Question]
4.

Column A		Column B
	The positive integers m and n leave remainders of 2 and 3, respectively, when divided by 6. $m > n$.	
The remainder when $m + n$ is divided by 6		The remainder when $m - n$ is divided by 6

[Quantitative Comparison Question]

5. Column A — The ratio of x to y is $3:4$, and the ratio of $x+7$ to $y+7$ is $4:5$. — Column B

$\dfrac{x+14}{y+14}$ $\dfrac{5}{6}$

[Quantitative Comparison Question]

6. Column A — The greatest possible number of points common to a triangle and a circle   Column B — 3

[Multiple-choice Question – Select One Answer Choice Only]

7. In the figure, what is the value of a?

(A) 30
(B) 45
(C) 60
(D) 72
(E) 90

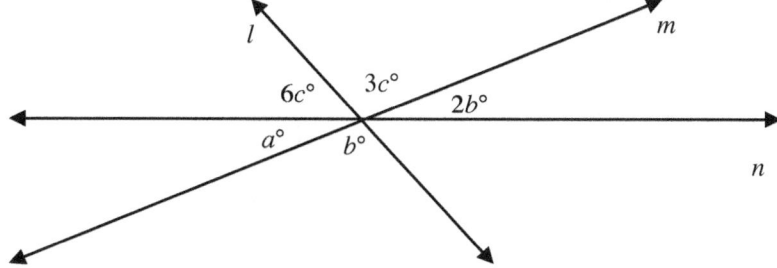

[Multiple-choice Question – Select One Answer Choice Only]

8. In the figure, the area of rectangle *EFGH* is 3 units greater than the area of rectangle *ABCD*. What is the value of *ab* if *a* + *b* = 8?

   (A) 9
   (B) 12
   (C) 15
   (D) 18
   (E) 21

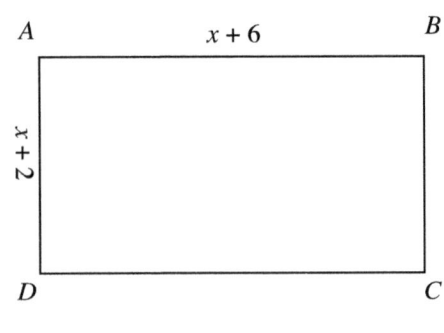
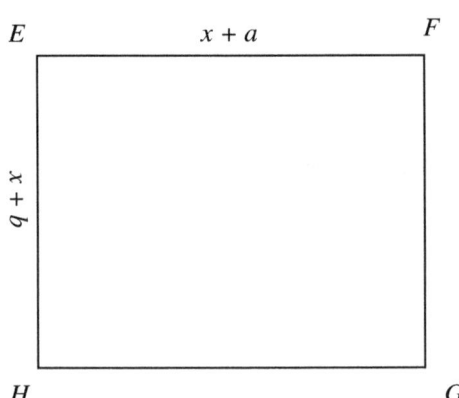

[Numeric Entry Question]
9. In the figure, ABCD is a rectangle. Points E and F cut the sides BC and CD of the rectangle respectively such that EC = 3, FC = 4, and AD = 12, and the areas of the crossed and the shaded regions in the figure are equal. What is the perimeter of rectangle ABCD?

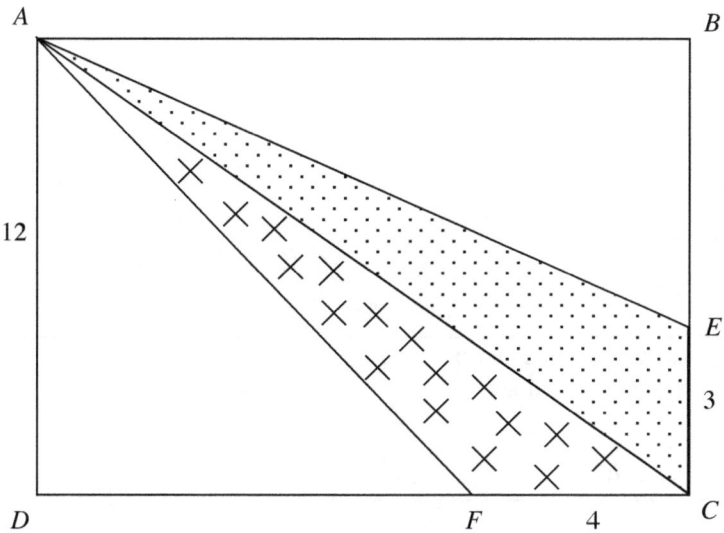

[Multiple-choice Question – Select One Answer Choice Only]
10. The length of a rectangle is increased by 25%. By what percentage should the width be decreased so that the area of the rectangle remains unchanged?

(A) 20
(B) 25
(C) 30
(D) 33.33
(E) 50

[Multiple-choice Question – Select One Answer Choice Only]
11. In the figure, AB is parallel to CD. What is the value of x ?

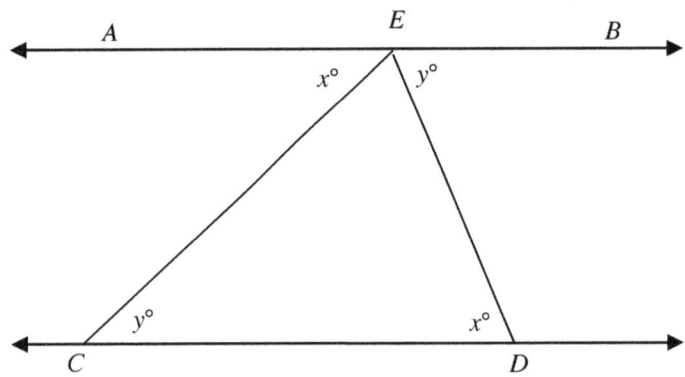

[Multiple-choice Question – Select One or More Answer Choices]
12. If $(x + 1)^2 - 2x > 2(x + 1) + 2$, then x cannot equal which of the following?

(A) −5
(B) −3
(C) 0
(D) 3
(E) 5

[Multiple-choice Question – Select One Answer Choice Only]

13. Jane gave three-fifths of the amount of money she had to Jack. Jane now has 200 dollars. How much did she give to Jack?

    (A) $80
    (B) $120
    (C) $200
    (D) $300
    (E) $500

[Multiple-choice Question – Select One Answer Choice Only]

14. If $(a+2)(a-3)(a+4) = 0$ and $a > 0$, then $a =$

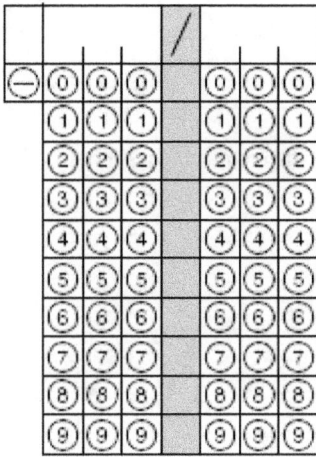

[Multiple-choice Question – Select One Answer Choice Only]
15. The five numbers 1056, 1095, 1098, 1100, and 1126 are represented on a number line by the points A, B, C, D, and E, respectively, as shown in the figure. Which one of the following points represents the average of the five numbers?

(A) Point A
(B) Point B
(C) Point C
(D) Point D
(E) Point E

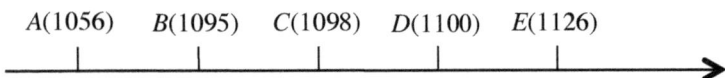

[Multiple-choice Question – Select One Answer Choice Only]
16. The arithmetic mean (average) of m and n is 50, and the arithmetic mean of p and q is 70. What is the arithmetic mean of m, n, p, and q?

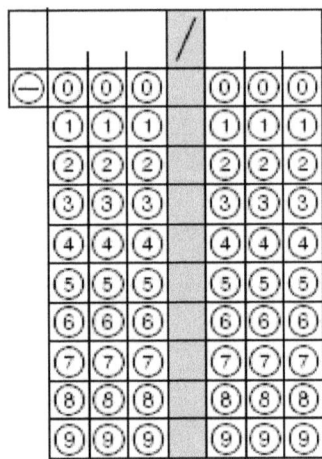

[Multiple-choice Question – Select One Answer Choice Only]
17. The ratio of two numbers is 10 and their difference is 18. What is the value of the smaller number?

    (A) 2
    (B) 5
    (C) 10
    (D) 21
    (E) 27

[Multiple-choice Question – Select One or More Answer Choices]
18. A precious stone when dropped breaks into pieces of equal size and weight. However, the stone is of a rare kind in which the value of the stone is always proportional to the square of its weight. On an exhibitory eve, the stone was dropped. Which of the following could be the price of each broken piece of stone?

    (A) One-half if the stone breaks into two.
    (B) One-fourth if the stone breaks into two.
    (C) One-third if the stone breaks into three.
    (D) One-ninth if the stone breaks into three.
    (E) $\dfrac{1}{\sqrt{3}}$ times if the stone breaks into three.

[Quantitative Comparison Question]

19. Column A	Column B
$\dfrac{\left(\left(\sqrt{7}\right)^x\right)^2}{\left(\sqrt{7}\right)^{11}}$	$\dfrac{7^x}{7^{11}}$

GRE Math Tests

[Multiple-choice Question – Select One or More Answer Choices]

20. If $\sqrt{3-2x} = 1$, then what is the value of $(3-2x) + (3-2x)^2$?

    (A)   0
    (B)   1
    (C)   2
    (D) $2x$
    (E) $x + 1$

USE THIS SPACE FOR SCRATCHWORK.

[Multiple-choice Question – Select One Answer Choice Only]

21. The cost of painting a wall increases by a fixed percentage each year. In 1970, the cost was $2,000; and in 1979, it was $3,600. What was the cost of painting in 1988?

    (A) $1,111
    (B) $2,111
    (C) $3,600
    (D) $6240
    (E) $6480

USE THIS SPACE FOR SCRATCHWORK.

[Multiple-choice Question – Select One Answer Choice Only]

22. How many coins of 0.5 dollars each and 0.7 dollars each together make exactly 4.6 dollars?

    (A)   1, 6
    (B)   2, 7
    (C)   3, 5
    (D)   4, 3
    (E)   5, 3

USE THIS SPACE FOR SCRATCHWORK.

[Multiple-choice Question – Select One Answer Choice Only]
23. A series has three numbers $a, ar,$ and ar^2. In the series, the first term is twice the second term. What is the ratio of the sum of the first two terms to the sum of the last two terms in the series?

    (A) 1 : 1
    (B) 1 : 2
    (C) 1 : 4
    (D) 2 : 1
    (E) 4 : 1

[Multiple-choice Question – Select One Answer Choice Only]
24. If x and y are two positive integers and $x + y = 5$, then what is the probability that x equals 1?

    (A) 1/2
    (B) 1/3
    (C) 1/4
    (D) 1/5
    (E) 1/6

[Multiple-choice Question – Select One Answer Choice Only]
25. Goodwin has 3 different colored pants and 2 different colored shirts. In how many ways can he choose a pair of pants and a shirt?

    (A) 2
    (B) 3
    (C) 5
    (D) 6
    (E) 12

GRE Math Tests

Answers and Solutions Test 13:

Question	Answer
1.	D
2.	B
3.	B
4.	C
5.	C
6.	A
7.	D
8.	C
9.	56
10.	A
11.	E
12.	C, D
13.	D
14.	3
15.	B
16.	60
17.	A
18.	A, C
19.	A
20.	C, D, E
21.	C
22.	E
23.	D
24.	C
25.	D

If you got 18/25 correct on this test, you are likely to get 750+ on the actual GRE by the time you complete all the tests in the book.

---

1. Suppose $x = 0$. Then $<x + 1> = <0 + 1> = <1> = 0$,* and $<x> + 1 = <0> + 1 = 0 + 1 = 1$. In this case, Column B is larger. Next, suppose $x = 1$. Then $<x + 1> = <1 + 1> = <2> = 10$, and $<x> + 1 = <1> + 1 = 0 + 1 = 1$. In this case, Column A is larger. The answer is (D).

---

2. We are given that x is divisible by 3 and 4. Hence, x must be a common multiple of 3 and 4. The least common multiple of 3 and 4 is 12. So, x is a multiple of 12. It need not be multiple of 24 or any other larger common multiples of 3 and 4. The answer is (B) only.

---

3. The consecutive integers $a, b,$ and c in the increasing order of size can be expressed as $a, a + 1, a + 2$, respectively. Substituting these expressions into the fractions yields

$$\frac{a}{5} - \frac{b}{6} = \frac{a}{5} - \frac{a+1}{6} \qquad \frac{b}{5} - \frac{c}{6} = \frac{a+1}{5} - \frac{a+2}{6}$$

Adding the fractions in each column yields

$$\frac{6a - 5a - 5}{30} = \frac{a-5}{30} \qquad \frac{6a + 6 - 5a - 10}{30} = \frac{a-4}{30}$$

---

* $<1> = 0$ because 0 is a multiple of 10: $0 = 0 \cdot 10$.

Test 13—Solutions

Multiplying both columns by 30 to clear fractions yields

$a - 5$ $\qquad\qquad\qquad\qquad\qquad\qquad\qquad\qquad\qquad\qquad$ $a - 4$

Adding 5 to both columns yields

a $\qquad\qquad\qquad\qquad\qquad\qquad\qquad\qquad\qquad\qquad\qquad\qquad$ $a + 1$

Finally, subtracting a from both columns yields

0 $\qquad\qquad\qquad\qquad\qquad\qquad\qquad\qquad\qquad\qquad\qquad\qquad\qquad$ 1

Hence, Column B is larger, and the answer is (B).

---

4. We are given that the numbers m and n, when divided by 6, leave remainders of 2 and 3, respectively. Hence, we can represent the numbers m and n as $6p + 2$ and $6q + 3$, respectively, where p and q are suitable integers.

Now, $m + n = (6p + 2) + (6q + 3) = 6p + 6q + 5 = 6(p + q) + 5$, so the remainder is 5. Hence, Column A equals 5.

Now, $m - n = (6p + 2) - (6q + 3) = 6p - 6q - 1 = 6(p - q) - 1$. Now, a remainder must be positive, so let's add 6 to this expression and compensate by subtracting 6:

$6(p - q) - 1 =$

$6(p - q) - 6 + 6 - 1 =$

$6(p - q) - 6 + 5 =$

$6(p - q - 1) + 5$

Thus, the remainder is 5, and Column B equals 5.

Since both columns equal 5, the answer is (C).

5. Forming the two ratios yields $\dfrac{x}{y} = \dfrac{3}{4}$ and $\dfrac{x+7}{y+7} = \dfrac{4}{5}$. Let's solve this system of equations by the substitution method. Multiplying the first equation by y yields $x = 3y/4$. Substituting this into the second equation yields $\dfrac{\frac{3y}{4} + 7}{y+7} = \dfrac{4}{5}$. Cross-multiplying yields

$$5(3y/4 + 7) = 4(y + 7)$$
$$15y/4 + 35 = 4y + 28$$
$$15y/4 - 4y = 28 - 35$$
$$-y/4 = -7$$
$$y = 28$$

and $x = \dfrac{3y}{4} = \dfrac{3 \cdot 28}{4} = 3 \cdot 7 = 21$.

Plugging these values for x and y into the expression in Column A, $\dfrac{x+14}{y+14}$, yields

$$\dfrac{21+14}{28+14} =$$
$$\dfrac{35}{42} =$$
$$\dfrac{5}{6} =$$
Column B

Hence, the answer is (C).

---

6. There are six possible points of intersection as shown in the diagram below:

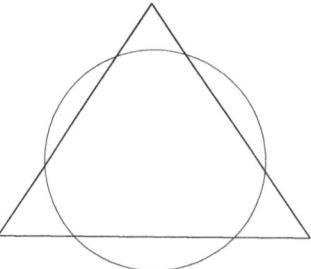

The answer is (A).

---

7. Equating vertical angles in the figure yields $a = 2b$ and $b = 3c$. From the first equation, we have $b = a/2$. Plugging this into the second equation yields $a/2 = 3c$, from which we can derive $c = a/6$. Since the angle made by a line is $180°$, we have for line l that $b + a + 6c = 180$. Replacing b with $a/2$ and c with $a/6$ in this equation yields $a/2 + a + 6(a/6) = a/2 + a + a = 180$. Summing the left-hand side yields $5a/2 = 180$, and multiplying both sides by $2/5$ yields $a = 180(2/5) = 72$. The answer is (D).

8. The formula for the area of a rectangle is *length×width*. Hence, the area of rectangle *ABCD* is $(x + 6)(x + 2) = x^2 + 8x + 12$, and the area of the rectangle *EFGH* is $(x + a)(x + b) = x^2 + (a + b)x + ab = x^2 + 8x + ab$ (given that $a + b = 8$). Now, we are given that the area of the rectangle *EFGH* is 3 units greater than the area of the rectangle *ABCD*. Hence, we have

$x^2 + 8x + ab = (x^2 + 8x + 12) + 3$
$ab = 12 + 3$ (by canceling x^2 and $8x$ from both sides)
$ab = 15$

The answer is (C).

---

9. The formula for the area of a triangle is $1/2 \times base \times height$. Hence, the area of $\triangle AFC$ (the crossed region) equals $1/2 \times FC \times AD$, and the area of $\triangle AEC$ (the shaded region) equals $1/2 \times EC \times AB$. We are given that the two areas are equal. Hence, we have the equation

$1/2 \times EC \times AB = 1/2 \times FC \times AD$

$1/2 \times 3 \times AB = 1/2 \times 4 \times 12$ from the figure, $EC = 3$, $FC = 4$, and $AD = 12$

$(3/2)AB = 24$

$AB = (2/3)(24) = 16$

Now, the perimeter of a rectangle is two times the sum of the lengths of any two adjacent sides of the rectangle. Hence, the perimeter of *ABCD* is $2(AB + AD) = 2(16 + 12) = 56$. Enter 56 in the grid.

---

10. Let *l* and *w* be the initial length and width of the rectangle, respectively. Then by the formula *Area of Rectangle = Length×Width*, the area of the rectangle = lw.

When the length is increased by 25%, the new length is $l(1 + 25/100)$. Now, let *x*% be the percentage by which the width of the new rectangle is decreased so that the area is unchanged. Then the new width should equal $w(1 - x/100)$. The area now is $l(1 + 25/100) \times w(1 - x/100)$, which equals lw (area remained unchanged).

Thus, we have the equation

$l(1 + 25/100) \times w(1 - x/100) = lw$

$(1 + 25/100) \times (1 - x/100) = 1$ canceling *l* and *w* from both sides

$(125/100)(1 - x/100) = 1$

$(1 - x/100) = 100/125$

$-x/100 = 100/125 - 1$

$-x/100 = 100/125 - 125/125$

$-x/100 = -25/125$

$x = 100 \times 25/125 = 100 \times 1/5 = 20$

The answer is (A).

11. Lines *AB* and *CD* are parallel (given) and cut by transversal *ED*. Hence, the alternate interior angles x and y are equal. Since $x = y$, $\triangle ECD$ is isosceles ($\angle C = \angle D$). Hence, angles x and y in $\triangle ECD$ could each range between 0° and 90°. No unique value for x is derivable. Hence, the answer is (E).

12. We are given the inequality

$(x + 1)^2 - 2x > 2(x + 1) + 2$
$x^2 + 2x + 1 - 2x > 2x + 2 + 2$
$x^2 + 1 > 2x + 4$
$x^2 - 2x + 1 > 4$ by subtracting $2x$ from both sides
$x^2 - 2(x)(1) + 1^2 > 4$ expressing the left side in the form $a^2 - 2ab + b^2$
$(x - 1)^2 > 4$ by using the formula $a^2 - 2ab + b^2 = (a - b)^2$

Square rooting both sides of the inequality yields two new inequalities: $x - 1 > 2$ or $x - 1 < -2$. Adding 1 to both sides of the solutions yields $x > 3$ and $x < -1$. Hence, x is either less than -1 or x is greater than 3. In either case, x does not equal 3 or 0. Hence, the answer is (C) and (D).

13. Let the original amount of money Jane had be x. Since she gave 3/5 of her money to Jack, she now has $1 - 3/5 = 2/5$ of the original amount. We are given that this 2/5 part equals 200 dollars. Hence, we have the equation $\frac{2}{5}x = 200$. Solving for x yields $x = 500$. Since she gave 3/5 of this amount to Jack, she gave him 300 ($= \frac{3}{5} \cdot 500$). The answer is (D).

14. We are given that $a > 0$ and $(a + 2)(a - 3)(a + 4) = 0$. Hence, the possible solutions are

$a + 2 = 0$; $a = -2$, a is not greater than 0, so reject.
$a - 3 = 0$; $a = 3$, a is greater than 0, so accept.
$a + 4 = 0$; $a = -4$, a is not greater than 0, so reject.

Enter 3 in the grid.

15. The average of the five numbers 1056, 1095, 1098, 1100, and 1126 is

$$\frac{1056 + 1095 + 1098 + 1100 + 1126}{5} =$$
$$\frac{5475}{5} =$$
$$1095$$

Hence, the answer is (B).

Test 13—Solutions

---

16. The arithmetic mean of m and n is 50. Hence, $(m + n)/2 = 50$. Multiplying the equation by 2 yields $m + n = 100$.

The arithmetic mean of p and q is 70. Hence, $(p + q)/2 = 70$. Multiplying the equation by 2 yields $p + q = 140$.

Now, the arithmetic mean of $m, n, p,$ and q is

$$\frac{m+n+p+q}{4} =$$
$$\frac{(m+n)+(p+q)}{4} =$$
$$\frac{100+140}{4} =$$
$$\frac{240}{4} =$$
$$60$$

Enter 60 in the grid.

---

17. Let x and y denote the numbers. Then $x/y = 10$ and $x - y = 18$. Solving the first equation for x and plugging it into the second equation yields

$$10y - y = 18$$
$$9y = 18$$
$$y = 2$$

Plugging this into the equation $x - y = 18$ yields $x = 20$. Hence, y is the smaller number. The answer is (A).

---

18. We are given that when dropped the stone breaks into pieces of equal size and weight, so if the stone has broken into n pieces, then the ratio of the value of the original stone to the value of each broken piece would be $1 : n^2$. Hence, the ratio for the original piece 1 to the total value of the n broken pieces is $n\,(1/n^2) = 1/n = 1 : n$.

Therefore, if the stone breaks into two, the net price is now one-half (1 : 2). Choose (A).

Also, if the stone breaks into three, the net price is now one-third (1 : 3). Choose (C).

The answer is (A) and (C).

19.

$$\frac{\left(\left(\sqrt{7}\right)^x\right)^2}{\left(\sqrt{7}\right)^{11}} = \qquad\qquad \frac{7^x}{7^{11}}$$

$$\frac{\left(\left(7^{\frac{1}{2}}\right)^x\right)^2}{\left(7^{\frac{1}{2}}\right)^{11}} = \qquad \text{since } \sqrt{7} = 7^{\frac{1}{2}}$$

$$\frac{7^{\frac{2x}{2}}}{7^{\frac{11}{2}}} =$$

$$\frac{7^x}{7^{\frac{11}{2}}}$$

Canceling 7^x from both columns yields

$$\frac{1}{7^{\frac{11}{2}}} = \qquad\qquad \frac{1}{7^{11}}$$

$$\frac{1}{\left(7^{11}\right)^{\frac{1}{2}}} =$$

$$\frac{1}{\sqrt{7^{11}}} =$$

Multiplying both columns by $7^{11} \cdot \sqrt{7^{11}}$ yields

$$7^{11} \qquad\qquad\qquad \sqrt{7^{11}}$$

Now, clearly, Column A is greater than Column B, and the answer is (A).

Method II:
In the numerator of the fraction in Column A, the square root and the square cancel each other. So, the numerator in Column A reduces to 7^x, which is the same as the numerator in Column B. Now, the denominator of the fraction in Column A is clearly smaller than the denominator of the fraction in Column B since $\sqrt{7} < 7$. Hence, Column A is larger: since all numerators and denominators are positive, the fraction with the same numerator and the smaller denominator has the larger value. In symbols, we have

$$\sqrt{7} < 7$$

$$\left(\sqrt{7}\right)^{11} < 7^{11}$$

$$\frac{1}{\left(\sqrt{7}\right)^{11}} > \frac{1}{7^{11}} \qquad \text{by reciprocating both sides of the inequality and reversing the direction of the inequality}$$

$$\frac{7^x}{\left(\sqrt{7}\right)^{11}} > \frac{7^x}{7^{11}} \qquad \text{by multiplying both sides of the inequality by } 7^x$$

Test 13—Solutions

20. We have $\sqrt{3-2x} = 1$. Squaring both sides of the equation yields $(3 - 2x) = 1$. Squaring both sides of the equation again yields $(3 - 2x)^2 = 1$. Hence, $(3 - 2x) + (3 - 2x)^2 = 1 + 1 = 2$. Select (C).

Adding $2x - 1$ to both sides of the equation $(3 - 2x) = 1$ yields $(3 - 2x) + (2x - 1) = 1 + 2x - 1$. Adding the like terms on both sides of this equation yields $2 = 2x$. Therefore, $2x = 2 =$ answer. Select (D) also.

Now, let's see whether we can create the expression $x + 1$ [Choice (E)] from the equation $2x = 2$. To that end divide both sides of the equation $2x = 2$ by 2, which gives $x = 1$. Now, add 1 to both sides of the equation $x = 1$ gives $x + 1 = 2 =$ answer. Select (E) also.

The answer is (C), (D), and (E).

---

21. Since the cost of painting increases by a fixed percentage each year and it increased $3,600/$2,000 = 1.8 times in the 9-year period from 1970 to 1979, it must increase by the same number of times in the period 1979 to 1988. Hence, the amount becomes $1.8 \times \$3,600 = \$6480$ by 1988. The answer is (E).

U23. Let r be the retail price. The list price is the price after a 20% discount on the retail price. Hence, it equals $r(1 - 20/100) = r(1 - 0.2) = 0.8r$.

The festival discount price is the price after a 30% discount on the list price. Hence, the festival discount price equals (list price)$(1 - 30/100) = (0.8r)(1 - 30/100) = (0.8r)(1 - 0.3) = (0.8r)(0.7) = 0.56r$.

Hence, the total discount offered is (Original Price − Price after discount)/Original Price × 100 = $(r - 0.56r)/r \times 100 = 0.44 \times 100 = 44\%$.

The answer is (C).

---

22. Let m coins of 0.5 dollars each and n coins of 0.7 dollars each add up to 4.6 dollars. Then, we have the equation $0.5m + 0.7n = 4.6$. Multiplying both sides by 10 to eliminate the decimals yields $5m + 7n = 46$. Since m is a positive integer, $5m$ is positive integer; and since n is a positive integer, $7n$ is a positive integer. Let $p = 5m$ and $q = 7n$. So, p is a multiple of 5 and q is a multiple of 7 and $p + q = 46$. Subtracting q from both sides yields $p = 46 - q$ [(a positive multiple of 5) equals 46 − (a positive multiple of 7)]. Let's seek such solution for p and q:

If $q = 7, p = 46 - 7 = 39$, not a multiple of 5. Reject.

If $q = 14, p = 46 - 14 = 32$, not a multiple of 5. Reject.

If $q = 21, p = 46 - 21 = 25$, a multiple of 5. Acceptable. So, $n = q/7 = 3$ and $m = p/5 = 5$.

The following checks are not actually required since we already have an acceptable solution.

If $q = 28, p = 46 - 28 = 18$, not a multiple of 5. Reject.

If $q = 35, p = 46 - 35 = 11$, not a multiple of 5. Reject.

If $q = 42, p = 46 - 42 = 4$, not a multiple of 5. Reject.

If $q \geq 49, p = 46 - 49 = -3$, not positive. Reject.

The answer is (E).

23. Since "the first term in the series is twice the second term," we have $a = 2(ar)$. Canceling a from both sides of the equation yields $1 = 2r$. Hence, $r = 1/2$.

Hence, the three numbers a, ar, and ar^2 become a, $a(1/2)$, and $a(1/2)^2$, or a, $a/2$, and $a/4$.

The sum of first two terms is $a + a/2$ and the sum of the last two terms is $a/2 + a/4$. Forming their ratio yields

$$\frac{a + \frac{a}{2}}{\frac{a}{2} + \frac{a}{4}} =$$

$$\frac{\frac{2a+a}{2}}{\frac{2a+a}{4}} =$$

$$\frac{\frac{3a}{2}}{\frac{3a}{4}} =$$

$$\left(\frac{3a}{2}\right)\left(\frac{4}{3a}\right) =$$

$$2 =$$

$$\frac{2}{1} \text{ or } 2:1$$

The answer is (D).

---

24. The possible positive integer solutions x and y of the equation $x + y = 5$ are $\{x, y\} = \{1, 4\}, \{2, 3\}, \{3, 2\}$, and $\{4, 1\}$. Each solution is equally probable. Exactly one of the 4 possible solutions has x equal to 1. Hence, the probability that x equals 1 is one in four ways, which equals 1/4. The answer is (C).

---

25. Pants can be selected from 3 in 3 ways, and a shirt can be selected from 2 in 2 ways. Hence, the pair can be selected in $3 \cdot 2 = 6$ ways. The answer is (D).

Test 14

GRE Math Tests

Questions: 24
Time: 45 minutes

[Multiple-choice Question – Select One or More Answer Choices]
1. If $(2x + 1)^2 = 100$, then which of the following COULD equal x?

   (A) $-11/2$
   (B) $-9/2$
   (C) $9/2$
   (D) $11/2$
   (E) $17/2$

[Multiple-choice Question – Select One or More Answer Choices]
2. The number m yields a remainder p when divided by 14 and a remainder q when divided by 7. If $p = q + 7$, then which of the following could be the value of m?

   (A) 45
   (B) 53
   (C) 72
   (D) 81
   (E) 85
   (F) 100

[Multiple-choice Question – Select One or More Answer Choices]
3. If n is a positive integer, which of the following numbers must have a remainder of 3 when divided by any one of the numbers 4, 5, and 6?

   (A) $12n + 3$
   (B) $24n + 3$
   (C) $60n + 3$
   (D) $80n + 2$
   (E) $120n + 3$

Test 14—Questions

[Multiple-choice Question – Select One Answer Choice only]
4. $a, b, c, d,$ and e are five consecutive integers in increasing order of size. Which one of the following expressions is not odd?

(A) $a + b + c$
(B) $ab + c$
(C) $ab + d$
(D) $ac + d$
(E) $ac + e$

[Quantitative Comparison Question]
5.   Column A   In the figure, ABCD and BECD   Column B
                                                          are parallelograms.

Area of parallelogram ABCD   Area of parallelogram BECD

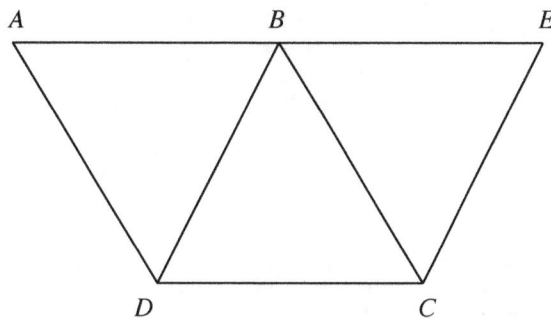

[Multiple-choice Question – Select One Answer Choice only]
6. $\triangle ABC$ is a right-angled isosceles triangle, and $\angle B$ is the right angle in the triangle. If AC measures $7\sqrt{2}$, then which one of the following would equal the lengths of AB and BC, respectively?

(A) 7, 7
(B) 9, 9
(C) 10, 10
(D) 11, 12
(E) 7, 12

[Multiple-choice Question – Select One Answer Choice only]
7. *A* and *B* are centers of two circles that touch each other externally, as shown in the figure. What is the area of the circle whose diameter is *AB* ?

   (A) 4π
   (B) $25\pi/4$
   (C) 9π
   (D) 16π
   (E) 25π

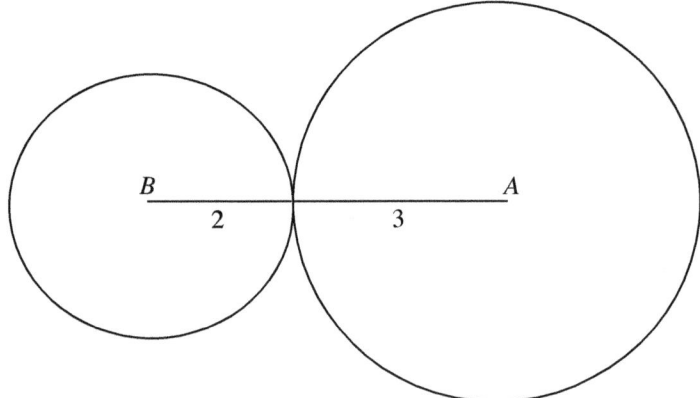

[Multiple-choice Question – Select One Answer Choice only]
8. In the figure, what is the value of *x* ?

   (A) 90
   (B) 95
   (C) 100
   (D) 105
   (E) 115

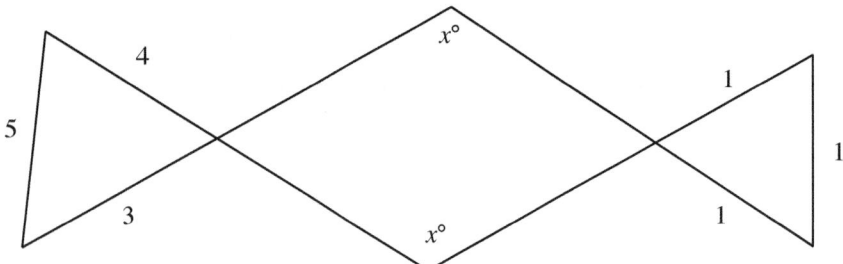

[Numeric Entry Question]
9. The area of the base of a tank is 100 sq. ft. It takes 20 seconds to fill the tank with water poured at rate of 25 cubic feet per second. What is the height in feet of the rectangular tank?

10. Column A

A circular park is enlarged uniformly such that it now occupies 21% more land.

Column B

The percentage increase in the radius of the park due to the enlargement

The percentage increase in the area of the park due to the enlargement

[Multiple-choice Question – Select One Answer Choice only]
11. What is the maximum number of 3 x 3 squares that can be formed from the squares in the 6 x 6 checker board shown?

(A) 4
(B) 6
(C) 12
(D) 16
(E) 24

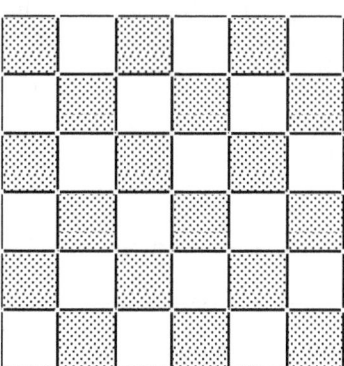

[Multiple-choice Question – Select One Answer Choice only]

12. What is the greatest prime factor of $(2^4)^2 - 1$?

    (A) 3
    (B) 5
    (C) 11
    (D) 17
    (E) 19

[Multiple-choice Question – Select One or More Answer Choices]

13. If $x + y = 750$, then which of the following additional details alone will determine the numeric value of x ?

    (A) $x + 2y = d$
    (B) $2x + 4y = 2d$
    (C) $2x + 2y = 1500$
    (D) $3x = 2250 - 3y$
    (E) $2x + y = 15$
    (F) $x + 2y = 2235$

[Multiple-choice Question – Select One Answer Choice only]

14. Which one of the following points in the figure is the median of the points $M, P, Q, R,$ and S ?

    (A) M
    (B) P
    (C) Q
    (D) R
    (E) S

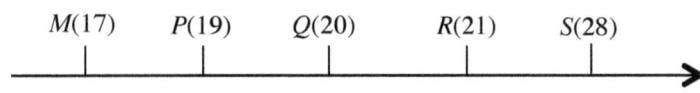

The figure is not drawn to scale.   x-axis

[Multiple-choice Question – Select One Answer Choice Only]
15. Which of the following points in the figure is either the median or the average of M, P, Q, R, and S?

   (A) M
   (B) P
   (C) Q
   (D) R
   (E) S

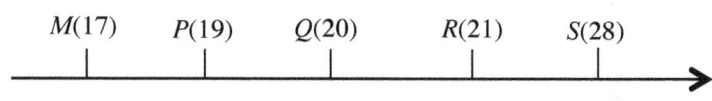

The figure is not drawn to scale. x-axis

[Multiple-choice Question – Select One Answer Choice Only]
16. The ratio of the sum of the reciprocals of x and y to the product of the reciprocals of x and y is 1 : 3. What is sum of the numbers x and y?

   (A) 1/3
   (B) 1/2
   (C) 1
   (D) 2
   (E) 4

17.   Column A $x = a, y = 2b, z = 3c.$ Column B
                               $x : y : z = 1 : 2 : 3.$

       $\dfrac{x+y+z}{a+b+c}$ 2

[Multiple-choice Question – Select One or More Answer Choices]

18. If $p = 216^{-1/3} + 243^{-2/5} + 256^{-1/4}$, then which of the following is an integer?

   (A) $p/19$
   (B) $p/36$
   (C) p
   (D) $19/p$
   (E) $36/p$
   (F) $19p$
   (G) $36p$

[Multiple-choice Question – Select One or More Answer Choices]

19. If $\dfrac{x+y}{x-y} = \dfrac{4}{3}$ and $x \ne 0$, then what percentage (to the nearest integer) of $x + 3y$ is $x - 3y$?

   (A) 20%
   (B) 25%
   (C) 30%
   (D) 35%
   (E) 40%
   (F) Same percentage $14y$ is of $5x$

[Multiple-choice Question – Select One Answer Choice Only]

20. The *list price* of a commodity is the price after a 20% discount on the retail price. The *festival discount price* on the commodity is the price after a 30% discount on the list price. Customers purchase commodities from stores at a festival discount price. What is the effective discount offered by the stores on the commodity on its retail price?

   (A) 20%
   (B) 30%
   (C) 44%
   (D) 50%
   (E) 56%

21.   Column A   12 trophies cost exactly 60 dollars.   Column B

If the cost of each trophy decreases by 1 dollar, $12 + x$ trophies cost exactly 60 dollars.

If the cost of each trophy increases by 1 dollar, $12 - y$ trophies cost exactly 60 dollars.

         x y

[Multiple-choice Question – Select One Answer Choice Only]

22. The sum of the first n terms of an arithmetic series whose nth term is n can be calculated by the formula $n(n + 1)/2$. Which one of the following equals the sum of the first eight terms in a series whose nth term is $2n$?

   (A) 24
   (B) 48
   (C) 56
   (D) 72
   (E) 96

[Multiple-choice Question – Select One Answer Choice Only]

23. A survey of n people in the town of Eros found that 50% of them prefer Brand A. Another survey of 100 people in the town of Angie found that 60% prefer Brand A. In total, 55% of all the people surveyed together prefer Brand A. What is the total number of people surveyed?

   (A) 50
   (B) 100
   (C) 150
   (D) 200
   (E) 250

[Multiple-choice Question – Select One Answer Choice Only]

24. A bowl contains one marble labeled 0, one marble labeled 1, one marble labeled 2, and one marble labeled 3. The bowl contains no other objects. If two marbles are drawn randomly without replacement, what is the probability that they will add up to 3?

(A) 1/12
(B) 1/8
(C) 1/6
(D) 1/4
(E) 1/3

Test 14—Solutions

Answers and Solutions Test 14:

Question	Answer
1.	A, C
2.	B, D
3.	C, E
4.	E
5.	C
6.	A
7.	B
8.	D
9.	5
10.	B
11.	D
12.	D
13.	E, F
14.	C
15.	C, D
16.	A
17.	C
18.	D, G
19.	E, F
20.	C
21.	A
22.	D
23.	D
24.	E

If you got 18/24 correct on this test, you are likely to get 750+ on the actual GRE **by the time you complete all the tests in the book.**

---

1. Choice (A): $(2x+1)^2 = \left(2\left[\dfrac{-11}{2}\right]+1\right)^2 = (-11+1)^2 = (-10)^2 = 100$.

Choice (C): $(2x+1)^2 = \left(2\left[\dfrac{9}{2}\right]+1\right)^2 = (9+1)^2 = (10)^2 = 100$.

Since these value of x satisfies the equation, the answer is (A) and (C). You can check that no other answer-choice satisfies the equation.

Method II (without substitution):
Square rooting both sides of the given equation $(2x + 1)^2 = 100$ yields two possible solutions: $2x + 1 = 10$ and $2x + 1 = -10$. Solving the first equation for x yields $x = 9/2$, and solving the second equation for x yields $x = -11/2$. We have the first and the second solutions in choices (A) and (C), so the answer is (A) and (C).

GRE Math Tests

2. Select the choice that satisfies the equation $p = q + 7$.

Choice (A): Suppose $m = 45$. Then $m/14 = 45/14 = 3 + 3/14$. So, the remainder is $p = 3$. Also, $m/7 = 45/7 = 6 + 3/7$. So, the remainder is $q = 3$. Here, $p \neq q + 7$. So, reject the choice.

Choice (B): Suppose $m = 53$. Then $m/14 = 53/14 = 3 + 11/14$. So, the remainder is $p = 11$. Also, $m/7 = 53/7 = 7 + 4/7$. So, the remainder is $q = 4$. Here, $p = q + 7$. So, select the choice.

Choice (C): Suppose $m = 72$. Then $m/14 = 72/14 = 5 + 2/14$. So, the remainder is $p = 2$. Now, $m/7 = 72/7 = 10 + 2/7$. So, the remainder is $q = 2$. Here, $p \neq q + 7$. So, reject the choice.

Choice (D): Suppose $m = 81$. Then $m/14 = 81/14 = 5 + 11/14$. So, the remainder is $p = 11$. Also, $m/7 = 81/7 = 11 + 4/7$. So, the remainder is $q = 4$. Here, $p = q + 7$. So, select the choice.

Choice (E): Suppose $m = 85$. Then $m/14 = 85/14 = 6 + 1/14$. So, the remainder is $p = 1$. Now, $m/7 = 85/7 = 12 + 1/7$. So, the remainder is $q = 1$. Here, $p \neq q + 7$. So, reject the choice.

Choice (F): Suppose $m = 100$. Then $m/14 = 100/14 = 7 + 2/14$. So, the remainder is $p = 2$. Now, $m/7 = 100/7 = 14 + 2/7$. So, the remainder is $q = 2$. Here, $p \neq q + 7$. So, reject the choice.

Hence, the answers are (B) and (D).

3. Let m be a number that has a remainder of 3 when divided by any of the numbers 4, 5, and 6. Then $m - 3$ must be exactly divisible by all three numbers. Hence, $m - 3$ must be a multiple of the Least Common Multiple of the numbers 4, 5, and 6. The LCM is $3 \cdot 4 \cdot 5 = 60$. Hence, we can suppose $m - 3 = 60p$, where p is a positive integer. Replacing p with n, we get $m - 3 = 60n$. So, $m = 60n + 3$. Choose choice (C). Choice (E) is also in the same format $120n + 3 = 60(2n) + 3$. Hence, choose (C) and (E).

We can also subtract 3 from each answer-choice, and the correct answer will be divisible by 60:

>Choice (A): If $n = 1$, then $(12n + 3) - 3 = 12n = 12$, not divisible by 60. Reject.
>Choice (B): If $n = 1$, then $(24n + 3) - 3 = 24n = 24$, not divisible by 60. Reject.
>Choice (C): If $n = 1$, then $(60n + 3) - 3 = 60n = 60$, not divisible by 60. Hence, correct.
>Choice (D): If $n = 1$, then $(80n + 2) - 3 = 80n - 1 = 79$, not divisible by 60. Reject.
>Choice (E): If $n = 1$, then $(120n + 3) - 3 = 120n = 120$, divisible by 60 for any integer n. Hence, correct.

4. Choice (A): $a + b + c$: Suppose a is an even number. Then b, the integer following a, must be odd, and c, the integer following b, must be even. Hence, $a + b + c$ = sum of two even numbers (a and c) and an odd number (b). Since the sum of any number of even numbers with an odd number is odd (For example, if $a = 4$, then $b = 5$, $c = 6$, and $a + b + c$ equals $4 + 5 + 6 = 15$ (odd)), $a + b + c$ is odd. Reject.

Choice (B): $ab + c$: At least one of every two consecutive positive integers a and b must be even. Hence, the product ab is an even number. Now, if c is odd (which happens when a is odd), $ab + c$ must be odd. For example, if $a = 3$, $b = 4$, and $c = 5$, then $ab + c$ must equal $12 + 5 = 17$, an odd number. Reject.

Choice (C): $ab + d$: We know that ab being the product of two consecutive numbers must be even. Hence, if d happens to be an odd number (it happens when b is odd), then the sum $ab + d$ is also odd. For example, if $a = 4$, then $b = 5$, $c = 6$, and $d = 7$, then $ab + d = 3 \cdot 5 + 7 = 15 + 7 = 23$, an odd number. Reject.

Choice (D): $ac + d$: Suppose a is odd. Then c must also be odd, being a number 2 more than a. Hence, ac is the product of two odd numbers and must therefore be odd. Now, d is the integer following c and must be even. Hence, $ac + d$ = odd + even = odd. For example, if $a = 3$, then $b = 3 + 1 = 4$, $c = 4 + 1 = 5$ (odd) and $d = 5 + 1 = 6$ (even) and $ac + d = 3 \cdot 5 + 6 = 21$, an odd number. Reject.

Choice (E): $ac + e$: Suppose a is an odd number. Then both c and e must also be odd. Now, ac is product of two odd numbers and therefore must be odd. Summing this with another odd number e yields an even number. For example, if $a = 1$, then c must equal 3, and e must equal 5 and $ac + e$ must equal $1 \cdot 3 + 5 = 8$, an even number. Now, suppose a is an even number. Then both c and e must also be even. Hence,

$$ac + e =$$

(product of two even numbers) + (an even number) =

(even number) + (even number) =

an even number

For example, if $a = 2$, then c must equal 4, and e must equal 6 and the expression $ac + e$ equals 14, an even number. Hence, in any case, $ac + e$ is even. Correct.

The answer is (E).

---

5. We know that a diagonal of a parallelogram cuts the parallelogram into two triangles of equal area. Since BD is a diagonal of parallelogram $ABCD$, the area of the parallelogram equals twice the area of either of the two equal triangles $\triangle ABD$ or $\triangle DBC$. Hence, Column A = 2(area of $\triangle DBC$).

Similarly, since BC is a diagonal of parallelogram $BECD$, the area of the parallelogram $BECD$ is 2(area of $\triangle DBC$).

Thus, both columns equal 2(area of $\triangle DBC$), and the answer is (C).

---

6. In a right-angled isosceles triangle, the sides of the right angle are equal. Now, in the given right-angled isosceles triangle ABC, $\angle B$ is given to be the right angle. Hence, the sides of the angle, AB and BC, are equal. Applying The Pythagorean Theorem to the triangle yields

$$AB^2 + BC^2 = AC^2$$
$$BC^2 + BC^2 = \left(7\sqrt{2}\right)^2 \quad \text{since } AB = BC$$
$$2(BC)^2 = 7^2 \times 2$$
$$BC^2 = 7^2$$
$$BC = 7 \quad \text{by square rooting both sides}$$

Hence, $AB = BC = 7$. The answer is (A).

---

7. Since the two circles touch each other, the distance between their centers, AB, equals the sum of the radii of the two circles, which is $2 + 3 = 5$. Hence, the area of a circle with diameter AB (or radius = $AB/2$) is

$$\pi \times radius^2 =$$

$$\pi(AB/2)^2 =$$

$$\pi(5/2)^2 =$$

$$25\pi/4$$

The answer is (B).

8. Let's name the vertices of the figure as shown below

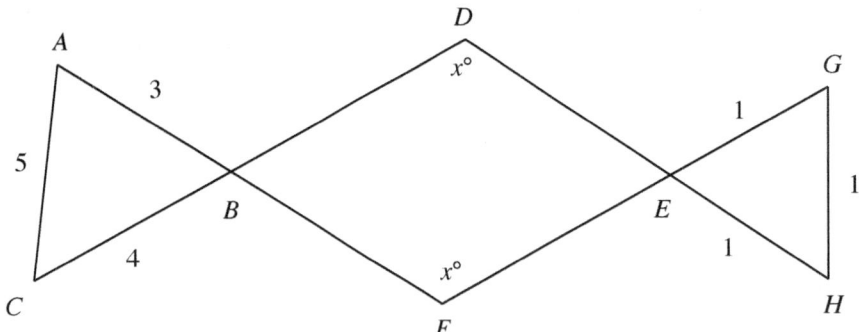

Now, in $\triangle ABC$, $AB = 3$, $BC = 4$, and $AC = 5$. Hence, AC^2 equals $5^2 = 25$, and $AB^2 + BC^2$ equals $3^2 + 4^2 = 9 + 16 = 25$. Hence, $AC^2 = 25 = AB^2 + BC^2$. From this, it is clear that the triangle satisfies The Pythagorean Theorem. Hence, by the theorem, ABC must be a right triangle, AC the longest side the hypotenuse, and the angle opposite the side, $\angle B$ must be the right angle. Hence, $\angle B = 90°$.

Now, in $\triangle EGH$, from the figure, we have each side (EG, GH, and EH) of the triangle measures 1 unit. Hence, all the sides of the triangle are equal, and the triangle is equilateral. Since in any equilateral triangle, each angle measures $60°$, we have $\angle E = 60°$.

Now, from the figure, we have

$\angle B$ in the quadrilateral = $\angle B$ in $\triangle ABC$ vertical angles are equal
$= 90°$ known result

Also, we have

$\angle E$ in the quadrilateral = $\angle E$ in $\triangle EGH$ vertical angles are equal
$= 60°$

Now, summing the angles of the quadrilateral to $360°$ yields $\angle B + \angle E + \angle D + \angle F = 360°$. Hence, we have $90 + 60 + x + x = 360$. Solving for x yields $2x = 210$ or $x = 210/2 = 105$. The answer is (D).

---

9. The formula for the volume of a rectangular tank is (*area of base*) ×*height*. Hence, we have

$$volume = (area\ of\ base) \times height$$

or

$$height = volume/(area\ of\ base)$$

We are given that it takes 20 seconds to fill the tank at the rate of 25 cubic feet per second. Hence, the volume of the tank = rate × time = 25 cubic feet × 20 seconds = 500 cubic feet. Using this in the equation for height yields *height* = *volume*/(*area of base*) = 500 cu. ft/100 sq. ft = 5 feet. Enter 5 in the grid.

10. Let the initial radius of the park be r, and let the radius after the enlargement be R. (Since the enlargement is uniform, the shape of the enlarged park is still circular.) By the formula for the area of a circle, the initial area of the park is πr^2, and the area after expansion is πR^2. Since the land occupied by the park is now 21% greater (given that the new area is 21% more), the new area is $(1 + 21/100)\pi r^2 = 1.21\pi r^2$. Equating this to πR^2 yields

$$\pi R^2 = 1.21\pi r^2$$
$$R^2 = 1.21 r^2 \quad \text{by canceling } \pi \text{ from both sides}$$
$$R = 1.1r \quad \text{by taking the square root of both sides}$$

Now, Column A equals the percentage increase in the radius:

$$\frac{\text{Final radius} - \text{Initial radius}}{\text{Initial radius}} \cdot 100 =$$
$$\frac{1.1r - r}{r} \cdot 100 =$$
$$\frac{0.1r}{r} \cdot 100 =$$
$$\frac{1}{10} \cdot 100 =$$
$$10\%$$

Column B equals the percentage increase in the area of the park due to the enlargement, which is same as the percentage increase in the area of the land, 21%. Hence, Column B is greater than Column A, and the answer is (B).

---

11. Clearly, there are more than four 3 x 3 squares in the checkerboard—eliminate (A). Next, eliminate (B) since it merely repeats a number from the problem, which is an eye-catcher for this hard problem. Further, eliminate (E) since it is the greatest, again this is an eye-catcher for this hard problem. This leaves choices (C) and (D). If you count carefully, you will find sixteen 3 x 3 squares in the checkerboard. The answer is (D).

---

12. $\left(2^4\right)^2 - 1 = (16)^2 - 1 = 256 - 1 = 255$. Since the question asks for the greatest prime factor, we eliminate 19, the greatest number. Now, we start with the next largest number and work our way up the list; the first number that divides into 255 evenly will be the answer. Dividing 17 into 255 gives

$$17\overline{)255} = 15$$

Hence, 17 is the largest prime factor of $\left(2^4\right)^2 - 1$. The answer is (D).

GRE Math Tests

13. Solving the given equation for y yields $y = 750 - x$. Now, let's substitute this into each answer-choice. The one that returns a numeric value for x is the answer.

Choice (A): $x + 2y = d$; $x + 2(750 - x) = d$; $x + 1500 - 2x = d$; $x = 1500 - d$; Since d is unknown, the value of x cannot be calculated. Reject.

Choice (B): $2x + 4y = 2d$; $2x + 4(750 - x) = 2d$; $2x + 3000 - 4x = 2d$; $-2x + 3000 = 2d$; $x = 1500 - d$; Since d is unknown, the value of x cannot be calculated. Reject.

Choice (C): $2x + 2y = 1500$; $2x + 2(750 - x) = 1500$; $2x + 1500 - 2x = 1500$; $0 = 0$, A known result. Hence, this is no additional information to derive the solution. Hence, the value of x cannot be calculated. Reject.

Choice (D): $3x = 2250 - 3y$; $3x = 2250 - 3(750 - x) = 3x$; $3x = 3x$; A known result. Hence, this is no additional information to derive the solution. Reject.

Choice (E): $2x + y = 15$; The slope of the equation is -2, different from -1 of the given equation. Hence, there will be a definite solution. Accept it. The calculation is in the brackets: [$2x + 750 - x = 15$; $x + 750 = 15$; $x = 15 - 750 = -735$.]

Choice (F): $x + 2y = 2235$; The slope of the equation is $-1/2$, different from -1 of the given equation. Hence, there will be a definite solution, needless to calculate. Accept it.

The answer is (E) and (F).

14. The definition of *median* is "When a set of numbers is arranged in order of size, the *median* is the middle number. If a set contains an even number of elements, then the median is the average of the two middle elements."

From the number line $M = 17$, $P = 19$, $Q = 20$, $R = 21$, and $S = 28$. The numbers arranged in order are 17, 19, 20, 21, and 28. The median is 20. Since $Q = 20$, the answer is (C).

15. We know that median is Q. Now, the average is $(17 + 19 + 20 + 21 + 28)/5 = 21 = R$. The answer is (C) and (D).

16. We are given that the ratio of the sum of the reciprocals of x and y to the product of the reciprocals of x and y is $1 : 3$. Writing this as an equation yields

$$\frac{\frac{1}{x} + \frac{1}{y}}{\frac{1}{x} \cdot \frac{1}{y}} = \frac{1}{3}$$

$$\frac{\frac{x+y}{xy}}{\frac{1}{xy}} = \frac{1}{3}$$

$$\frac{x+y}{xy} \cdot \frac{xy}{1} = \frac{1}{3}$$

$$x + y = \frac{1}{3} \qquad \text{by canceling } xy \text{ from the numerator and denominator}$$

The answer is (A).

Test 14—Solutions

17. We are given the equations $x = a$, $y = 2b$, $z = 3c$, and the proportion $x : y : z = 1 : 2 : 3$. Substituting the first three equations into the last equation yields $a : 2b : 3c = 1 : 2 : 3$. Forming the resultant ratio yields $a/1 = 2b/2 = 3c/3$. Simplifying the equation yields $a = b = c$. Thus, we have that both a and b equal c. Hence, from the given equations, we have $x = a = c$, $y = 2b = 2c$, and $z = 3c$.

Now, Column A $= \dfrac{x + y + z}{a + b + c}$

$= \dfrac{c + 2c + 3c}{c + c + c}$ because $x = a, y = 2c, z = 3c,$ and $a = b = c$

$= \dfrac{6c}{3c} = 2 =$ Column B.

Hence, the answer is (C).

---

18. Simplifying the given equation yields

$p = 216^{-1/3} + 243^{-2/5} + 256^{-1/4}$
$= (6^3)^{-1/3} + (3^5)^{-2/5} + (4^4)^{-1/4}$ because $216 = 6^3$, $243 = 3^5$, and $256 = 4^4$
$= 6^{3(-1/3)} + 3^{5(-2/5)} + 4^{4(-1/4)}$
$= 6^{-1} + 3^{-2} + 4^{-1}$
$= \dfrac{1}{6} + \dfrac{1}{9} + \dfrac{1}{4}$
$= \dfrac{6 + 4 + 9}{36}$
$= \dfrac{19}{36}$

Now,

    Choice (A): $p/19 = (19/36)/19 = 1/36$, not an integer. Reject.
    Choice (B): $p/36 = (19/36)/36 = 19/36^2$, not an integer. Reject.
    Choice (C): $p = 19/36$, not an integer. Reject.
    Choice (D): $19/p = 19/(19/36) = 19 \cdot 36/19 = 36$, an integer. Accept.
    Choice (E): $36/p = 36/(19/36) = 36^2/19$, not an integer. Reject.
    Choice (F): $19p = 19(19/36) = 19^2/36$, not an integer. Reject.
    Choice (G): $36p = 36(19/36) = 19$, an integer. Accept.

The answer is (D) and (G).

19. Dividing both the numerator and the denominator of the given equation $\frac{x+y}{x-y} = \frac{4}{3}$ by y yields

$\frac{\frac{x}{y}+1}{\frac{x}{y}-1} = \frac{4}{3}$. Cross-multiplying this equation yields $\frac{3x}{y} + 3 = \frac{4x}{y} - 4$. Solving for x/y yields $x/y = 7$.

Now, the percentage of $x + 3y$ the expression $x - 3y$ makes is $\frac{x-3y}{x+3y} \cdot 100$. Dividing both the numerator and the denominator of the expression by y yields

$$\frac{\frac{x}{y}-3}{\frac{x}{y}+3} \cdot 100 = \frac{7-3}{7+3} \cdot 100 = \frac{4}{10} \cdot 100 = 40\%$$

Choice (E) is correct.

Now, let's evaluate choice (F), "The percentage $14y$ is of $5x$." Let p be the percentage. Then, $14y$ is $p\%$ of $5x$. Therefore, we have

$$\frac{p}{100} \cdot 5x = 14y$$

$$= \frac{14y}{5x} \cdot 100$$

$$= \frac{14 \cdot 100}{5} \cdot \frac{y}{x}$$

$$= \frac{14 \cdot 100}{5} \cdot \frac{1}{\frac{x}{y}}$$

$$= \frac{14 \cdot 100}{5} \cdot \frac{1}{7}$$

$$= 40$$

Therefore, (F) is also a correct answer.

The correct answers are (E) and (F).

---

20. Let r be the retail price. The list price is the price after a 20% discount on the retail price. Hence, it equals $r(1 - 20/100) = r(1 - 0.2) = 0.8r$.

The festival discount price is the price after a 30% discount on the list price. Hence, the festival discount price equals (list price)$(1 - 30/100) = (0.8r)(1 - 30/100) = (0.8r)(1 - 0.3) = (0.8r)(0.7) = 0.56r$.

Hence, the total discount offered is (Original Price − Price after discount)/Original Price × 100 = $(r - 0.56r)/r \times 100 = 0.44 \times 100 = 44\%$.

The answer is (C).

Test 14—Solutions

21. 12 trophies cost 60 dollars, so each trophy costs 60/12 = 5 dollars.

If the price decreases by 1, the new price is 5 − 1 = 4 dollars. Hence, 60 dollars can now buy 15 (= 60/4) trophies. Equating this to 12 + x yields 12 + x = 15, or x = 3.

If the price increases by 1, the new price is 5 + 1 = 6 dollars. Hence, 60 dollars can now buy 10 (= 60/6) trophies. Equating this to 12 − y yields 12 − y = 10, or y = 2.

Now, Column A equals x = 3, and Column B equals y = 2. Hence, Column A > Column B, and the answer is (A).

---

22. The sum of the first n terms of an arithmetic series whose nth term is n is $n(n + 1)/2$. Hence, we have

$$1 + 2 + 3 + \ldots + n = n(n + 1)/2$$

Multiplying each side by 2 yields

$$2 + 4 + 6 + \ldots + 2n = 2n(n + 1)/2 = n(n + 1)$$

Hence, the sum to 8 terms equals $n(n + 1) = 8(8 + 1) = 8(9) = 72$. The answer is (D).

---

23. 50% of n people from Eros prefer brand A. 50% of n is $50/100 \times n = n/2$.

60% of 100 people from Angie prefer brand A. 60% of 100 is $60/100 \times 100 = 60$.

Of the total $n + 100$ people surveyed, $n/2 + 60$ prefer brand A. Given that this is 55%, we have

$$\frac{\frac{n}{2} + 60}{n + 100} \times 100 = 55$$

Solving the equation yields

$$\frac{\frac{n}{2} + 60}{n + 100} \times 100 = 55$$
$$\frac{n}{2} + 60 = \frac{55}{100}(n + 100)$$
$$\frac{n}{2} + 60 = \frac{11}{20}n + 55$$
$0 = 11n/20 − n/2 + 55 − 60$ subtracting $n/2$ and 60 from both sides
$0 = n/20 − 5$
$5 = n/20$ adding 5 to both sides
$n = 20 \times 5 = 100$ multiplying both sides by 20

Hence, the total number of people surveyed is $n + 100 = 100 + 100 = 200$. The answer is (D).

24. The following list shows all 12 ways of selecting the two marbles:

$$\begin{array}{llll}(0,1) & (1,0) & (2,0) & \mathbf{(3,0)} \\ (0,2) & \mathbf{(1,2)} & \mathbf{(2,1)} & (3,1) \\ \mathbf{(0,3)} & (1,3) & (2,3) & (3,2)\end{array}$$

The four pairs in bold are the only ones whose sum is 3. Hence, the probability that two randomly drawn marbles will have a sum of 3 is

$$4/12 = 1/3$$

The answer is (E).

Test 15

GRE Math Tests

Questions: 24
Time: 45 minutes

[Quantitative Comparison Question]
1.   Column A $y \neq 0$ Column B

      x/y xy

USE THIS SPACE FOR SCRATCHWORK.

[Quantitative Comparison Question]
2.   Column A   For any positive integer n, $n!$ Column B
                          denotes the product of all the
                          integers from 1 through n.

      $1!(10-1)!$ $2!(10-2)!$

USE THIS SPACE FOR SCRATCHWORK.

[Multiple-choice Question – Select One Answer Choice Only]
3.   A housing subdivision contains only two types of homes: ranch-style homes and townhomes. There are twice as many townhomes as ranch-style homes. There are 3 times as many townhomes with pools than without pools. What is the probability that a home selected at random from the subdivision will be a townhome with a pool?

   (A) 1/6
   (B) 1/5
   (C) 1/4
   (D) 1/3
   (E) 1/2

USE THIS SPACE FOR SCRATCHWORK.

[Quantitative Comparison Question]
4. Column A   Column B

The difference between two   179.5°
angles of a triangle

[Quantitative Comparison Question]
5. Column A   Column B

Volume of a cylinder with   Volume of a cone with a
a height of 10   height of 10

[Multiple-choice Question – Select One or More Answer Choices]
6. In the figure, if $AB = 8$, $BC = 6$, $AE = 10$, $BD = 16$ and the two given triangles are similar triangles, then DE could equal which of the following?

   (A) 12
   (B) 13
   (C) 14.4
   (D) 15.8
   (E) 18

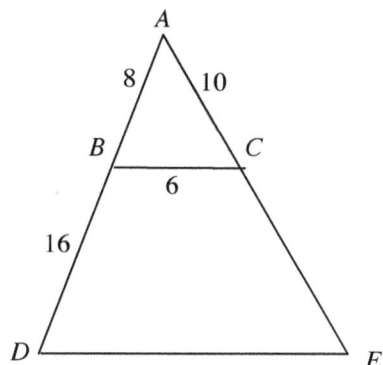

[Multiple-choice Question – Select One Answer Choice Only]
7. In the figure, ABCD is a parallelogram. Which one of the following is true?

    (A) $x < y$
    (B) $x > q$
    (C) $x > p$
    (D) $y > p$
    (E) $y > q$

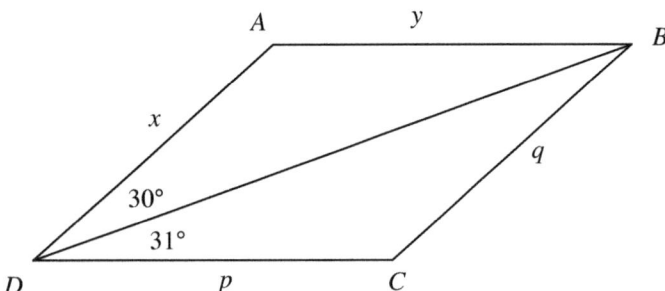

[Multiple-choice Question – Select One Answer Choice Only]
8. In the figure, ABCD is a rectangle. The area of quadrilateral EBFD is one-half the area of the rectangle ABCD. Which one of the following is the value of AD ?

    (A) 5
    (B) 6
    (C) 7
    (D) 12
    (E) 15

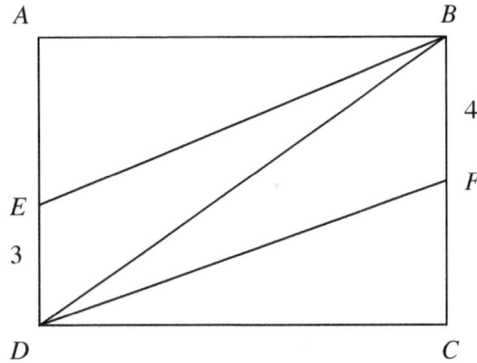

[Quantitative Comparison Question]

9.   Column A   The perimeter of rectangle *ABCD*   Column B
                              is 5/2 times as long as the side *AB*.

      Length of side *AB*   Length of side *BC*

[Multiple-choice Question – Select One Answer Choice Only]

10. The distance between cities A and B is 120 miles. A car travels from A to B at 60 miles per hour and returns from B to A along the same route at 40 miles per hour. What is the average speed for the round trip?

   (A) 48
   (B) 50
   (C) 52
   (D) 56
   (E) 58

[Quantitative Comparison Question]

11.   Column A $x = 1/y$ Column B

       $x + 1 + 1/x$ $y + 1 + 1/y$

[Multiple-choice Question – Select One or More Answer Choices]

12. Which two of the following numbers are closest to 1?

   (A) $\dfrac{3}{3 + 0.3}$
   (B) $\dfrac{3}{3 + 0.3^2}$
   (C) $\dfrac{3}{3 - 0.3}$
   (D) $\dfrac{3}{3 - 0.3^2}$
   (E) $\dfrac{3}{3 + 0.33}$

[Quantitative Comparison Question]

13. Column A

$7x + 3y = 12$
$3x + 7y = 8$

Column B

$x - y$

1

[Multiple-choice Question – Select One Answer Choice Only]

14. If $x + y = 7$ and $x^2 + y^2 = 25$, then which one of the following equals the value of $x^3 + y^3$?

(A) 7
(B) 25
(C) 35
(D) 65
(E) 91

[Quantitative Comparison Question]

15. Column A

The mean monthly rainfall for the 8 months

The monthly rainfall (in inches) for the first eight months of the year was 2, 4, 4, 5, 7, 9, 10, 11.

Column B

The median of the rainfall for the 8 months

[Quantitative Comparison Question]

16.
Column A		Column B
	A certain recipe requires 3/2 cups of sugar and makes 2-dozen cookies.	
The number of cups of sugar required for the same recipe to make 30 cookies		2

[Numeric Entry]

17. The cost of production of a certain instrument is directly proportional to the number of units produced. The cost of production for 300 units is $300. What is the cost of production for 270 units?

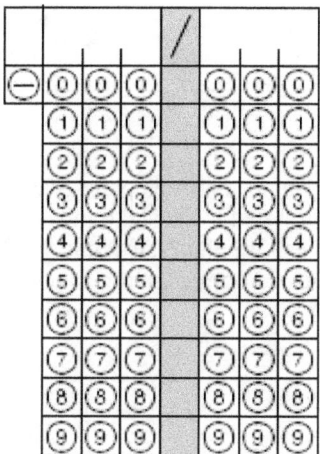

[Multiple-choice Question – Select One Answer Choice Only]

18. If a sprinter takes 30 steps in 9 seconds, how many steps does he take in 54 seconds?

    (A)   130
    (B)   170
    (C)   173
    (D)   180
    (E)   200

[Multiple-choice Question – Select One or More Answer Choices]
19. If 9/100 of *x* is 9, then which of the following are true?

   (A) *y* percent of *x* is *y*
   (B) 1/4 of *x* is 0.25
   (C) *x* is 120% of 80
   (D) *x* is 9 percent of 90
   (E) 90 percent of *x* is 9
   (F) *x* percent of *x* is *x*
   (G) 1/*x* is 1/*x* percent of *x*

[Quantitative Comparison Question]
20.   Column A   Column B

Distance between point *A* and a   Distance between point *B* and a
point that is located 8 miles   point that is located 6 miles
East of point *P*, if Point *P* is   West of point *Q*, if Point *Q* is
located 6 miles North of point   located 8 miles South of point
A   *B*

[Multiple-choice Question – Select One Answer Choice Only]
21. Hose A can fill a tank in 5 minutes, and Hose B can fill the same tank in 6 minutes. How many tanks would Hose B fill in the time Hose A fills 6 tanks?

   (A) 3
   (B) 4
   (C) 5
   (D) 5.5
   (E) 6

[Multiple-choice Question – Select One Answer Choice Only]
22. A worker is hired for 7 days. Each day, he is paid 10 dollars more than what he is paid for the preceding day of work. The total amount he was paid in the first 4 days of work equaled the total amount he was paid in the last 3 days. What was his starting pay?

   (A) 90
   (B) 138
   (C) 153
   (D) 160
   (E) 163

[Multiple-choice Question – Select One Answer Only]
23. Thirty airmail and 40 ordinary envelopes are the only envelopes in a bag. Thirty-five envelopes in the bag are unstamped, and 5/7 of the unstamped envelopes are airmail letters. What is the probability that an envelope picked randomly from the bag is an ordinary airmail envelope?

   (A) 1/7
   (B) 1/3
   (C) 5/14
   (D) 17/38
   (E) 23/70

[Multiple-choice Question – Select One Answer Choice Only]
24. There are 3 doors to a lecture room. In how many ways can a lecturer enter and leave the room?

   (A) 1
   (B) 3
   (C) 6
   (D) 9
   (E) 12

Answers and Solutions Test 15:

Question	Answer
1.	D
2.	A
3.	E
4.	D
5.	D
6.	C, E
7.	C
8.	C
9.	A
10.	A
11.	C
12.	B, D
13.	C
14.	E
15.	A
16.	B
17.	270
18.	D
19.	D, F, G
20.	C
21.	C
22.	A
23.	A
24.	D

If you got 18/24 correct on this test, you are likely to get 750+ on the actual GRE **by the time you complete all the tests in the book.**

---

1. If $x = y = 1$, then both columns equal 1. If $x = y = 2$, then $x/y = 1$ and $xy = 4$. In this case, the columns are unequal. The answer is (D).

Test 15—Solutions

2. First, subtracting the numbers in the parentheses yields

Column A	Column B
$1! \cdot 9!$	$2! \cdot 8!$

As defined, $2! = 2 \cdot 1 = 2$ and $8! = 8 \cdot 7 \cdot 6 \cdot 5 \cdot 4 \cdot 3 \cdot 2 \cdot 1$. Hence, in Column B, we have

$$2! \cdot 8! = 2(8 \cdot 7 \cdot 6 \cdot 5 \cdot 4 \cdot 3 \cdot 2 \cdot 1)$$

Similarly, 1! equals the product of the integers from 1 through 1. Hence, $1! = 1$. Also, $9! = 9 \cdot 8 \cdot 7 \cdot 6 \cdot 5 \cdot 4 \cdot 3 \cdot 2 \cdot 1$. Hence, in Column A, we have

$$1! \cdot 9! = 1(9 \cdot 8 \cdot 7 \cdot 6 \cdot 5 \cdot 4 \cdot 3 \cdot 2 \cdot 1) = 9 \cdot 8 \cdot 7 \cdot 6 \cdot 5 \cdot 4 \cdot 3 \cdot 2 \cdot 1$$

Now, rewriting $9 \cdot 8 \cdot 7 \cdot 6 \cdot 5 \cdot 4 \cdot 3 \cdot 2 \cdot 1$ as $9(8 \cdot 7 \cdot 6 \cdot 5 \cdot 4 \cdot 3 \cdot 2 \cdot 1)$ yields

Column A	Column B
$9(8 \cdot 7 \cdot 6 \cdot 5 \cdot 4 \cdot 3 \cdot 2 \cdot 1)$	$2(8 \cdot 7 \cdot 6 \cdot 5 \cdot 4 \cdot 3 \cdot 2 \cdot 1)$

Finally, dividing both columns by $8 \cdot 7 \cdot 6 \cdot 5 \cdot 4 \cdot 3 \cdot 2 \cdot 1$ yields

Column A	Column B
9	2

The answer is (A).

---

3. Since there are twice as many townhomes as ranch-style homes, the probability of selecting a townhome is 2/3.[*] Now, "there are 3 times as many townhomes with pools than without pools." So the probability that a townhome will have a pool is 3/4. Hence, the probability of selecting a townhome with a pool is

$$\frac{2}{3} \cdot \frac{3}{4} = \frac{1}{2}$$

The answer is (E).

---

4. Suppose the angles of the triangle measure $179.8°$, $0.1°$, and $0.1°$. Then the difference between the first two angles of the triangle is $179.8 - 0.1 = 179.7 >$ Column B and the difference between the last two angles is $0.1 - 0.1 = 0 <$ Column B.

Hence, we have a double case, and the answer is (D).

---

[*] Caution: Were you tempted to choose 1/2 for the probability because there are "twice" as many townhomes? One-half (= 50%) would be the probability if there were an equal number of townhomes and ranch-style homes. Remember the probability of selecting a townhome is not the ratio of townhomes to ranch-style homes, but the ratio of townhomes to the total number of homes. To see this more clearly, suppose there are 3 homes in the subdivision. Then 2 would be townhomes and 1 would be a ranch-style home. So the ratio of townhomes to total homes would be 2/3.

5. Since we are not given the radius of the cylinder, we can make the cylinder very narrow or very broad by taking the radius to very small or very large. The same can be done with the cone. Hence, we have a double case, and the answer is (D).

6. Corresponding sides of similar triangles are proportional.

If $ABC \approx ADE$, then

$$\frac{AB}{BC} = \frac{AD}{DE}$$

$$\frac{8}{6} = \frac{8+16}{DE}$$

$$DE = \frac{6 \times 24}{8} = 18$$

If $ABC \approx AED$, then

$$\frac{AB}{AE} = \frac{BC}{ED} = \frac{AC}{AD}$$

$$\frac{8}{10+CE} = \frac{6}{ED} = \frac{10}{8+16}$$

$$ED = \frac{24}{10} \times 6 = 14.4$$

The answers are (C) and (E).

7. Since $ABCD$ is a parallelogram, opposite sides are equal. So, $x = q$ and $y = p$. Now, line BD is a transversal cutting opposite sides AB and DC in the parallelogram. So, the alternate interior angles $\angle ABD$ and $\angle BDC$ both equal 31°. Hence, in $\triangle ABD$, $\angle B$ (which equals 31°) is greater than $\angle D$ (which equals 30°, from the figure). Since the sides opposite greater angles in a triangle are greater, we have $x > y$. But, $y = p$ (we know). Hence, $x > p$, and the answer is (C).

8. From the figure, it is clear that the area of quadrilateral $EBFD$ equals the sum of the areas of the triangles $\triangle EBD$ and $\triangle DBF$. Hence, the area of the quadrilateral $EBFD$

= area of $\triangle EBD$ + area of $\triangle DBF$

= 1/2 × ED × AB + 1/2 × BF × DC   area of a triangle equals 1/2 × *base* × *height*

= 1/2 × 3 × AB + 1/2 × 4 × DC   from the figure, ED = 3 and BF = 4

= (3/2)AB + 2DC

= (3/2)AB + 2AB   opposite sides AB and DC must be equal

= (7/2)AB

Now, the formula for the area of a rectangle is *length* × *width*. Hence, the area of the rectangle $ABCD$ equals $AD \times AB$. Since we are given that the area of quadrilateral $EBFD$ is half the area of the rectangle $ABCD$, we have

$$\frac{1}{2}(AD \times AB) = \frac{7}{2}AB$$

$AD \times AB = 7AB$

$AD = 7$ (by canceling AB from both sides)

The answer is (C).

9. The ordering *ABCD* indicates that *AB* and *BC* are adjacent sides of a rectangle with common vertex *B* (see figure below).

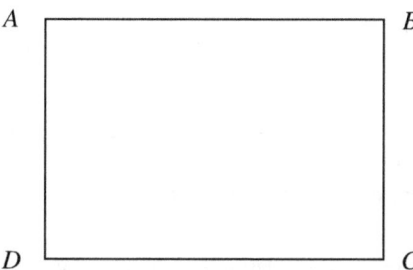

Remember that the perimeter of a rectangle is equal to twice its length plus twice its width. Hence,

$P = 2AB + 2BC$

We are given that the perimeter is 5/2 times the length of side *AB*. Replacing the left-hand side of the equation with $\frac{5}{2}AB$ yields

$\frac{5}{2}AB = 2AB + 2BC$
$5AB = 4AB + 4BC$ (by multiplying the equation by 2)
$AB = 4BC$ (by subtracting 4*AB* from both sides)

Thus, the length of side *AB* is four times as long as side *BC*. Hence, side *AB* is greater than side *BC*. The answer is (A).

---

10. We can eliminate 50 (the mere average of 40 and 60) since that would be too elementary. Now, the average must be closer to 40 than to 60 because the car travels for a longer time at 40 mph. But 48 is the only number given that is closer to 40 than to 60. The answer is (A).

It's instructive to also calculate the answer. *Average Speed* = $\frac{Total\ Distance}{Total\ Time}$. Now, a car traveling at 40 mph will cover 120 miles in 3 hours. And a car traveling at 60 mph will cover the same 120 miles in 2 hours. So, the total traveling time is 5 hours. Hence, for the round trip, the average speed is

$$\frac{120+120}{5} = 48$$

---

11. Substituting 1/*y* for *x* in Column A yields

$$\frac{1}{y} + 1 + \frac{1}{\frac{1}{y}} =$$

$$\frac{1}{y} + 1 + y =$$

Column B

The answer is (C).

12. Let's subtract 1 from each answer-choice. The answer-choice that has the lowest positive value should be closest to 1.

Choice (A): $\dfrac{3}{3+0.3} - 1 = \dfrac{3-(3+0.3)}{3+0.3} = \dfrac{-0.3}{3+0.3} = \dfrac{-0.3}{3.3} = \dfrac{-1}{11}$. Hence, Choice (A) is 1/11 units away from 1.

Choice (B): $\dfrac{3}{3+0.3^2} - 1 = \dfrac{3-(3+0.3^2)}{3+0.3^2} = \dfrac{-0.3^2}{3+0.3^2} = \dfrac{-0.09}{3+0.09} = \dfrac{-0.09}{3.09} = \dfrac{-9}{309}$. Hence, Choice (B) is 9/309 units away from 1. Since 9/309 is less than 1/11, Choice (B) is closer than Choice (A).

Choice (C): $\dfrac{3}{3-0.3} - 1 = \dfrac{3-(3-0.3)}{3-0.3} = \dfrac{0.3}{2.7} = \dfrac{1}{9}$. Hence, Choice (C) is 1/9 units away from 1. Clearly, this is greater than 9/309. Hence, Choice (A), the second closest is closer than Choice (C). Hence, eliminate choice (C).

Choice (D): $\dfrac{3}{3-0.3^2} - 1 = \dfrac{3-(3-0.3^2)}{3-0.3^2} = \dfrac{0.3^2}{3-0.3^2} = \dfrac{0.09}{2.91} = \dfrac{9}{291}$. Hence, Choice (D) is 9/291 units away from 1. This is less than 1/11. Hence, Choice (D) is closer than Choice (A). Hence, eliminate choice (A).

Choice (E): $\dfrac{3}{3+0.33} - 1 = \dfrac{3-(3+0.33)}{3+0.33} = \dfrac{-0.33}{3.33} = \dfrac{-33}{333}$. So, Choice (E) is 33/333 units away from 1. This is greater than 9/291, the second closest number. Hence, eliminate choice (E).

The answer is (B) and (D).

13. We are given the two equations:

$$7x + 3y = 12$$
$$3x + 7y = 8$$

Subtracting the bottom equation from the top equation yields

$$(7x + 3y) - (3x + 7y) = 12 - 8$$
$$7x + 3y - 3x - 7y = 4$$
$$4x - 4y = 4$$
$$4(x - y) = 4$$
$$x - y = 1$$

Hence, Column A equals 1; and since Column B also equals 1, the answer is (C).

14. Solving the top equation for y yields $y = 7 - x$. Substituting this into the bottom equation yields

$$x^2 + (7 - x)^2 = 25$$
$$x^2 + 49 - 14x + x^2 = 25$$
$$2x^2 - 14x + 24 = 0$$
$$x^2 - 7x + 12 = 0$$
$$(x - 3)(x - 4) = 0$$
$$x - 3 = 0 \text{ or } x - 4 = 0$$
$$x = 3 \text{ or } x = 4$$

If $x = 3$, then $y = 7 - 3 = 4$. If $x = 4$, then $y = 7 - 4 = 3$. In either case, $x^3 + y^3 = 3^3 + 4^3 = 27 + 64 = 91$. The answer is (E).

15. Column A: The mean rainfall for the 8 months is the sum of the eight rainfall measurements divided by 8:

$$\frac{2+4+4+5+7+9+10+11}{8} = 6.5$$

Column B: When a set of numbers is arranged in order of size, the *median* is the middle number. If a set contains an even number of elements, then the median is the average of the two middle elements. The average of 5 and 7 is 6, which is the median of the set. Hence, Column A is greater than Column B, and the answer is (A).

---

16. This problem can be solved by setting up a proportion between the number of cookies and the number of cups of sugar required to make the corresponding number of cookies. Since there are 12 items in a dozen, 2-dozen cookies is $2 \times 12 = 24$ cookies. Since 3/2 cups are required to make the 24 cookies, we have the proportion

$$\frac{24 \text{ cookies}}{\frac{3}{2} \text{ cookies}} = \frac{30 \text{ cookies}}{x \text{ cups}}$$

$24x = 30 \cdot 3/2 = 45$ by Cross-multiplying
$x = 45/24$

Hence, Column A equals 45/24, which is less than 2 (= Column B). Hence, Column B is greater, and the answer is (B).

---

17. The cost of production is proportional to the number of units produced. Hence, we have the equation

The Cost of Production = $k \times$ *Quantity*
(where k is a constant)

We are given that 300 units cost 300 dollars. Putting this in the proportionality equation yields

$$300 = k \times 300$$

Solving the equation for k yields $k = 300/300 = 1$. Hence, the *Cost of Production* of 270 units equals

$$k \times 270 = 1 \times 270 = 270$$

Enter 270 in the grid.

---

18. This is a direct proportion: as the time increases so does the number of steps that the sprinter takes. Setting ratios equal yields

$$\frac{30}{9} = \frac{x}{54}$$

$$\frac{30 \cdot 54}{9} = x$$

$$180 = x$$

The answer is (D).

---

19. We are given that 9/100 of x is 9. Now, 9/100 of x can be expressed as 9% of x. Hence, 9% of x is 9. Hence, 25 percent of x must equal 25. The choice (A) is correct. Similarly, x% of x must be x and $1/x$ percent of x must be $1/x$. Hence, choice (F) and (G) are also correct. Note, this is so because x is 100 so n% of 100 is n.

Select (D), (F), and (G).

20. Column A:
First, place point A arbitrarily. Then locate point P 6 miles North of point A, and then locate a new point 8 miles East of P. Name the new point M. Now, Column A equals AM. The map drawn should look like this:

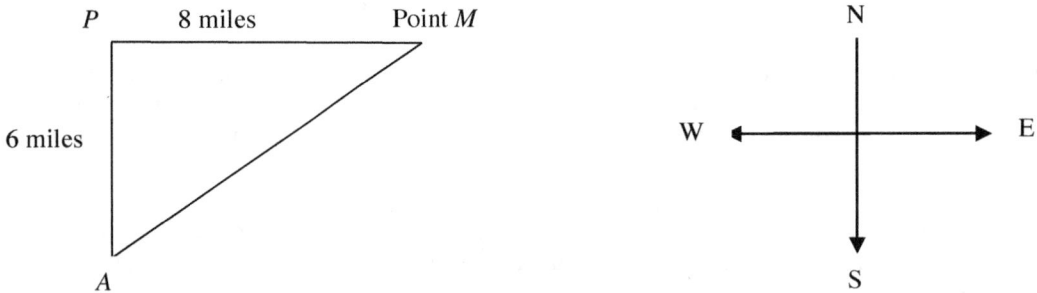

Since the angle between the standard directions East and South is 90°, the three points A, P and M form a right triangle, with right angle at P. So, AM is the hypotenuse. By The Pythagorean Theorem, the hypotenuse equals the square root of the sum of the squares of the other two sides. Hence,

$$AM = \sqrt{AP^2 + PM^2}$$
$$= \sqrt{6^2 + 8^2}$$
$$= \sqrt{36 + 64}$$
$$= \sqrt{100}$$
$$= 10$$

Column B:
Similarly, place point B arbitrarily. Then locate point Q 8 miles South of it, and locate a new point 6 miles West of the point Q. Name the new point N. Now, Column B equals BN. The map should look like this:

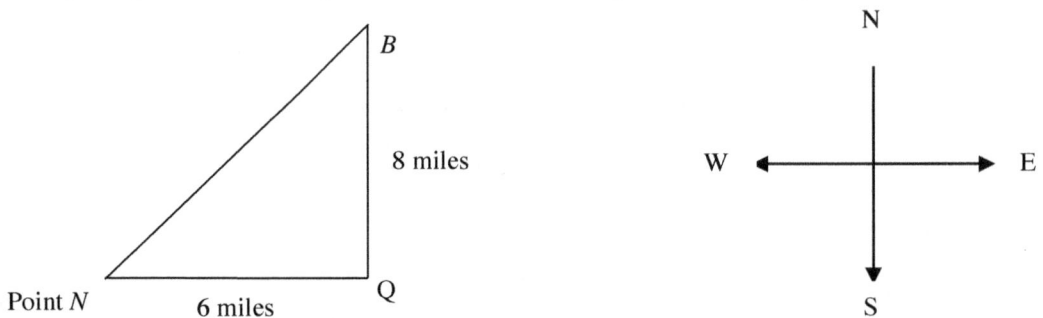

Again, since the angle between standard directions is 90°, we have a right triangle BQN, with right angle at Q, and, by The Pythagorean Theorem mentioned above, the hypotenuse BN equals

$$\sqrt{BQ^2 + QN^2}$$
$$= \sqrt{8^2 + 6^2}$$
$$= \sqrt{64 + 36}$$
$$= \sqrt{100}$$
$$= 10$$

Since both columns equal 10, the answer is (C).

Test 15—Solutions

---

21. Hose A takes 5 minutes to fill one tank. To fill 6 tanks, it takes 6 · 5 = 30 minutes. Hose B takes 6 minutes to fill one tank. Hence, in the 30 minutes, it would fill 30/6 = 5 tanks. The answer is (C).

---

22. This problem can be solved with a series. Let the payments for the 7 continuous days be $a_1, a_2, a_3, ..., a_7$. Since each day's pay was 10 dollars more than the previous day's pay, the rule for the series is $a_{n+1} = a_n + 10$.

By the rule, let the payments for each day be listed as

a_1
$a_2 = a_1 + 10$
$a_3 = a_2 + 10 = (a_1 + 10) + 10 = a_1 + 20$
$a_4 = a_3 + 10 = (a_1 + 20) + 10 = a_1 + 30$
$a_5 = a_4 + 10 = (a_1 + 30) + 10 = a_1 + 40$
$a_6 = a_5 + 10 = (a_1 + 40) + 10 = a_1 + 50$
$a_7 = a_6 + 10 = (a_1 + 50) + 10 = a_1 + 60$

We are given that the net pay for the first 4 days equals the net pay for the last 3 days.

The net pay for the first 4 days is $a_1 + (a_1 + 10) + (a_1 + 20) + (a_1 + 30) = 4a_1 + 60$.

The net pay for the last (next) 3 days is $(a_1 + 40) + (a_1 + 50) + (a_1 + 60) = 3a_1 + 150$.

Equating the two expressions yields

$4a_1 + 60 = 3a_1 + 150$
$a_1 = 90$

The answer is (A).

Method II: Let's repeat the solution without all the sequence notation. Let P be the first day's pay. Then the second day's pay is $P + 10$, and the third day's pay is $P + 20$, etc.

The net pay for the first 4 days is $P + (P + 10) + (P + 20) + (P + 30) = 4P + 60$.

The net pay for the last (next) 3 days is $(P + 40) + (P + 50) + (P + 60) = 3P + 150$.

Equating the two expressions yields

$4P + 60 = 3P + 150$
$P = 90$

The answer is (A).

---

23. We have that 30 airmail and 40 ordinary envelopes are the only envelopes in the bag. Hence, the total number of envelopes is 30 + 40 = 70. We also have that 35 envelopes in the bag are unstamped, and 5/7 of these envelopes are airmail letters. Now, 5/7 × 35 = 25. So the remaining 35 – 25 = 10 are ordinary unstamped envelopes. Hence, the probability of picking such an envelope from the bag is

(Number of unstamped ordinary envelopes) / (Total number of envelopes) =

10/70 =

1/7

The answer is (A).

24. Recognizing the Problem:

1) Is it a permutation or a combination problem?
Here, order is important. Suppose A, B, and C are the three doors. Entering by door A and leaving by door B is not the same way as entering by door B and leaving by door A. Hence, AB ≠ BA implies the problem is a *permutation* (order is important).

2) Are repetitions allowed?
Since the lecturer can enter and exit through the same door, *repetition* is allowed.

3) Are there any indistinguishable objects in the base set?
Doors are different. They are not indistinguishable, so *no indistinguishable objects*.

Hence, we have a permutation problem, with repetition allowed and no indistinguishable objects.

Method I:
The lecturer can enter the room in 3 ways and exit in 3 ways. So, in total, the lecturer can enter and leave the room in 9 (= 3 · 3) ways. The answer is (D). This problem allows repetition: the lecturer can enter by a door and exit by the same door.

Method II:
Let the 3 doors be A, B, and C. We must choose 2 doors: one to enter and one to exit. This can be done in 6 ways: {A, A}, {A, B}, {B, B}, {B, C}, {C, C}, and {C, A}. Now, the order of the elements is important because entering by A and leaving by B is not same as entering by B and leaving by A. Let's permute the combinations, which yields

$$A - A$$
$$A - B \text{ and } B - A$$
$$B - B$$
$$B - C \text{ and } C - B$$
$$C - C$$
$$C - A \text{ and } A - C$$

The total is 9, and the answer is (D).

Test 16

Questions: 24
Time: 45 minutes

[Multiple-choice Question – Select One or More Answer Choices]
1. In a series of five consecutive even integers with middle integer n, the difference between the greatest integer and the smallest integer in the series is

   (A) $n - 5$
   (B) 8 if the order is increasing and 10 if the order is decreasing.
   (C) 8 if the order is decreasing and 10 if the order is increasing.
   (D) 8 whether the order is decreasing or increasing.
   (E) 8 whether the integers are positive or negative.

Quantitative Comparison Question]
2. Column A   Column B
    The number of multiples of 3 $\dfrac{729 - 102}{3}$
    between 102 and 729, inclusive

[Multiple-choice Question – Select One or More Answer Choices]
3. If each of the three nonzero numbers a, b, and c is divisible by 3, then abc must be divisible by which of the following the numbers?

   (A) 9
   (B) 27
   (C) 81
   (D) 121
   (E) 159

[Numeric-Entry]
4. The remainder when $m + n$ is divided by 12 is 8, and the remainder when $m - n$ is divided by 12 is 6. If $m > n$, then what is the remainder when mn divided by 6?

Quantitative Comparison Question]
5. Column A

Percentage of wheat lost in the first three days

A tank is filled with x pounds of wheat. The tank has a hole at the bottom and each day 1% of the wheat is lost from the tank through the hole.

Column B

3%

Quantitative Comparison Question]
6. Column A $x > 0$ Column B
    $x^3 + 1$ $x^4 + 1$

[Multiple-choice Question – Select One Answer Choice Only]
7. In the figure, ABC is a right triangle. What is the value of y ?

    (A) 20
    (B) 30
    (C) 50
    (D) 70
    (E) 90

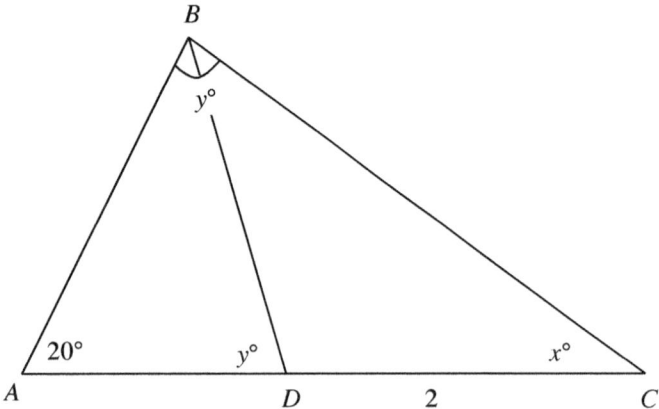

[Multiple-choice Question – Select One Answer Choice Only]
8. In the figure, ABCD is a parallelogram, what is the value of b ?

    (A) 46
    (B) 48
    (C) 72
    (D) 84
    (E) 96

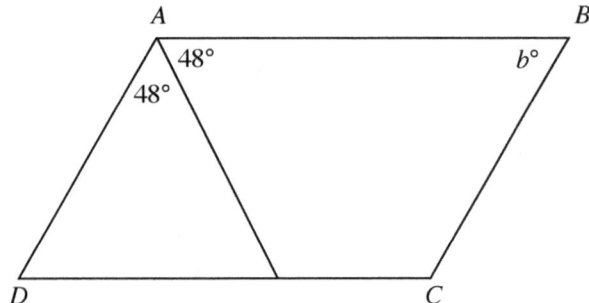

[Multiple-choice Question – Select One Answer Choice Only]

9. In the figure, *ABC* and *ADC* are right triangles. Which of the following could be the lengths of *AD* and *DC*, respectively?

   (I) $\sqrt{3}$ and $\sqrt{4}$
   (II) 4 and 6
   (III) 1 and $\sqrt{24}$
   (IV) 1 and $\sqrt{26}$

   (A) I and II only
   (B) II and III only
   (C) III and IV only
   (D) IV and I only
   (E) I, II and III only

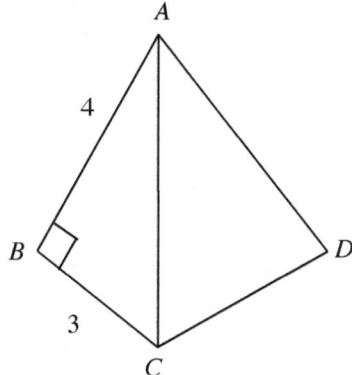

Figure not drawn to scale

[Numeric Entry Question]

10. In the rectangular coordinate system shown, *ABCD* is a parallelogram. If the coordinates of the points *A*, *B*, *C*, and *D* are $(0, 2)$, (a, b), $(a, 2)$, and $(0, 0)$, respectively, then $b =$

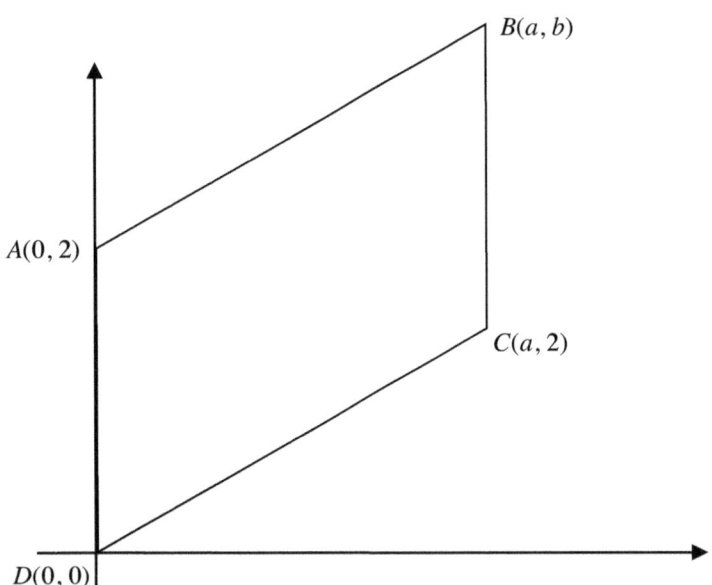

[Numeric Entry Question]
11. In the figure, *ABCD* and *ABEC* are parallelograms. The area of the quadrilateral *ABED* is 6. What is the area of the parallelogram *ABCD* ?

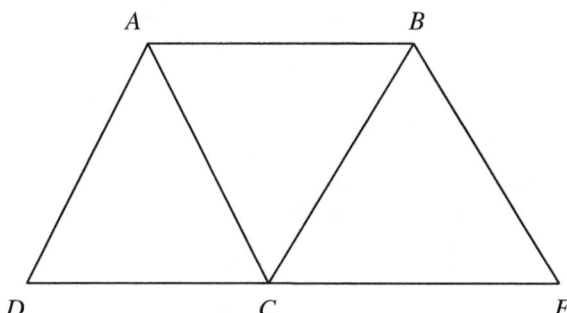

Quantitative Comparison Question]

12.   Column A　　　　In the two figures shown, line *l*　　　Column B
　　　　　　　　　　represents the function *f* and line
　　　　　　　　　　m represents the function *g*.

　　　　f(10)　　　　　　　　　　　　　　　　　　　　　*g*(10)

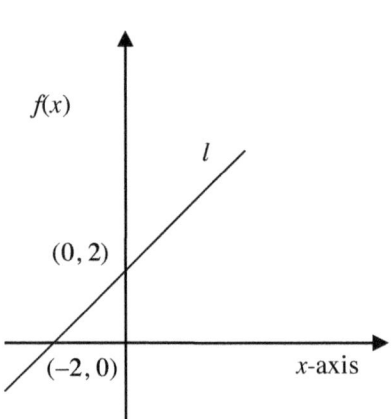
Fig. 1

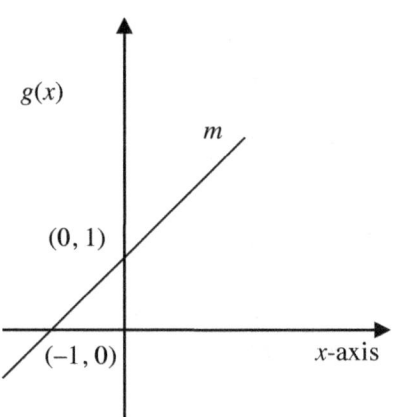
Fig. 2

[Multiple-choice Question – Select One Answer Choice Only]

13.   If $x + 3$ is positive, then which one of the following must be positive?

(A) $x - 3$
(B) $(x - 3)(x - 4)$
(C) $(x - 3)(x + 3)$
(D) $(x - 3)(x + 4)$
(E) $(x + 3)(x + 6)$

[Multiple-choice Question – Select One Answer Choice Only]

14. If $a, b,$ and c are three different numbers and $\dfrac{x}{b-c} = \dfrac{y}{c-a} = \dfrac{z}{a-b}$, then what is the value of $ax + by + cz$?

   (A) 0
   (B) 1
   (C) 2
   (D) 3
   (E) 4

Quantitative Comparison Question]

15.

Column A	Column B
The last digit in the average of the numbers 13 and 23	The last digit in the average of the numbers 113 and 123

[Multiple-choice Question – Select One Answer Choice Only]

16. If $a, b,$ and c are three different numbers and $ax : by : cz = 1 : 2 : -3$, then $ax + by + cz =$

   (A) 0
   (B) 1/2
   (C) 3
   (D) 6
   (E) 9

[Quantitative Comparison Question]

17.   Column A   A precious stone was accidentally   Column B
                                 dropped and broke into 3 pieces of
                                 equal weight. The stone is of a
                                 rare type such that the stone is
                                 always proportional to the square
                                 of its weight.

The value of the 3 broken   The value of the original stone
pieces together

[Multiple-choice Question – Select One Answer Choice Only]

18. If  $\sqrt[m]{27} = 3^{3m}$ and $4^m > 1$, then what is the value of m ?

   (A) -1
   (B) $-1/4$
   (C) 0
   (D) $1/4$
   (E) 1

[Multiple-choice Question – Select One or More Answer Choices]

19. If b equals 10% of a and c equals 20% of b, then which of the following equals 30% of c ?

   (A) 0.0006% of a
   (B) 0.006% of a
   (C) 0.06% of a
   (D) 0.6% of a
   (E) 6% of b

Test 16—Questions

[Multiple-choice Question – Select One Answer Choice Only]
20. In a market, a dozen eggs cost as much as a pound of rice, and a half-liter of kerosene costs as much as 8 eggs. If the cost of each pound of rice is $0.33, then how many cents does a liter of kerosene cost? [One dollar has 100 cents.]

   (A) 0.33
   (B) 0.44
   (C) 0.55
   (D) 44
   (E) 55

USE THIS SPACE FOR SCRATCHWORK.

Quantitative Comparison Question]
21.   Column A   A piece of string 35 cm long is cut   Column B
                                        into three smaller pieces of
                                        different lengths along the length
                                        of the string. The length of the
                                        longest piece is three times the
                                        length of the shortest piece.
       The length of the medium-size   15
       piece

USE THIS SPACE FOR SCRATCHWORK.

[Multiple-choice Question – Select One Answer Choice Only]
22. There are 750 male and female participants in a meeting. Half the female participants and one-quarter of the male participants are Democrats. How many participants could be both female and Democrat?

   (A) 75
   (B) 100
   (C) 125
   (D) 175
   (E) 225

USE THIS SPACE FOR SCRATCHWORK.

Quantitative Comparison Question]

23.   Column A   In a box of 5 eggs, 2 are rotten.   Column B
        The probability that one egg   The probability that two eggs
        chosen at random from the box   chosen at random from the box
        is rotten   are rotten

[Multiple-choice Question – Select One Answer Choice Only]

24. In a small town, 16 people own Fords and 11 people own Toyotas. If exactly 15 people own only one of the two types of cars, how many people own both types of cars.

    (A) 2
    (B) 6
    (C) 7
    (D) 12
    (E) 14

Answers and Solutions Test 16:

Question	Answer
1.	D, E
2.	A
3.	A, B
4.	1
5.	D
6.	B
7.	E
8.	D
9.	C
10.	4
11.	4
12.	A
13.	E
14.	A
15.	C
16.	A
17.	A
18.	E
19.	D, E
20.	D
21.	B
22.	C
23.	A
24.	B

If you got 18/24 correct on this test, you are likely to get 750+ on the actual GRE by the time you complete all the tests in the book.

---

1. Let $n-4, n-2, n, n+2$, and $n+4$, be the five consecutive even integers. The smallest number is $n-4$, and the largest is $n+4$. The difference between them is $n+4-(n-4) = 4+4 = 8$. The value is a constant that is independent of the order (decreasing or increasing) of the sequence or the value of the middle number n. Hence, Choose (D) and (E).

---

2. The numbers 102 and 729 are themselves multiples of 3. Also, a multiple of 3 exists once in every three consecutive integers. Counting the multiples of 3 starting with 1 for 102, 2 (= 1 + (105 − 102)/3 = 1 + 1 = 2) for 105, 3 (= 1 + (108 − 102)/3 = 1 + 2 = 3) for 108, and so on, the count we get for 729 equals 1 + (729 − 102)/3 = 1 + Column B. Hence, Column A is greater than Column B by 1 unit. Hence, the answer is (A).

GRE Math Tests

3. Since each one of the three numbers a, b, and c is divisible by 3, the numbers can be represented as $3p$, $3q$, and $3r$, respectively, where p, q, and r are integers. The product of the three numbers is

$$3p \cdot 3q \cdot 3r = 27(pqr)$$

Since p, q, and r are integers, pqr is an integer and therefore abc is divisible by 27. The answer is (A) and (B).

Note: any number divisible by 27 must also be divisible by the factors of 27 (9 is a factor of 27).

---

4. Since the remainder when $m + n$ is divided by 12 is 8, $m + n = 12p + 8$; and since the remainder when $m - n$ is divided by 12 is 6, $m - n = 12q + 6$. Here, p and q are integers. Adding the two equations yields

$$2m = 12p + 12q + 14$$

Solving for m by dividing both sides of the equation by 2 yields

$$m = 6p + 6q + 7 =$$

$$m = 6p + 6q + (6 + 1) = \qquad \text{(by rewriting 7 as 6 + 1)}$$

$$6(p + q + 1) + 1 =$$

$$6r + 1, \text{ where } r \text{ is a positive integer equaling } p + q + 1$$

Now, let's subtract the equations $m + n = 12p + 8$ and $m - n = 12q + 6$. This yields

$$2n = (12p + 8) - (12q + 6) = 12(p - q) + 2$$

Solving for n by dividing by 2 yields $n = 6(p - q) + 1 = 6t + 1$, where t is an integer equaling $p - q$.

Hence, we have

$$mn = (6r + 1)(6t + 1)$$

$$= 36rt + 6r + 6t + 1$$

$$= 6(6rt + r + t) + 1 \qquad \text{by factoring out 6}$$

Hence, the remainder is 1. Enter 1 in the grid.

---

5. Intuitively, one expects $x^4 + 1$ to be larger than $x^3 + 1$. But this is a hard problem, so we can reject (B) as the answer. Now, if $x = 1$, then the expressions are equal. However, for any other value of x, the expressions are unequal. Hence, the answer is (D).

Test 16—Solutions

6. We are given that each day 1% of the remaining wheat in the tank is lost. The initial content in the tank is x pounds. By the end of the first day, the content remaining is

(Initial content)(1 − Loss percent/100) = $x(1 − 1/100) = 0.99x$

Similarly, the content remaining after the end of the second day is $0.99x(1 − 1/100) = (0.99)(0.99)x$; and by the end of the third day, the content remaining is $(0.99)(0.99)(0.99)x$. Hence, Column A, which equals the net loss percentage in the three consecutive days, is

$$\frac{\text{Initial content} - \text{Final content}}{\text{Final content}} \cdot 100 =$$

$$\frac{x - 0.99 \cdot 0.99 \cdot 0.99 x}{x} \cdot 100 =$$

$100 − 99(0.99)(0.99) \approx 2.9$, which is less than 3

Hence, Column A < Column B and the answer is (B).

---

7. In the given right triangle, $\triangle ABC$, $\angle A$ is 20°. Hence, $\angle A$ is not the right angle in the triangle. Hence, either of the other two angles, $\angle C$ or $\angle B$, must be right angled.

Now, $\angle BDA$ (= $\angle ABC = y°$, from the figure) is an exterior angle to $\triangle BCD$ and therefore equals the sum of the remote interior angles $\angle C$ and $\angle DBC$. Clearly, the sum is larger than $\angle C$ and therefore if $\angle C$ is a right angle, $\angle BDA$ (= $\angle ABC$) must be larger than a right angle, so $\angle BDA$, hence, $\angle ABC$ must be obtuse. But a triangle cannot accommodate a right angle and an obtuse angle simultaneously because the angle sum of the triangle would be greater than 180°. So, $\angle C$ is not a right angle and therefore the other angle, $\angle B$, is a right angle. Hence, $y° = \angle B = 90°$. The answer is (E).

---

8. From the figure, $\angle A = 48 + 48 = 96$. Since the sum of any two adjacent angles of a parallelogram equals 180°, we have $\angle A + b = 180$ or $96 + b = 180$. Solving for b yields $b = 180 − 96 = 84$. The answer is (D).

---

9. From the figure, we have that $\angle B$ is a right angle in $\triangle ABC$. Applying The Pythagorean Theorem to the triangle yields $AC^2 = AB^2 + BC^2 = 4^2 + 3^2 = 25$. Hence, $AC = \sqrt{25} = 5$.

Now, we are given that $\triangle ADC$ is a right-angled triangle. But, we are not given which one of the three angles of the triangle is right-angled. We have two possibilities: either the common side of the two triangles, AC, is the hypotenuse of the triangle, or it is not.

In the case AC is the hypotenuse of the triangle, we have by The Pythagorean Theorem,

$AC^2 = AD^2 + DC^2$
$5^2 = AD^2 + DC^2$

This equation is satisfied by III since $5^2 = 1^2 + \left(\sqrt{24}\right)^2$. Hence, III is possible.

In the case AC is not the hypotenuse of the triangle and, say, DC is the hypotenuse, then by applying The Pythagorean Theorem to the triangle, we have

$AD^2 + AC^2 = DC^2$
$AD^2 + 5^2 = DC^2$

This equation is satisfied by IV: $5^2 + 1^2 = \left(\sqrt{26}\right)^2$.

Hence, we conclude that III and IV are possible. The two are available in choice (C). Hence, the answer is (C).

10. In the figure, points $A(0, 2)$ and $D(0, 0)$ have the same *x*-coordinate (which is 0). Hence, the two lines must be on the same vertical line in the coordinate system.

Similarly, the *x*-coordinates of points B and C are the same (both equal a). Hence, the points are on the same vertical line in the coordinate system.

Now, if $ABCD$ is a parallelogram, then the opposite sides must be equal. Hence, AD must equal BC.

Since AD and BC are vertical lines, AD = *y*-coordinate difference of the points A and D, which equals $2 - 0 = 2$, and BC = the *y*-coordinate difference of the points B and C, which equals $b - 2$. Equating AD and BC yields $b - 2 = 2$, or $b = 4$. Enter 4 in the grid.

---

11. We know that a diagonal of a parallelogram divides the parallelogram into two triangles of equal area. Since AC is a diagonal of parallelogram $ABCD$, the area of $\triangle ACD$ = the area of $\triangle ABC$; and since BC is a diagonal of parallelogram $ABEC$, the area of $\triangle CBE$ = the area of $\triangle ABC$. Hence, the areas of triangles ACD, ABC, and CBE, which form the total quadrilateral $ABED$, are equal. Since $ABCD$ forms only two triangles, ACD and ABC, of the three triangles, the area of the parallelogram equals two thirds of the area of the quadrilateral $ABED$. This equals $2/3 \times 6 = 4$. Enter 4 in the grid.

---

12. If the two graphs in figures 1 and 2 were represented on a single coordinate plane, the figure would look like this:

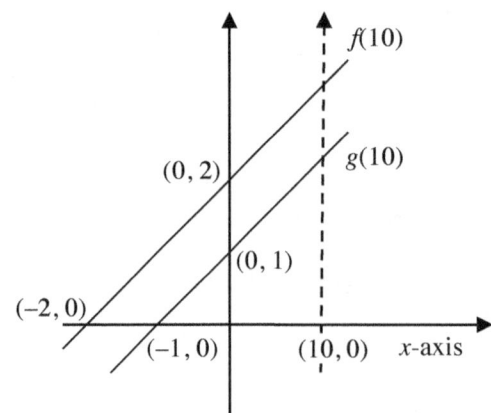

We know that the slope of a line through any two points (x_1, y_1) and (x_2, y_2) is given by $\frac{y_2 - y_1}{x_2 - x_1}$. Now, $(-1, 0)$ and $(0, 1)$ are two points on the line $g(x)$. Hence, the slope of the line is $\frac{1 - 0}{0 - (-1)} = \frac{1}{1} = 1$. Similarly, $(-2, 0)$ and $(0, 2)$ are two points on the line $f(x)$. Hence, the slope of the line is $\frac{2 - 0}{0 - (-2)} = \frac{2}{2} = 1$. Since the lines have same slope (1), they must be parallel.

Now, the point above the origin ($x = 0$) on the line $f(x)$ is $(0, 2)$ and is 2 units above the origin. But, the corresponding point (above origin, i.e., above $x = 0$) on $g(x)$ is $(0, 1)$ and is only 1 unit above the origin. This combined with the fact that the lines $f(x)$ and $g(x)$ are parallel shows that each point on $f(x)$ for a given value of x is above the corresponding point on $g(x)$. Hence, for any given value of x, $f(x)$ is greater than $g(x)$. Hence, $f(10)$ must be greater than $g(10)$. Hence, Column A > Column B, and the answer is (A).

Test 16—Solutions

13. We are given $x + 3 > 0$. Adding 3 to both sides of this inequality yields $x + 6 > 3$. Hence, $x + 6 > 0$. Now, the product of two positive numbers is positive, so $(x + 3)(x + 6) > 0$. The answer is (E).

Regarding the other answer-choices: We are given that $x + 3 > 0$. Subtracting 3 from both sides yields $x > -3$. If x equals 2, $x - 3$ is negative. Hence, reject choice (A).

$(x - 3)(x - 4)$ is negative for the values of x between 3 and 4 [For example, when x equals 3.5, the expression is negative]. The known inequality $x > -3$ allows the values to be in this range. Hence, the expression can be negative. Reject choice (B).

$(x - 3)(x + 3)$ is negative for the values of x between 3 and -3 [For example, when x equals 0, the expression is negative]. The known inequality $x > -3$ allows the values to be in this range. Hence, the expression can be negative. Reject choice (C).

$(x - 3)(x + 4)$ is negative for the values of x between 3 and -4 [For example, when x equals 0, the expression is negative]. The known inequality $x > -3$ allows the values to be in this range. Hence, the expression can be negative. Reject choice (D).

$(x + 3)(x + 6)$ is negative only for the values of x between -3 and -6 [For example, when x equals -4, the expression is negative]. But the known inequality $x > -3$ does not allow the values to be in this range. Hence, the expression cannot be negative. Hence, accept choice (E).

---

14. Let each expression in the equation equal k. Then we have $\dfrac{x}{b-c} = \dfrac{y}{c-a} = \dfrac{z}{a-b} = k$. This reduces to

$$x = (b - c)k$$
$$y = (c - a)k$$
$$z = (a - b)k$$

Now, $ax + by + cz$ equals

$$a(b - c)k + b(c - a)k + c(a - b)k =$$

$$k(ab - ac + bc - ba + ca - cb) =$$

$$k \times 0 =$$

$$0$$

The answer is (A).

---

15. The average of 13 and 23 is $(13 + 23)/2 = 36/2 = 18$, so the last digit is 8.

The average of 113 and 123 is $(113 + 123)/2 = 236/2 = 118$, so the last digit is also 8.

Hence, Column A equals Column B, and the answer is (C).

---

16. We are given that $ax : by : cz = 1 : 2 : -3$. Let $ax = t$, $by = 2t$, and $cz = -3t$ (such that $ax : by : cz = 1 : 2 : -3$). Then $ax + by + cz = t + 2t - 3t = 0$. The answer is (A).

GRE Math Tests

17. Forming the ratio yields $\frac{a+6}{b+6} = \frac{5}{6}$. Multiplying both sides of the equation by $6(b+6)$ yields

$6(a+6) = 5(b+6)$
$6a + 36 = 5b + 30$
$6a = 5b - 6$
$a = 5b/6 - 1$
$0 < 5b/6 - 1$ since a is positive
$1 < 5b/6$
$6/5 < b$
$1.2 < b$
$1 < 1.2 < b$ since $1 < 1.2$
Column B $< 1.2 <$ Column A

Hence, the answer is (A).

18. We have

$\sqrt[m]{27} = 3^{3m}$

$\sqrt[m]{3^3} = 3^{3m}$ By replacing 27 with 3^3
$(3^3)^{1/m} = 3^{3m}$ Since by definition $\sqrt[m]{a} = a^{1/m}$
$3^{3/m} = 3^{3m}$ Since $(x^a)^b$ equals x^{ab}
$3/m = 3m$ By equating the powers on both sides
$m^2 = 1$ By multiplying both sides by $m/3$
$m = \pm 1$ By square rooting both sides

We have $4^m > 1$. If $m = -1$, then $4^m = 4^{-1} = 1/4 = 0.25$, which is not greater than 1. Hence, m must equal the other value 1. Here, $4^m = 4^1 = 4$, which is greater than 1. Hence, $m = 1$. The answer is (E).

19. $b = 10\%$ of $a = (10/100)a = 0.1a$.

$c = 20\%$ of $b = (20/100)b = 0.2b = (0.2)(0.1a)$.

Now, 30% of $c = (30/100)c = 0.3c = (0.3)(0.2)(0.1a) = 0.006a = 0.6\%a$.

The choice (D) is correct.

We know that $b = 0.1a$. Multiplying both sides by 10 yields $10b = a$. Therefore, the answer $0.6\%a$ also equals 0.6% of $10b = 0.6 \times 10 \%$ of $b = 6\%$ of b. Hence, choice (E) is also correct.

The answers are (D) and (E).

20. One pound of rice costs 0.33 dollars. A dozen eggs cost as much as one pound of rice, and a dozen has 12 items. Hence, 12 eggs cost 0.33 dollars.

Now, since half a liter of kerosene costs as much as 8 eggs, one liter must cost 2 times the cost of 8 eggs, which equals the cost of 16 eggs.

Now, suppose 16 eggs cost x dollars. We know that 12 eggs cost 0.33 dollars. So, forming the proportion yields

$$\frac{0.33 \text{ dollars}}{12 \text{ eggs}} = \frac{x \text{ dollars}}{16 \text{ eggs}}$$

$$x = 16 \times \frac{0.33}{12} = 4 \times \frac{0.33}{3} = 4 \times 0.11$$

$$= 0.44 \text{ dollars} = 0.44 \ (100 \text{ cents}) \qquad \text{since one dollar has 100 cents}$$

$$= 44 \text{ cents}$$

The answer is (D).

21. The string is cut into three along its length. Let l be the length of the smallest piece. Then the length of the longest piece is $3l$, and the total length of the three pieces is 35 cm. The length of the longest and shortest pieces together is $l + 3l = 4l$. Hence, the length of the third piece (medium-size piece) must be $35 - 4l$. Arranging the lengths of the three pieces in increasing order of length yields the following inequality:

$l < 35 - 4l < 3l$	
$5l < 35 < 7l$	by adding $4l$ to each part of the inequality
$5l < 35$ and $35 < 7l$	by separating into two inequalities
$l < 7$ and $5 < l$	by dividing first inequality by 5 and the second inequality by 7
$5 < l < 7$	combining the two inequalities
$20 < 4l < 28$	multiplying each part by 4
$-20 > -4l > -28$	by multiplying the inequalities by -1 and flipping the directions of the inequalities
$35 - 20 > 35 - 4l > 35 - 28$	adding 35 to each part
$15 > 35 - 4l > 7$	
$15 >$ The length of the medium-size piece > 7	
Column B $>$ Column A > 7	

Hence, the answer is (B).

Method II:

Had the length of the medium-size piece been greater than or equal to 15, the length of the longest-size piece would be greater than 15 and the length of the smallest piece, which equals 1/3 the length of the longest piece, would be greater than $15/3 = 5$. Hence, the sum of the three lengths exceeds $15 + 15 + 5$ (= 35). Since this is impossible, our assumption that the length of the medium-sized piece is greater than or equal to 15 is false. Hence, it is less than 15 and therefore Column A is less than Column B. The answer is (B).

22. Let m be the number of male participants and f be the number of female participants in the meeting. The total number of participants is given as 750. Hence, we have

$$m + f = 750$$

Now, we have that half the female participants and one-quarter of the male participants are Democrats. Let d equal the number of the Democrats. Then we have the equation

$$f/2 + m/4 = d$$

Now, we have that one-third of the total participants are Democrats. Hence, we have the equation

$$d = 750/3 = 250$$

Solving the three equations yields the solution $f = 250$, $m = 500$, and $d = 250$. The number of female democratic participants equals half the female participants equals $250/2 = 125$. The answer is (C).

---

23. Since 2 of the 5 eggs are rotten, the chance of selecting a rotten egg the first time is 2/5. For the second selection, there is only one rotten egg, out of the 4 remaining eggs. Hence, there is a 1/4 chance of selecting a rotten egg again. Hence, the probability of selecting 2 rotten eggs in a row is $2/5 \times 1/4 = 1/10$. Since $2/5 > 1/10$, Column A is greater than Column B. The answer is (A).

---

24. Let x be the number of people who own both types of cars. Then the number of people who own only Fords is $16 - x$, and the number of people who own only Toyotas is $11 - x$. Adding these two expressions gives the number of people who own only one of the two types of cars, which we are are told is 15:

$$(16 - x) + (11 - x) = 15$$

Adding like terms yields $27 - 2x = 15$. Subtracting 27 from both sides of the equation yields $-2x = -12$. Finally, divide both sides of the equation by -2 yields $x = 6$. The answer is (B).

Test 17

GRE Math Tests

Questions: 24
Time: 45 minutes

Quantitative Comparison Question]
1.　　　　　Column A　　　　　$0 < x < 2$　　　　　Column B

　　　　　　　x^2　　　　　　　　　　　　　　　　　$\sqrt{x}$

USE THIS SPACE FOR SCRATCHWORK.

[Multiple-choice Question – Select One or More Answer Choices]
2.　This question is based on the following system of equations:

$x + 2y + z = 7$
$x + y + 2z = 7$

Which of the following could be the solution of the given system of equations?

(A)　$x = 1, y = 2$
(B)　$x = 3, y = 2$
(C)　$x = 1, y = 3$
(D)　$y = 1, z = 2$
(E)　$y = 5, z = 5$

USE THIS SPACE FOR SCRATCHWORK.

Quantitative Comparison Question]
3.　　　Column A　　　a and b are the digits of a two-　　　Column B
　　　　　　　　　　　digit number ab, and $b = a + 3$.

　The positive two-digit number　　　　　　　　　　　The positive two-digit number
　　　　　　ab　　　　　　　　　　　　　　　　　　　　　　　　ba

USE THIS SPACE FOR SCRATCHWORK.

Quantitative Comparison Question]

4.

Column A		Column B
	n is a positive integer	
n		The sum of two integers whose product is n

USE THIS SPACE FOR SCRATCHWORK.

[Multiple-choice Question – Select One Answer Choice only]

5. If p is the circumference of the circle Q and the area of the circle is 25π, what is the value of p ?

(A) 25
(B) 10π
(C) 35
(D) 15π
(E) 25π

USE THIS SPACE FOR SCRATCHWORK.

[Numeric Entry Question]

6. A, B, C, and D are points on a line such that point B bisects line AC and point A bisects line CD. The ratio of AB to CD = ☐ : ☐

USE THIS SPACE FOR SCRATCHWORK.

[Multiple-choice Question – Select One Answer Choice only]
7. In the figure, if line CE bisects ∠ACB, then x =

   (A) 45
   (B) 50
   (C) 55
   (D) 65
   (E) 70

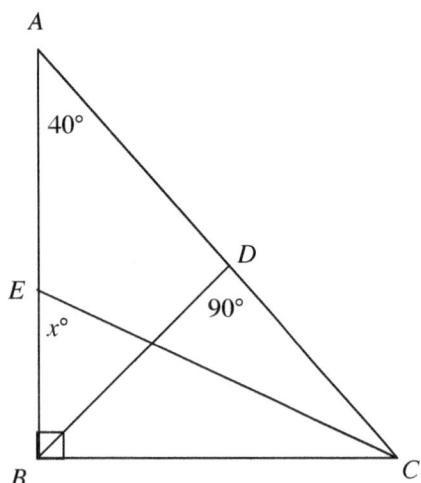

Quantitative Comparison Question]

8. | Column A | The width and length of a rectangle are 6 inches and 12 inches, respectively. | Column B |
|---|---|---|
| The area of the rectangle when its length is decreased by 4 inches and its width is not changed | | The area of the rectangle when its width is decreased by 4 inches and its length is not changed |

[Multiple-choice Question – Select One or More Answer Choices]
9. In the figure, O is the center of the circle of radius 3, and ABCD is a square. If PC = 3 and the side BC of the square is a tangent to the circle, then what is the area of the square ABCD ?

(A) 25
(B) 27
(C) 36
(D) 42
(E) 56

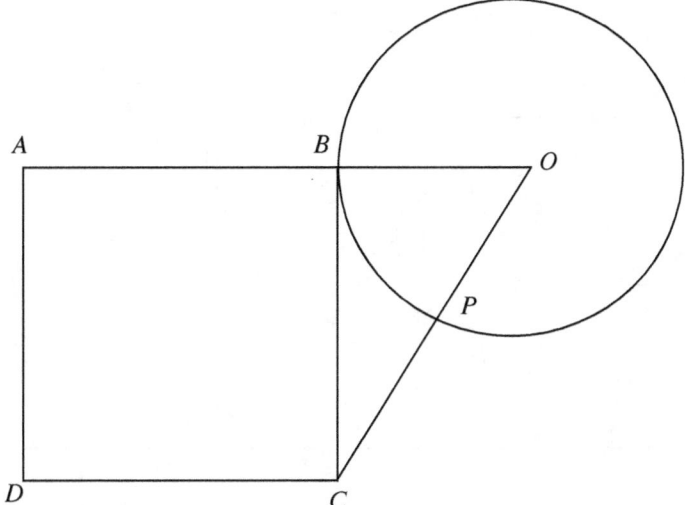

GRE Math Tests

Quantitative Comparison Question]
10.   Column A   In the figure, P and Q are centers   Column B
                                of the two circles of radii 3 and 4,
                                respectively. A and B are the
                                points at which a common
                                tangent touches each circle.
              AB   PQ

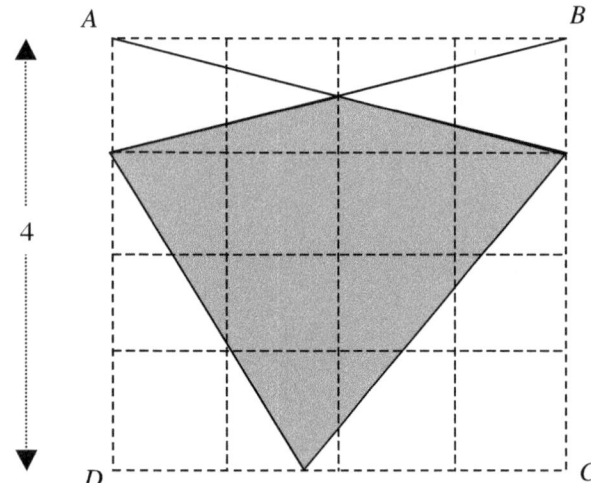

[Multiple-choice Question – Select One Answer Choice only]
11. In the figure, the horizontal and vertical lines divide the square *ABCD* into 16 equal squares as shown. What is the area of the shaded region?

(A) 4
(B) 4.5
(C) 5
(D) 6.5
(E) 7

328

[Multiple-choice Question – Select One Answer Choice Only]
12. If $x + z > y + z$, then which of the following must be true?

   (I) $x - z > y - z$
   (II) $xz > yz$
   (III) $x/z > y/z$

   (A) I only
   (B) II only
   (C) III only
   (D) I and II only
   (E) II and III only

[Multiple-choice Question – Select One or More Answer Choice only]
13. If $|yz - zx| = 3$ and $|zx - xy| = 4$, then $xy - yz =$

   (A) –7
   (B) –1
   (C) 0
   (D) 1
   (E) 7

[Multiple-choice Question – Select One or More Answer Choices]
14. In quadrilateral $ABCD$,

   $\angle A$ is greater than the average of the other three angles by at least 20°.
   $\angle B$ is greater than the average of the other three angles by at least 15°.
   $\angle C$ is exactly 10° more than the average of the other three angles.

   What could $\angle D$ be ?

   (A) 45
   (B) 55
   (C) 59
   (D) 60
   (E) 66

GRE Math Tests

Quantitative Comparison Question]
15.   Column A $a : b = 2 : 3$.   Column B
                                 a is positive.

          $\dfrac{a+5}{b+5}$ 1

USE THIS SPACE FOR SCRATCHWORK.

Quantitative Comparison Question]
16.   Column A   A spirit and water solution is sold   Column B
                                in a market. The cost per liter of
                                the solution is directly
                                proportional to the part (fraction)
                                of spirit (by volume) the solution
                                has.

                                A solution of 1 liter of spirit and 1
                                liter of water costs x dollars.

                                A solution of 1 liter of spirit and 2
                                liters of water costs y dollars.
          x y

USE THIS SPACE FOR SCRATCHWORK.

[Multiple-choice Question – Select One or More Answer Choices]
17. If r is negative, which of the following must also be negative?

   (A) r^2
   (B) r^3
   (C) r^4
   (D) $r^2 + r^3$
   (E) $r^2 \cdot r^3$

USE THIS SPACE FOR SCRATCHWORK.

Test 17 – Questions

[Multiple-choice Question – Select One Answer Choice only]
18. If 80 percent of the number a is 80, then how much is 20 percent of the number a ?

   (A) 20
   (B) 40
   (C) 50
   (D) 60
   (E) 80

[Multiple-choice Question – Select One Answer Choice only]
19. One ton has 2000 pounds, and one pound has 16 ounces. How many packets containing wheat weighing 16 pounds and 4 ounces each would totally fill a gunny bag of capacity 13 tons?

   (A) 1600
   (B) 1700
   (C) 2350
   (D) 2500
   (E) 8000

[Quantitative Comparison Question]
20. Column A — A train takes 15 seconds to cross a bridge at 50 mph, and at the same speed takes 10 seconds to cross the same bridge when the train's length is halved. — Column B

   Length of the bridge

   Original length of the train

331

[Multiple-choice Question – Select One Answer Choice only]
21. The sequence of numbers a, ar, ar^2, and ar^3 are in geometric progression. The sum of the first four terms in the series is 5 times the sum of first two terms and $r \neq -1$ and $a \neq 0$. How many times larger is the fourth term than the second term?

   (A) 1
   (B) 2
   (C) 4
   (D) 5
   (E) 6

[Multiple-choice Question – Select One Answer Choice only]
22. In a factory, there are workers, executives and clerks. 59% of the employees are workers, 460 are executives, and the remaining 360 employees are clerks. How many employees are there in the factory?

   (A) 1500
   (B) 2000
   (C) 2500
   (D) 3000
   (E) 3500

[Multiple-choice Question – Select One Answer Choice only]
23. A national math examination has 4 statistics problems. The distribution of the number of students who answered the questions correctly is shown in the chart. If 400 students took the exam and each question was worth 25 points, then what is the average score of the students taking the exam?

Question Number	Number of students who solved the question
1	200
2	304
3	350
4	250

(A) 1 point
(B) 25 points
(C) 26 points
(D) 69 points
(E) 263.5 points

[Multiple-choice Question – Select One Answer Choice only]
24. On average, a sharpshooter hits the target once every 3 shots. What is the probability that he will not hit the target until 4th shot?

(A) 1
(B) 8/81
(C) 16/81
(D) 65/81
(E) 71/81

GRE Math Tests

Answers and Solutions Test 17:

Question	Answer
1.	D
2.	A, E
3.	B
4.	D
5.	B
6.	1/4 or 0.25
7.	D
8.	A
9.	B
10.	B
11.	E
12.	A
13.	A, B, D, E
14.	A, B
15.	B
16.	C
17.	B, E
18.	A
19.	A
20.	B
21.	C
22.	B
23.	D
24.	B

If you got 18/24 correct on this test, you are likely to get 750+ on the actual GRE **by the time you complete all the tests in the book.**

1. If $x = 1$, then $x^2 = 1^2 = 1 \neq \sqrt{1} = \sqrt{x}$. In this case, the columns are equal.

If $x = 1/2$, then $x^2 = \left(\frac{1}{2}\right)^2 = \frac{1}{4} \neq \sqrt{\frac{1}{2}} = \sqrt{x}$. In this case, the columns are not equal and therefore the answer is (D).

2. There are different ways of checking whether a substitution is consistent. A different method is used for each of the choices below to explain them—though you can use any method for any of the choices. Note all the methods below mean the same and the objective is to choose the substitutions that are consistent with the system and eliminate the ones that are not.

Choice (A): $x = 1, y = 2$:
$x + 2y + z = 7$
$1 + 2(2) + z = 7$ by substitution
$z = 7 - 4 - 1 = 2$ subtracting 4 and 1 from both sides

$x + y + 2z = 7$
$1 + 2 + 2z = 7$ by substitution
$2z = 7 - 2 - 1 = 4$ subtracting 2 and 1 from both sides
$z = 4/2 = 2$ dividing both sides by 2

The values of z derived according to the equations of the system are consistent. So, we can say $x = 1, y = 2$, and $z = 3$ are solutions. Accept.

    Choice (B): $x = 3, y = 2$:
    Assume the choice is the solution.
    Let's derive the value of z from the first equation in the system:
        $x + 2y + z = 7$
        $3 + 2(2) + z = 7$ by substitution
        $z = 7 - 4 - 3$ subtracting 4 and 3 from both sides
        $z = 0$
    Then substitute this result in the second equation:
        $x + y + 2z = 7$
        $3 + 2 + 2(0) = 7$
        $5 = 7$, an absurd result and therefore the substation not consistent with the system.
    Reject the choice.

Choice (C): $x = 1, y = 3$:
    Assuming the choice is the solution.
    Let's derive the value of z from the second equation in the system:
        $x + y + 2z = 7$
        $1 + 3 + 2z = 7$ by substitution
        $2z = 7 - 3 - 1$ by subtracting 3 and 1 from both sides
        $2z = 3$
        $z = 3/2 = 1.5$

    Substituting the result into the second equation yields
        $x + 2y + z = 7$
        $1 + 2(3) + 1.5 = 7$ by substitution
        $5.5 = 7$
    Inconsistent. Reject.

Choice (D): $y = 1, z = 2$;
        $x + 2y + z = 7$
        $x + 2(1) + 2 = 7$ by substitution
        $x + 2 + 2 = 7$
        $x = 7 - 2 - 2$ subtracting 2 and 2 from both sides
        $x = 3$
    Break to second equation.
        $x + y + 2z = 7$
        $x + 1 + 2(2) = 7$
        $x = 7 - 4 - 1$
        $x = 2$
    The system is inconsistent. Reject the choice.

Choice (E): $y = 5, z = 5$:
$$x + 2y + z = 7$$
$$x + 2(5) + 5 = 7$$
$$x + 10 + 5 = 7$$
$$x = 7 - 10 - 5$$
$$x = -8$$
Break to second equation.
$$x + y + 2z = 7$$
$$-8 + 5 + 2(5) = 7$$
$$-8 + 5 + 10 = 7$$
$$7 = 7$$

True. Hence, solution is consistent. Accept as a possibility. Choose the solutions (A) and (E).

3. Since $b = a + 3$, the digit b is greater than the digit a.

In a two-digit number, the leftmost digit (the tens-digit) is the "more significant" digit. Since $b > a$, the number ba (which has b in tens-digit position) is greater than the number ab (which has a in tens-digit position). For example, 58 satisfies the equation $b = a + 3$, and interchanging its digits gives 85, which is greater than 58. Hence, Column B is greater than Column A, and the answer is (B).

Method II:
Since $b = a + 3$, the digit b is greater than the digit a and $a - b = -3$.

Column A: $ab = 10a + b$

Column B: $ba = 10b + a$

Column A – Column B = $10a + b - (10b + a) = 9a - 9b = 9(a - b) = 9(-3) = -27$.

Hence, Column A is 27 units less than Column B. The answer is (B).

4. Suppose $n = 6$. Now, the factors of 6 are 6 and 1, and 2 and 3.

In the first case, factors 6 and 1 sum to 7 (= 6 + 1), which is greater than 6 (= Column A). Here, Column B is greater.

In the second case, the factors 2 and 3 sum to 5 (= 2 + 3), which is less than 6 (= Column A). Here, Column A is greater. This is a double case, and therefore the answer is (D).

Test 17 — Solutions

5. The area of the circle Q is $\pi \cdot radius^2 = 25\pi$. Solving the equation for the radius yields $radius = \sqrt{\frac{25\pi}{\pi}} = 5$.

Now, the circumference of the circle Q is $2\pi \cdot radius = 2\pi \cdot 5 = 10\pi$.

The answer is (B).

6. Drawing the figure given in the question, we get

```
   _____
   D         A    B    C
```

$$AB = BC$$
$$CA = DA$$

Suppose AB equal 1 unit. Since point B bisects line segment AC, AB equals half AC. Hence, AC equals twice $AB = 2(AB) = 2(1\text{ unit}) = 2$ units. Again, since point A bisects line segment DC, DC equals twice $AC = 2(AC) = 2(2\text{ units}) = 4$ units. Hence, $AB/DC = 1\text{ unit}/4\text{ units} = 1/4$. Enter 1 in the first box and 4 in the second box. You can also choose to enter 2 in the first box and 8 in the second box. A calculator program that runs in the background attached to the screen evaluates your ratio to the most reduced form and checks for the correctness. For the paper-based test, the evaluation is done manually by the evaluator.

7. Summing the angles of $\triangle ABC$ to $180°$ yields

$$\angle A + \angle B + \angle C = 180$$
$$40 + 90 + \angle C = 180 \qquad \text{by substituting known values}$$
$$\angle C = 180 - (40 + 90) = 50$$

We are given that CE bisects $\angle ACB$. Hence, $\angle ECB = \angle ACE = $ one half of the full angle $\angle ACB$, which equals $1/2 \cdot 50 = 25$. Now, summing the angles of $\triangle ECB$ to $180°$ yields

$$\angle BEC + \angle ECB + \angle CBE = 180$$
$$x + 25 + 90 = 180$$
$$x = 180 - 25 - 90 = 65$$

The answer is (D).

8. The formula for the area of a rectangle is $length \times width$. The original length and width of the rectangle are 12 inches and 6 inches, respectively.

Column A: After the length of the rectangle is decreased by 4 inches, the new length becomes $12 - 4 = 8$ inches and the new area becomes $8 \cdot 6 = 48$.

Column B: After the width of the rectangle is decreased by 4 inches, the new width becomes $6 - 4 = 2$ inches and the new area becomes $12 \cdot 2 = 24$.

Since Column A is greater than Column B, the answer is (A).

GRE Math Tests

9. We are given that the radius of the circle is 3 units and BC, which is tangent to the circle, is a side of the square ABCD.

Since we have that PC = 3 (given), OC equals

$$[OP \text{ (radius of circle)}] + [PC \text{ (= 3 units, given)}] =$$

$$\text{radius} + 3 =$$

$$3 + 3 =$$

$$6$$

Now, since BC is tangent to the circle (given), $\angle OBC = 90°$, a right triangle. Hence, applying The Pythagorean Theorem to the triangle yields

$$OC^2 = BC^2 + OB^2$$

$$6^2 = BC^2 + 3^2$$

$$BC^2 = 6^2 - 3^2 = 36 - 9 = 27$$

Since the area of a square is (*side length*)^2, the area of square ABCD is $BC^2 = 27$. The answer is (B).

10. Since AB is tangent to both circles, $\angle BAQ = 90°$ and $\angle ABP = 90°$. Hence, AQ is parallel to BP. Now, draw a line through point P and parallel to AB as shown in the figure.

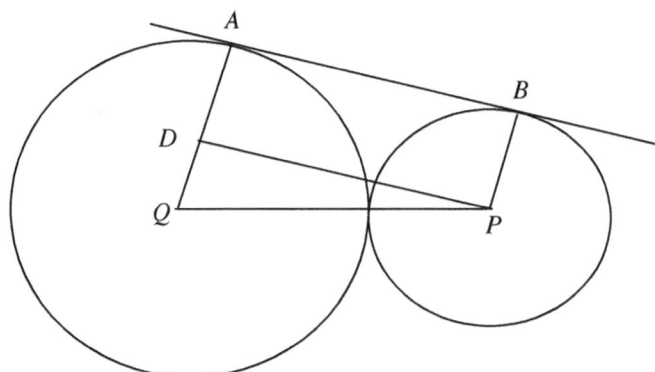

Then ABPD must be a rectangle. Hence, AB = DP. Also, $\angle PDQ$ equals the corresponding angle, $\angle BAD$. So, both equal 90°. Since a right angle is the greatest angle in a triangle, the side opposite the angle is the longest. PQ is the side opposite the right angle $\angle PDQ$ in $\triangle PQD$. Hence, PQ is greater than the other side DP. Hence, PQ is also greater than AB, which equals DP (we know). Hence, Column B is greater than Column A, and the answer is (B).

11. Since each side of the larger square measures 4 units and is divided by the four horizontal lines, the distance between any two adjacent horizontal lines must be 1 unit (= 4 units/4). Similarly, the larger square is divided into four vertical lines in the figure and any two adjacent lines are separated by 1 unit (= 4 units/4).

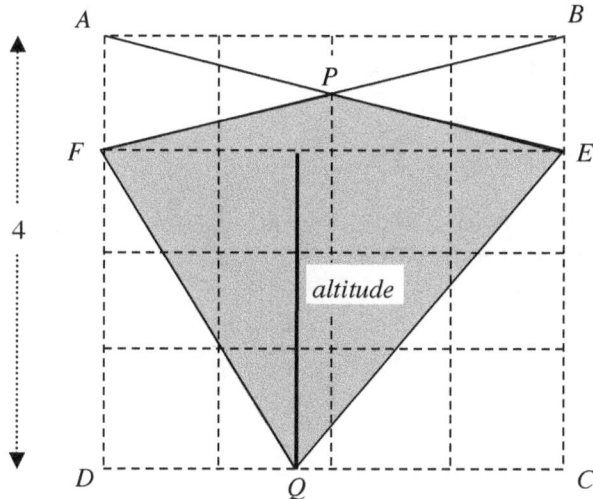

The shaded region can be divided into two sub-regions by the line FE, forming $\triangle FPE$ and $\triangle FQE$.

Now, the line-segment FE is spread across the two opposite sides of the larger square and measures 4 units. The altitude to it, shown in the bottom triangle, is spread vertically across the three horizontal segments and therefore measures 3 units.

Hence, the area of the lower triangle = $1/2 \times base \times height = 1/2 \times 4 \times 3 = 6$.

The shaded region above the line FE is one of the four sections formed by the two diagonals of the rectangle $ABEF$ and therefore its area must equal one-fourth the area of rectangle $ABEF$. The area of rectangle $ABEF$ is $length \times width = FE \times AF = 4 \times 1 = 4$. Hence, the area of the shaded region above the line FE is $1/4 \times 4 = 1$.

Summing the areas of the two shaded regions yields the area of the total shaded region: $1 + 6 = 7$. The answer is (E).

12. Canceling z from both sides of the inequality $x + z > y + z$ yields $x > y$. Adding $-z$ to both sides yields $x - z > y - z$. Hence, I is true.

If z is negative, multiplying the inequality $x > y$ by z would flip the direction of the inequality resulting in the inequality $xz < yz$. Hence, II may not be true.

If z is negative, dividing the inequality $x > y$ by z would flip the direction of the inequality resulting in the inequality $x/z < y/z$. Hence, III may not be true.

Since we are asked for statements that MUST be true, the answer is (A)

13. The equations $|yz - zx| = 3$ and $|zx - xy| = 4$ yield two pairs of cases:

$$yz - zx = 3 \text{ or } yz - zx = -3 \text{ and}$$
$$zx - xy = 4 \text{ or } zx - xy = -4.$$

Adding the two given equations yields

$$(yz - zx) + (zx - xy) = (3 \text{ or } -3) + (4 \text{ or } -4)$$
$$yz - zx + zx - xy = (3 + 4) \text{ or } (-3 + 4) \text{ or } (3 - 4) \text{ or } (-3 - 4)$$
$$yz - xy = 7 \text{ or } 1 \text{ or } -1 \text{ or } -7$$

The possible answers are (A), (B), (D), and (E). Choose them.

14. Let $a, b, c,$ and d be the respective angles of the quadrilateral $ABCD$. Sum of angles is $a + b + c + d = 360$. So, $b + c + d = 360 - a$, $a + c + d = 360 - b$, and $a + b + d = 360 - c$.

We are also given the following

$$a \geq (b + c + d)/3 + 20$$
$$b \geq (a + c + d)/3 + 15$$
$$c = (a + b + d)/3 + 10$$

Adding the inequalities yields

$$a + b + c \geq (b + c + d)/3 + 20 + (a + c + d)/3 + 15 + (a + b + d)/3 + 10$$
$$\geq (b + c + d)/3 + (a + c + d)/3 + (a + b + d)/3 + 20 + 15 + 10$$
$$\geq (b + c + d + a + c + d + a + b + d)/3 + 45$$
$$\geq (2b + 2c + 3d + 2a)/3 + 45$$
$$3a + 3b + 3c \geq (2b + 2c + 3d + 2a) + 135$$
$$3a + 3b + 3c \geq 2b + 2c + 3d + 2a + 135$$
$$a + b + c \geq 3d + 135$$
$$360 - d \geq 3d + 135$$
$$360 \geq 3d + 135 + d$$
$$360 \geq 4d + 135$$
$$360 - 135 \geq 4d$$
$$56.25 \geq d$$

Choose (A) and (B) since they fit in the range.

15. Forming the ratio yields $a/b = 2/3$. Solving for b yields $b = 3a/2$. Since a is positive (given), this equation tells us that b is also positive. Hence, $b + 5$ is positive, and we can safely multiply both columns by $b + 5$. This yields

$a + 5$ $b + 5$

Subtracting 5 from both columns yields

a b

Since both a and b are positive and b is one-and-a-half times larger than a ($b = 3a/2$), Column B is larger than Column A. The answer is (B).

16. Since the cost of *each* liter of the spirit water solution is proportionally to the part (fraction) of spirit the solution has, the cost per liter can be expressed as kf, where f is the fraction (part of) of pure spirit the solution has.

Now, m liters of the solution containing n liters of the spirit should cost

$$m \times \text{cost of each liter} =$$
$$m \times kf =$$
$$m \times k(n/m) =$$
$$kn$$

Hence, the solution is only priced for the content of spirit the solution has (n here). Hence, the cost of the two samples given in the problem must be equal since both have exactly 1 liter of the spirit. Hence, x equals y and therefore Column A equals Column B. The answer is (C).

17. If r is negative, then r^2 and r^4 must be positive since they have even exponents [reject (A) and (C)]; and r^3 and $r^2 \cdot r^3 = r^5$ must be negative since they have odd exponents [accept (B) and (E)]. If r lies between -1 and 0, then r^2 is numerically greater than r^3 and $r^2 + r^3$ is positive. Reject choice (D).

The answer is (B) and (E).

18. We have that 80 is 80 percent of a. Now, 80 percent of a is $80/100 \times a$. Equating the two yields $80/100 \times a = 80$. Solving the equation for a yields $a = 100/80 \times 80 = 100$. Now, 20 percent of a is $20/100 \times 100 = 20$. The answer is (A).

19. One ton has 2000 pounds. The capacity of the gunny bag is 13 tons. Hence, its capacity in pounds would equal 13×2000 pounds.

One pound has 16 ounces. We are given the capacity of each packet is 16 pounds and 4 ounces. Converting it into pounds yields 16 pounds + 4/16 pounds = 16 1/4 pounds = $(16 \times 4 + 1)/4$ pounds = 65/4 pounds.

Hence, the number of packets required to fill the gunny bag equals

$$(\text{Capacity of the gunny bag}) \div (\text{Capacity of the each packet}) =$$
$$13 \times 2000 \text{ pounds} \div (65/4) \text{ pounds} =$$
$$13 \times 2000 \times 4/65 =$$
$$2000 \times 4/5 =$$
$$1600$$

The answer is (A).

GRE Math Tests

20. Let the length of the train be l and the length of the bridge be b. We are given that the train crosses the bridge in 15 seconds, and it crosses the same bridge in 10 seconds when its length is halved. In each case, the speed is the same. Hence, let's derive expressions for speed for each case and equate. The distance traveled by the train in crossing the bridge is

$$(length\ of\ bridge) + (length\ of\ train)$$

In the first case, distance traveled is $b + l$, and, in the second case, with train length halved, distance traveled is $b + l/2$. In the first case, time taken is 15 seconds, and in the second case, time taken is 10 seconds. By the formula *Speed = Distance/Time*, the speed in the first case is $(b + l)/15$ and in the second case is $(b + l/2)/10$. Since the speed in both cases is the same, we have

$$\frac{b+l}{15} = \frac{b+l/2}{10}$$

$$\frac{b+l}{3} = \frac{b+l/2}{2}$$

$$2(b + l) = 3(b + l/2)$$

$$2b + 2l = 3b + 3l/2$$

$$l/2 = b$$

$$l = 2b$$

Hence, the length of the train is twice the length of the bridge. So, Column B is greater, and the answer is (B).

21. In the given progression, the sum of first two terms is $a + ar$, and the sum of first four terms is $a + ar + ar^2 + ar^3$. Since "the sum of the first four terms in the series is 5 times the sum of the first two terms," we have

$a + ar + ar^2 + ar^3 = 5(a + ar)$
$a + ar + ar^2 + ar^3 = 5a + 5ar$ by distributing the 5
$1 + r + r^2 + r^3 = 5 + 5r$ by dividing every term by $a \neq 0$
$-4 - 4r + r^2 + r^3 = 0$ by subtracting 5 and $5r$ from both sides
$-4(1 + r) + r^2(1 + r) = 0$ by factoring -4 from the first two terms and factoring r^2 from the second two terms
$(1 + r)(-4 + r^2) = 0$ by factoring out the common factor $1 + r$
$1 + r = 0$ or $-4 + r^2 = 0$
$-1 = r$ or $4 = r^2$
$\pm 2 = r$ by taking the square root of both sides of $4 = r^2$ and discarding $-1 = r$ because we are given that $r \neq -1$

Now, to see how many times the fourth term is greater than the second term, we divide them. For example, to see how many times 6 is greater than 2, we divide 6 by 2: $6/2 = 3$. Hence, 6 is 3 times larger than 2. Dividing the fourth term by the second term in the sequence gives $ar^3/ar = r^2 = 2^2 = 4$. Hence, the fourth term is 4 times larger than the second term, and the answer is (C).

22. We are given that that 59% of the employees E are workers. Since the factory consists of only workers, executives, and clerks, the remaining $100 - 59 = 41\%$ of the employees must include only executives and clerks. Since we are given that the number of executives is 460 and the number of clerks is 360, which sum to $460 + 360 = 820$, we have the equation $(41/100)E = 820$, or $E = 100/41 \times 820 = 2000$. The answer is (B).

23. The average score of the students is equal to the net score of all the students divided by the number of students. The number of students is 400 (given). Now, let's calculate the net score. Each question carries 25 points, the first question is solved by 200 students, the second one by 304 students, the third one by 350 students, and the fourth one by 200 students. Hence, the net score of all the students is

$$200 \times 25 + 304 \times 25 + 350 \times 25 + 250 \times 25 =$$

$$25(200 + 304 + 350 + 250) =$$

$$25(1104)$$

Hence, the average score equals

$$25(1104)/400 =$$

$$1104/16 =$$

$$69$$

The answer is (D).

24. The sharpshooter hits the target once in every 3 shots. Hence, the probability of hitting the target is 1/3. The probability of not hitting the target is $1 - 1/3 = 2/3$.

He will not hit the target on the first, second, and third shots, but he will hit it on the fourth shot. The probability of this is

$$\frac{2}{3} \cdot \frac{2}{3} \cdot \frac{2}{3} \cdot \frac{1}{3} = \frac{8}{81}$$

The answer is (B).

Test 18

GRE Math Tests

> **Questions: 24**
> **Time: 45 minutes**

[Multiple-choice Question – Select One Choice Only]
1. Which one of the following is the solution of the system of equations given?

$$x + 2y = 7$$
$$x + y = 4$$

(A) $x = 3, y = 2$
(B) $x = 2, y = 3$
(C) $x = 1, y = 3$
(D) $x = 3, y = 1$
(E) $x = 7, y = 1$

[Multiple-choice Question – Select One or More Answer Choices]
2. The last digit of the positive even number n equals the last digit of n^2. Which of the following could be n?

(A) 10
(B) 14
(C) 15
(D) 16
(E) 17

[Multiple-choice Question – Select One Answer Choice Only]
3. How many positive five-digit numbers can be formed with the digits 0, 3, and 5?

(A) 14
(B) 15
(C) 108
(D) 162
(E) 243

Quantitative Comparison Question]

4. **Column A**
The sum of the positive integers from 1 through n can be calculated by the formula
$$\frac{n(n+1)}{2}$$
Column B

The sum of the multiples of 6 between 0 and 100

The sum of the multiples of 8 between 0 and 100

Quantitative Comparison Question]

5. **Column A**
The average of three numbers if the greatest is 20

Column B
The average of three numbers if the greatest is 2

[Multiple-choice Question – Select One Choice only]

6. The side length of a square inscribed in a circle is 2. What is the area of the circle?

   (A) π
   (B) $\sqrt{2}\pi$
   (C) 2π
   (D) $2\sqrt{2}\pi$
   (E) π^2

[Multiple-choice Question – Select One Choice Only]
7. In the figure, ∠P =

   (A) 15°
   (B) 30°
   (C) 35°
   (D) 40°
   (E) 50°

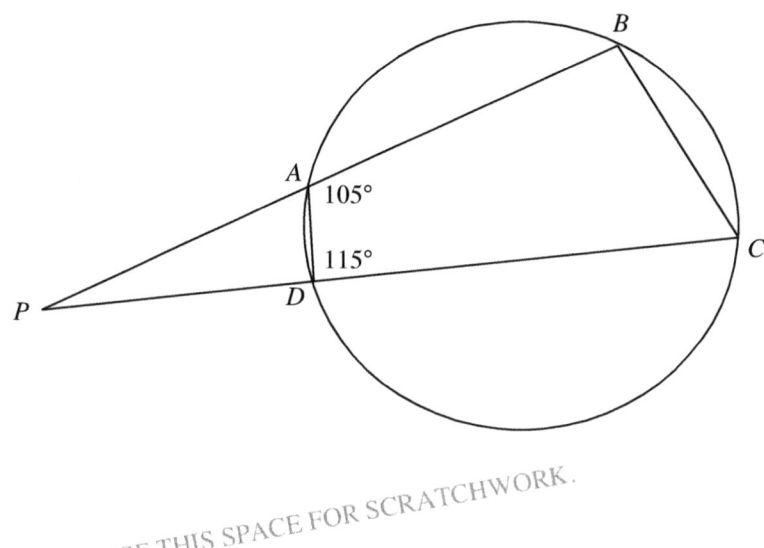

[Multiple-choice Question – Select One or More Answer Choices]
8. In the figure, lines *l* and *k* are parallel. Which of the following must be true?

   (A) $a < b$
   (B) $a = 2b + 30$
   (C) $a = b + 10$
   (D) $a \geq b$
   (E) $a > b$

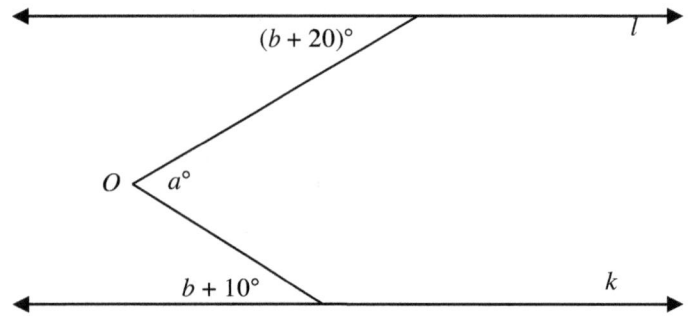

The following figure is for problems 9 and 10. Lines *l* and *k* are parallel and *a* is an acute angle, that is, *a* is less than 90 degrees in measure.

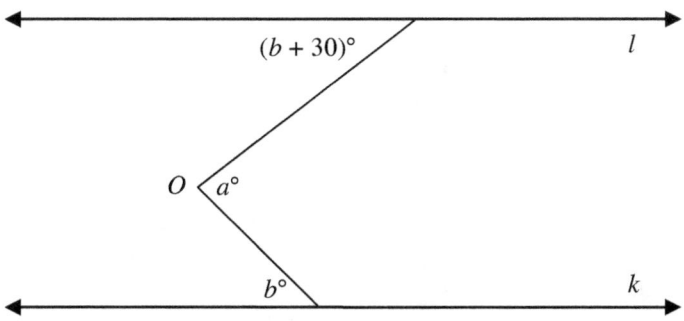

[Multiple-choice Question – Select One or More Answer Choices]
9. Which of the following must be true?

   (A) $b > 10$
   (B) $b > 15$
   (C) $b < 20$
   (D) $b < 30$
   (E) $b > 45$

[Multiple-choice Question – Select One or More Answer Choices]
10. Which of the following could be true?

   (A) $b > 10$
   (B) $b > 15$
   (C) $b < 20$
   (D) $b < 30$
   (E) $b > 45$

[Multiple-choice Question – Select One Answer Choice Only]
11. In a triangle with sides of lengths 3, 4, and 5, the smallest angle is 36.87°. In the figure, O is the center of the circle of radius 5. A and B are two points on the circle, and the distance between the points is 6. What is the value of x ?

(A) 36.87
(B) 45
(C) 53.13
(D) 116.86
(E) 126.86

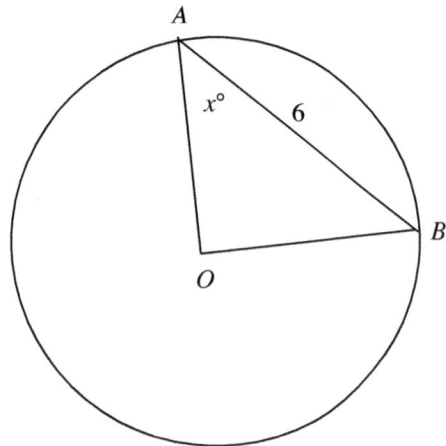

Quantitative Comparison Question]

12. Column A — In the figure, $\triangle ABC$ is inscribed in the circle. The triangle does not contain the center of the circle O. — Column B

x — 90

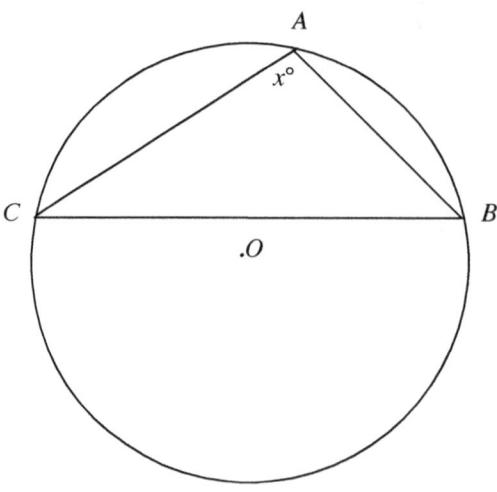

[Multiple-choice Question – Select One or More Answer Choices]
13. The slope of the line $2x + y = 3$ is NOT the same as the slope of which of the following lines?

   (A) $2x + y = 5$
   (B) $x + y/2 = 3$
   (C) $x = -y/2 - 3$
   (D) $y = 7 - 2x$
   (E) $x + 2y = 9$

[Multiple-choice Question – Select One or More Answer Choices]
14. If $|x| + x = 4$, then which of the following is odd?

   (A) $x^2 + 3x$
   (B) $x^2 + 3x + 2$
   (C) $x^2 + 4x$
   (D) $x^2 + 4x + 2$
   (E) $x^2 + 4x + 3$

Quantitative Comparison Question]
15.   Column A $(x - 2y)(x + 2y) = 5$ Column B
                               $(2x - y)(2x + y) = 35$

       $2x^2 - y^2$ $x^2 - 2y^2$

GRE Math Tests

[Multiple-choice Question – Select One Answer Choice Only]
16. In 2012, the arithmetic mean of the annual incomes of Jack and Jill was $3800. The arithmetic mean of the annual incomes of Jill and Jess was $4800, and the arithmetic mean of the annual incomes of Jess and Jack was $5800. What is the arithmetic mean of the incomes of the three?

    (A) $4000
    (B) $4200
    (C) $4400
    (D) $4800
    (E) $5000

Quantitative Comparison Question]
17.

Column A		Column B
	The savings from a person's income is the difference between his income and his expenditure. The ratio of Marc's income to his boss's is 3 : 4. Their respective expenditures ratio is 1 : 2.	
The savings from income of Marc		The savings from income of Marc's Boss

Quantitative Comparison Question]
18.

Column A		Column B
	Item A cost John $90 and item B cost him $100.	
Percentage of profit that John earned by selling item A with a profit of $10		Percentage of profit that John earned by selling item B with a profit of $10

Quantitative Comparison Question]

19.   Column A

James purchased Medicine A for x dollars, which included a sales tax of 5%. Kate was charged 5% for sales tax on x dollars that Medicine B costs.

Column B

Sales tax paid by James

Sales tax paid by Kate

[Multiple-choice Question – Select One or More Answer Choices]

20. A driver plans that he should travel at 80 mph for the first half (either by distance or by time) of a trip and at 40 mph for the second half of the trip. Which of the following could be the average speed of the car during the entire trip?

(A) 50
(B) 60
(C) 80
(D) 80/3
(E) 160/3

[Multiple-choice Question – Select One Answer Choice Only]
21. For how many integers n between 5 and 20, inclusive, is the sum of $3n$, $9n$, and $11n$ greater than 200?

   (A)   4
   (B)   8
   (C)   12
   (D)   16
   (E)   20

[Multiple-choice Question – Select One Answer Choice Only]
22. What is the probability that the product of two integers (not necessarily different integers) randomly selected from the numbers 1 through 20, inclusive, is odd?

   (A)   0
   (B)   1/4
   (C)   1/2
   (D)   2/3
   (E)   3/4

[Multiple-choice Question – Select One or More Answer Choices]
23. Removing which of the following numbers from the set $S = \{1, 2, 3, 4, 5, 6\}$ would move the median of the set S to the right on the number line?

   (A)   1
   (B)   2
   (C)   3
   (D)   4
   (E)   5
   (F)   6

[Multiple-choice Question – Select One Answer Choice Only]

24. On average, a sharpshooter hits the target once every 3 shots. What is the probability that he will hit the target in 4 shots?

    (A) 1
    (B) 1/81
    (C) 1/3
    (D) 65/81
    (E) 71/81

USE THIS SPACE FOR SCRATCHWORK.

GRE Math Tests

Answers and Solutions Test 18:

Question	Answer
1.	C
2.	A, D
3.	D
4.	A
5.	D
6.	C
7.	D
8.	E
9.	D
10.	A, B, C, D
11.	C
12.	A
13.	E
14.	E
15.	A
16.	D
17.	D
18.	A
19.	B
20.	B, E
21.	C
22.	B
23.	A, B, C
24.	D

If you got 18/24 correct on this test, you are likely to get 750+ on the actual GRE **by the time you complete all the tests in the book.**

---

1. The given system of equations is $x + 2y = 7$ and $x + y = 4$. Now, just substitute each answer-choice into the two equations and see which one works (start checking with the easier equation, $x + y = 4$):

   Choice (A): $x = 3$, $y = 2$: Here, $x + y = 3 + 2 = 5 \neq 4$. Reject.

   Choice (B): $x = 2$, $y = 3$: Here, $x + y = 2 + 3 = 5 \neq 4$. Reject.

   Choice (C): $x = 1$, $y = 3$: Here, $x + y = 1 + 3 = 4 = 4$, and $x + 2y = 1 + 2(3) = 7$. Correct.

   Choice (D): $x = 3$, $y = 1$: Here, $x + y = 3 + 1 = 4$, but $x + 2y = 3 + 2(1) = 5 \neq 7$. Reject.

   Choice (E): $x = 7$, $y = 1$: Here, $x + y = 7 + 1 = 8 \neq 4$. Reject.

The answer is (C).

Test 18 — Solutions

Method II (without substitution):
In the system of equations, subtracting the bottom equation from the top one yields $(x + 2y) - (x + y) = 7 - 4$, or $y = 3$. Substituting this result in the bottom equation yields $x + 3 = 4$. Solving the equation for x yields $x = 1$.

The answer is (C).

---

2. Numbers ending with the digits 0, 1, 5, or 6 will have their squares also ending with the same digit.

For example,

> 10 ends with 0, and 10^2 (= 100) also ends with 0.
> 11 ends with 1, and 11^2 (= 121) also ends with 1.
> 15 ends with 5, and 15^2 (= 225) also ends with 5.
> 16 ends with 6, and 16^2 (= 256) also ends with 6.

Among the four numbers 0, 1, 5, or 6, even numbers only end with 0 or 6. Choices (A) and (D) have such numbers. The answer is (A) and (D).

---

3. Let the digits of the five-digit positive number be represented by 5 compartments:

Each of the last four compartments can be filled in 3 ways (with any one of the numbers 0, 3 and 5).

The first compartment can be filled in only 2 ways (with only 3 and 5, not with 0, because placing 0 in the first compartment would yield a number with fewer than 5 digits).

3	0	0	0	0
5	3	3	3	3
	5	5	5	5

Hence, the total number of ways the five compartments can be filled in is $2 \cdot 3 \cdot 3 \cdot 3 \cdot 3 = 162$. The answer is (D).

---

4. The sum of the multiples of 6 between 0 and 100 equals

$6 + 12 + 18 + \ldots + 96 = 6(1) + 6(2) + 6(3) + \ldots + 6(16) = 6(1 + 2 + 3 + \ldots + 16) = 6\left(\dfrac{16(16+1)}{2}\right) = 3 \cdot 16 \cdot 17$.

The sum of multiples of 8 between 0 and 100 equals

$$8 + 16 + 24 + \ldots + 96 =$$
$$8(1) + 8(2) + 8(3) + \ldots + 8(12) =$$
$$8(1 + 2 + 3 + \ldots + 12) =$$
$$8\left(\dfrac{12(12+1)}{2}\right) = 4 \cdot 12 \cdot 13$$

Since $3 \cdot 16 \cdot 17$ (= 48 · 17) is clearly greater than $4 \cdot 12 \cdot 13$ (= 48 · 13), Column A is greater than Column B and the answer is (A).

5. At first glance, Column A appears larger than Column B. However, the problem does not exclude negative numbers. Suppose the three numbers in Column A are –20, 0, and 20 and that the three numbers in Column B are 0, 1, and 2. Then the average for Column A would be $\frac{-20+0+20}{3} = \frac{0}{3} = 0$, and the average for Column B would be $\frac{0+1+2}{3} = \frac{3}{3} = 1$. In this case, Column B is larger. Clearly, there are also numbers for which Column A would be larger. Hence, the answer is (D).

6.
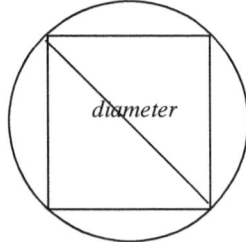

The diagonal of a square inscribed in a circle is a diameter of the circle. The formula for the diagonal of a square is $\sqrt{2} \times side$. Hence, the diameter of the circle inscribing the square of side length 2 is $\sqrt{2} \times 2 = 2\sqrt{2}$. Since *radius = diameter*/2, the radius of the circle is $\frac{2\sqrt{2}}{2} = \sqrt{2}$. Hence, the area of the circle is $\pi \cdot radius^2 = \pi(\sqrt{2})^2 = 2\pi$. The answer is (C).

---

7. Since the angle in a line is 180°, we have ∠*PAD* + ∠*DAB* = 180, or ∠*PAD* + 105 = 180 (from the figure, ∠*DAB* = 105°). Solving for ∠*PAD* yields ∠*PAD* = 75.

Since the angle in a line is 180°, we have ∠*ADP* + ∠*ADC* = 180, or ∠*ADP* + 115 = 180 (from the figure, ∠*ADC* = 115°). Solving for ∠*ADP* yields ∠*ADP* = 65.

Now, summing the angles of the triangle *PAD* to 180° yields

∠*P* + ∠*PAD* + ∠*ADP* = 180

∠*P* + 75 + 65 = 180

∠*P* = 40

The answer is (D).

8. Draw line m passing through O and parallel to both line l and line k.

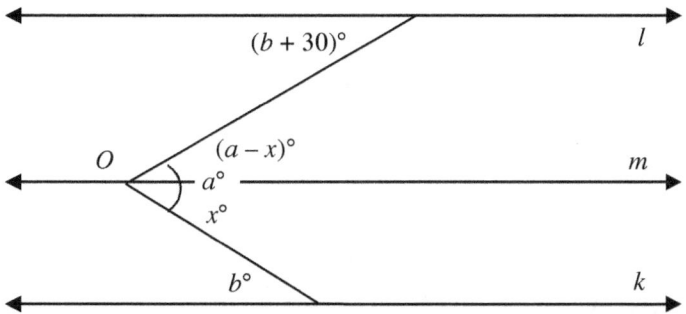

Now, observe that angle x is only part of angle a, and $x = b$ since they are alternate interior angles. Since x is only part of angle a, $a > x$ and $a > b$. The answer is (E).

---

9. Draw a line parallel to both of the lines l and k and passing through O.

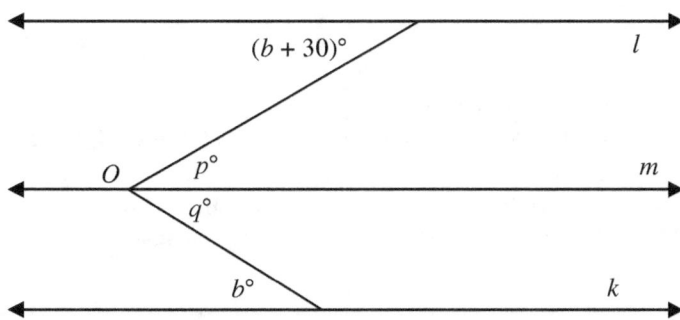

We are given that a is an acute angle. Hence, $a < 90$. Since angles p and $b + 30$ are alternate interior angles, they are equal. Hence, $p = b + 30$. Similarly, angles q and b are alternate interior angles. Hence, $q = b$. Since angle a is the sum of its sub-angles p and q, $a = p + q = (b + 30) + b = 2b + 30$. Solving this equation for b yields $b = (a - 30)/2 = a/2 - 15$. Now, dividing both sides of the inequality $a < 90$ by 2 yields $a/2 < 45$. Also, subtracting 15 from both sides of the inequality yields $a/2 - 15 < 30$. Since $a/2 - 15 = b$, we have $b < 30$. The answer is (D).

---

10. From the previous question, all the choices could be true except (E). We know that $b < 30$. The answer is (A), (B), (C), and (D).

11. We are given that the smallest angle of any triangle of side lengths 3, 4, and 5 is 36.87°. The smallest angle is the angle opposite the smallest side, which measures 3.

Now, in $\triangle AOB$, $OA = OB$ = radius of the circle = 5 (given). Hence, the angles opposite the two sides in the triangle are equal and therefore the triangle is isosceles. Just as in any isosceles triangle, the altitude on the third side AB must divide the side equally. Say, the altitude cuts AB at J. Then we have $AJ = JB$, and both equal $AB/2 = 3$.

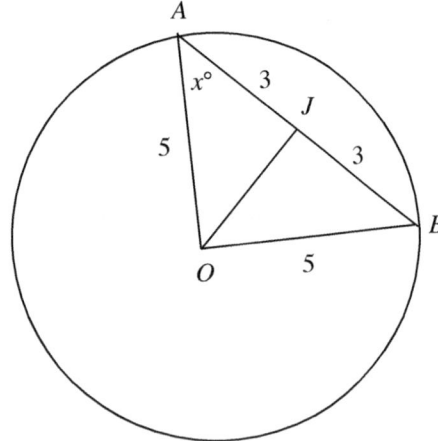

Since the altitude is a perpendicular, applying The Pythagorean Theorem to $\triangle AOJ$ yields $AJ^2 + JO^2 = OA^2$; $3^2 + JO^2 = 5^2$; $JO^2 = 5^2 - 3^2 = (25 - 9) = 16 = 4^2$. Square rooting both sides yields $JO = 4$. Hence, in $\triangle AOJ$, $AJ = 3$, $JO = 4$, and $AO = 5$. Hence, the smallest angle, the angle opposite the smallest side, $\angle AOJ$ equals 36.87°. Summing the angles of $\triangle AOJ$ to 180° yields $x + \angle AJO + \angle JOA = 180$; $x + 90 + 36.87 = 180$. Solving this equation yields $x = 180 - (90 + 36.87) = 180 - 126.87 = 53.13$. The answer is (C).

---

12. A chord makes an acute angle on the circle to the side containing the center of the circle and makes an obtuse angle to the other side. In the figure, BC is a chord and does not have a center to the side of point A. Hence, BC makes an obtuse angle at point A on the circle. Hence, $\angle A$, which equals $x°$, is obtuse and therefore is greater than 90°. Hence, Column A is greater than Column B, and the answer is (A).

---

13. The slope of a line expressed as $y = mx + b$ is m. Expressing the given line $2x + y = 3$ in that format yields $y = -2x + 3$. Hence, the slope is -2, the coefficient of x. Let's express each line in the form $y = mx + b$ and pick the line whose slope is not -2.

Choice (A): $2x + y = 5$; $y = -2x + 5$, slope is -2. Reject.
Choice (B): $x + y/2 = 3$; $y = -2x + 6$, slope is -2. Reject.
Choice (C): $x = -y/2 - 3$; $y = -2x - 6$, slope is -2. Reject.
Choice (D): $y = 7 - 2x$; $y = -2x + 7$, slope is -2. Reject.
Choice (E): $x + 2y = 9$; $y = -\frac{1}{2}x + \frac{9}{2}$, slope is $-\frac{1}{2} \neq -2$. Accept the choice.

The answer is (E).

Test 18 — Solutions

14. We are given that $|x| + x = 4$. If x is negative or zero, then $|x|$ equals $-x$ and $|x| + x$ equals $-x + x = 0$. This conflicts with the given equation, so x is not negative nor 0. Hence, x is positive and therefore $|x|$ equals x. Putting this in the given equation yields

$$|x| + x = 4$$
$$x + x = 4$$
$$2x = 4$$
$$x = 2$$

Now, select the answer-choice that results in an odd number when $x = 2$.

Choice (A): $x^2 + 3x = 2^2 + 3(2) = 4 + 6 = 10$, an even number. Reject.
Choice (B): $x^2 + 3x + 2 = 2^2 + 3(2) + 2 = 4 + 6 + 2 = 12$, an even number. Reject.
Choice (C): $x^2 + 4x = 2^2 + 4(2) = 4 + 8 = 12$, an even number. Reject.
Choice (D): $x^2 + 4x + 2 = 2^2 + 4(2) + 2 = 4 + 8 + 2 = 14$, an even number. Reject.
Choice (E): $x^2 + 4x + 3 = 2^2 + 4(2) + 3 = 4 + 8 + 3 = 15$, an odd number. Accept.

The answer is (E).

---

15. We are given the two equations:

$$(x - 2y)(x + 2y) = 5$$
$$(2x - y)(2x + y) = 35$$

Applying the Difference of Squares formula, $(a + b)(a - b) = a^2 - b^2$, to the left-hand sides of each equation yields

$$x^2 - (2y)^2 = 5$$
$$(2x)^2 - y^2 = 35$$

Simplifying these two equations yields

$$x^2 - 4y^2 = 5$$
$$4x^2 - y^2 = 35$$

Subtracting the bottom equation from the top one yields

$$(x^2 - 4y^2) - (4x^2 - y^2) = 5 - 35$$
$$-3x^2 - 3y^2 = -30$$
$$-3(x^2 + y^2) = -30$$
$$x^2 + y^2 = -30/-3 = 10$$

Now, Column A – Column B = $(2x^2 - y^2) - (x^2 - 2y^2) = x^2 + y^2 = 10$. Hence, Column A is 10 units greater than Column B. The answer is (A).

GRE Math Tests

16. Let a, b, and c be the annual incomes of Jack, Jill, and Jess, respectively.

Now, we are given that

The arithmetic mean of the annual incomes of Jack and Jill was $3800. Hence, $(a + b)/2 = 3800$. Multiplying by 2 yields $a + b = 2 \times 3800 = 7600$.

The arithmetic mean of the annual incomes of Jill and Jess was $4800. Hence, $(b + c)/2 = 4800$. Multiplying by 2 yields $b + c = 2 \times 4800 = 9600$.

The arithmetic mean of the annual incomes of Jess and Jack was $5800. Hence, $(c + a)/2 = 5800$. Multiplying by 2 yields $c + a = 2 \times 5800 = 11{,}600$.

Summing these three equations yields

$$(a + b) + (b + c) + (c + a) = 7600 + 9600 + 11{,}600$$

$$2a + 2b + 2c = 28{,}800$$

$$a + b + c = 14{,}400$$

The average of the incomes of the three equals the sum of the incomes divided by 3:

$$(a + b + c)/3 =$$
$$14{,}400/3 =$$
$$4800$$

The answer is (D).

---

17. Suppose Marc's income is $300 and his boss's income is $400 (incomes match the given ratio, 3 : 4). Also, suppose Marc's expenditure is $100 and his boss's expenditure is $200 (expenditures match the given ratio, 1 : 2). Then the savings from income of Marc is $300 − 100 = 200$ and that of his boss is $400 − 200 = 200$. In this case, Marc's savings equals his boss's savings.

Now, suppose the expenditures are different: Marc's expenditure is $150 and his boss's expenditure is $300 (expenditures match the given ratio, 1 : 2). In this case, Marc's savings is $300 − 150 = 150$, and his boss's savings is $400 − 300 = 100$ (savings are not equal). Hence, we have a double case, and the answer is (D).

18. The formula for the profit percentage is $\dfrac{\text{Profit}}{\text{Cost}} \cdot 100$.

Hence, Column A equals $10/90 \cdot 100 = 100/9\% = 11.1\%$, and Column B equals $10/100 \cdot 100 = 10\%$.

Hence, Column A is greater than Column B, and the answer is (A).

---

19. Let the original cost of Medicine A be a dollars. 5% tax on this equals $(5/100)a = a/20$. So, the total cost including sales tax is $a + a/20 = 21a/20$. This equals x dollars (what medicine A cost James). Hence, we have $21a/20 = x$. Solving for a yields $a = 20x/21$. Sales tax on A is $\dfrac{a}{20} = \dfrac{\tfrac{20x}{21}}{20} = \dfrac{x}{21} =$ Column A.

Now, Kate was charged sales tax on Medicine B. The charge was 5% of the cost of the medicine (x). 5% of x is $(5/100)x = x/20 =$ Column B. Since $1/20 > 1/21$, $x/20 > x/21$ and Column B > Column A. The answer is (B).

Test 18 — Solutions

20. Case I (Half by time). Let t be the entire time of the trip.

We have that the car traveled at 80 mph for $t/2$ hours and at 40 mph for the remaining $t/2$ hours. Remember that *Distance = Speed × Time*. Hence, the net distance traveled during the two periods equals $80 \times t/2 + 40 \times t/2$. Now, remember that

$$\textit{Average Speed} =$$
$$\frac{\textit{Net Distance}}{\textit{Time Taken}} =$$
$$\frac{80 \times \dfrac{t}{2} + 40 \times \dfrac{t}{2}}{t} =$$
$$80 \times \frac{1}{2} + 40 \times \frac{1}{2} =$$
$$40 + 20 =$$
$$60$$

Select choice (B).

Case II (Half by distance).

Let d be the entire time of the trip.

We have that the car traveled at 80 mph for $d/2$ hours and at 40 mph for the remaining $d/2$ hours.

Remember that *Distance = Speed × Time*. Hence, the net time taken to travel the two half distances equals $(d/2)/80 + (d/2)/40$. Now, the average speed

$$\textit{Average Speed} =$$
Total distance / Total time =
$$d / [(d/2)/80 + (d/2)/40] =$$
$$1 / [(1/2)/80 + (1/2)/40] =$$
$$2 / [2(1/2)/80 + 2(1/2)/40] =$$
$$2 / [1/80 + 1/40] =$$
$$2 \times 80 / [80 \times 1/80 + 80 \times 1/40] =$$
$$160 / [1 + 2] =$$
$$160 / 3$$

Select choice (E). The answers to select are (B) and (E).

21. The sum of $3n, 9n,$ and $11n$ is $23n$. Since this is to be greater than 200, we get the inequality $23n > 200$. From this, we get $n > 200/23 \approx 8.7$. Since n is an integer, $n > 8$. Now, we are given that $5 \leq n \leq 20$. Hence, the values for n are 9 through 20, a total of 12 numbers. The answer is (C).

22. The product of two integers is odd when both integers are themselves odd. Hence, the probability of the product being odd equals the probability of both numbers being odd. Since there is one odd number in every two numbers (there are 10 odd numbers in the 20 numbers 1 through 20, inclusive), the probability of a number being odd is 1/2. The probability of both numbers being odd (independent case) is

$$1/2 \times 1/2 = 1/4$$

The answer is (B).

23. The median of the $S = $ set $\{1, 2, 3, 4, 5, 6\}$ is half the sum of the middle numbers:

$$(3 + 4)/2 =$$

$$7/2 =$$

$$3.5$$

To move the median to the right on the number line, remove at least one of the elements on the left of the median on the number line.

For example, if we remove 3 from the set $S = $ set $\{1, 2, 4, 5, 6\}$, then median is 4, the median increased.

The correct answers are the numbers to the left of 3.5 on the number line, the choices (A), (B), and (C).

---

24. The sharpshooter hits the target once in 3 shots. Hence, the probability of hitting the target is 1/3. The probability of not hitting the target is $1 - 1/3 = 2/3$.

Now, (the probability of not hitting the target even once in 4 shots) + (the probability of hitting at least once in 4 shots) equals 1, because these are the only possible cases.

Hence, the probability of hitting the target at least once in 4 shots is

$$1 - \text{(the probability of not hitting even once in 4 shots)}$$

The probability of not hitting in the 4 chances is $\frac{2}{3} \cdot \frac{2}{3} \cdot \frac{2}{3} \cdot \frac{2}{3} = \frac{16}{81}$. Now, $1 - 16/81 = 65/81$. The answer is (D).

This methodology is similar to Model 2. You might try analyzing why. Clue: The numerators of $\frac{2}{3} \cdot \frac{2}{3} \cdot \frac{2}{3} \cdot \frac{2}{3} = \frac{16}{81}$ are the number of ways of doing the specific jobs, and the denominators are the number of ways of doing all possible jobs.

Test 19

GRE Math Tests

Questions: 24
Time: 45 minutes

Quantitative Comparison Question]
1.   Column A   A function * is defined for all even positive integers n as the number of even factors of n other than n itself.   Column B

          *(48)   *(122)

[Multiple-choice Question – Select One or More Answer Choices]
2. If a and b are integers, and $x = 2 \cdot 3 \cdot 7 \cdot a$, and $y = 2 \cdot 2 \cdot 8 \cdot b$, and the values of both x and y lie between 120 and 130 (not including the two), then $a - b$ could be

    (A)   –2
    (B)   –1
    (C)   0
    (D)   1
    (E)   2

[Numeric Entry Question]
3. The positive integers m and n leave remainders of 2 and 3, respectively, when divided by 6. $m > n$. What is the remainder when $m - n$ is divided by 6?

[Multiple-choice Question – Select One Answer Choice Only]
4. In the figure, triangles ABC and ABD are right triangles. What is the value of x ?

(A) 20
(B) 30
(C) 50
(D) 70
(E) 90

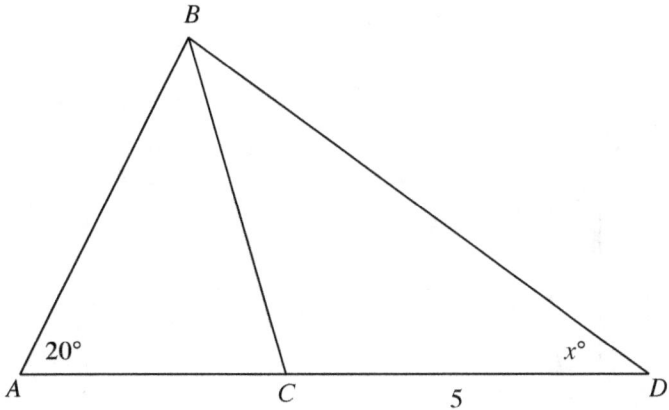

[Multiple-choice Question – Select One Answer Choice Only]
5. Which of the following must be true?

(I) The area of triangle P.
(II) The area of triangle Q.
(III) The area of triangle R.

(A) I = II = III
(B) I < II < III
(C) I > II < III
(D) III < I < II
(E) III > I > II

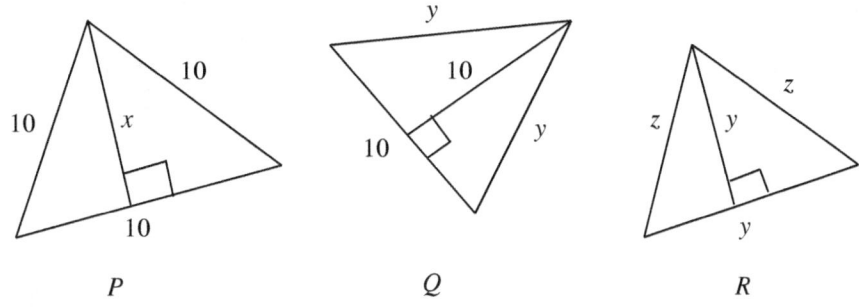

GRE Math Tests

[Multiple-choice Question – Select One Answer Choice Only]
6. In the figure, ABCD and PQRS are rectangles inscribed in the circle shown in the figure. If AB = 5, AD = 3, and QR = 4, then what is the value of *l* ?

  (A) 3
  (B) 4
  (C) 5
  (D) $\sqrt{15}$
  (E) $3\sqrt{2}$

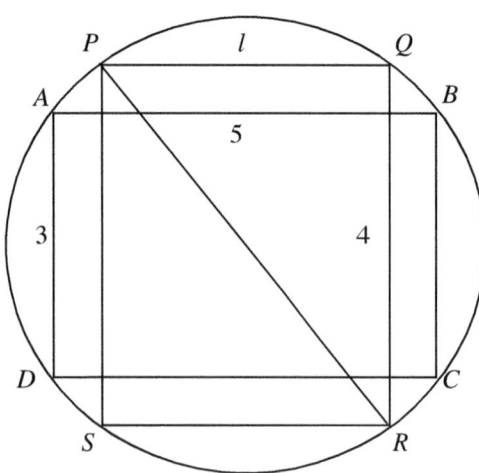

[Quantitative Comparison Question]
7. Column A | Column B
    The clockwise angle made by the hour hand and the minute hand at 12:15pm | The counterclockwise angle made by the hour hand and the minute hand at 12:45pm

Quantitative Comparison Question]

8.   Column A   Column B
      The slope of a line passing   The slope of a line passing
      through (–3, –4) and the origin   through (–5, –6) and the origin

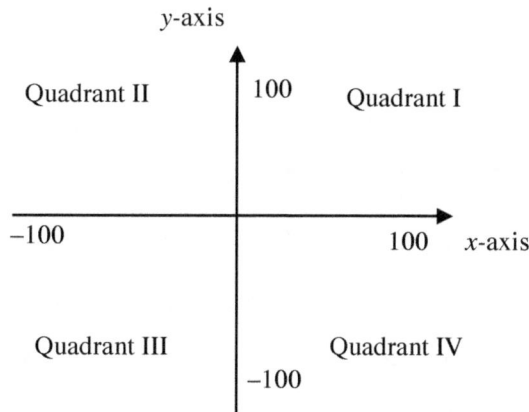

[Multiple-choice Question – Select One or More Answer Choices]

9.   If $a = 3 + b$, which of the following must be true?

   (I) $a > b + 2.5$
   (II) $a < b + 2.5$
   (III) $a > 2 + b$

   (A) I
   (B) II
   (C) III
   (D) I and II
   (E) I and III

GRE Math Tests

[Multiple-choice Question – Select One Answer Choice Only]
10. If $x^2 - y^2 = 16$ and $x + y > x - y$, then which one of the following could $x - y$ equal?

 (A) 3
 (B) 4
 (C) 5
 (D) 6
 (E) 7

Quantitative Comparison Question]
11. Column A $\quad\quad\quad x = 1/y \quad\quad\quad$ Column B

$\dfrac{x^2+1}{x}$ $\quad\quad\quad\quad\quad\quad\quad\quad\quad\quad\quad\quad \dfrac{y^2+1}{y}$

[Multiple-choice Question – Select One Answer Choice Only]
12. If a, b, and c are not equal to 0 or 1 and if $a^x = b$, $b^y = c$, and $c^z = a$, then $xyz =$

 (A) 0
 (B) 1
 (C) 2
 (D) a
 (E) abc

Quantitative Comparison Question]

13.

Column A		Column B
	The average ages of the players on team A and team B, are 20 and 30 years, respectively. The average age of the players of the teams together is 26.	
The number of players on team A		The number of players on team B

[Multiple-choice Question – Select One or More Answer Choices]

14. If $(x - 3)(x + 2) = 0$, then $x =$

    (A)   −3
    (B)   −2
    (C)   0
    (D)   2
    (E)   3

[Multiple-choice Question – Select One or More Answer Choices]

15. The ratio of the numbers a, b, c, d, and e is −6, −2, 1, 3, 5. Which of the three following choices sum to 0 ?

    (A) a
    (B) b
    (C) c
    (D) d
    (E) e

[Multiple-choice Question – Select One or More Answer Choices]
16. Kelvin spends at least $3 a day and Steve spends at least $5 a day. Kelvin spends one additional dollar for each additional dollar spent by Steve. They spend in integer numbers of dollars. Which of the following could be the ratio of the expenditures of Kelvin and Steve ?

   (A) 3/5
   (B) 51/52
   (C) 92/155
   (D) 44/75
   (E) 156/155

[Multiple-choice Question – Select One Answer Choice Only]
17. If $x = 10^{1.4}$, $y = 10^{0.7}$, and $x^z = y^3$, then what is the value of z ?

   (A) 0.5
   (B) 0.66
   (C) 1.5
   (D) 2
   (E) 3

Quantitative Comparison Question]
18. Column A   3/11 of a number is 23.   Column B
     3/5 of the number   60% of the number

[Multiple-choice Question – Select One or More Answer Choices]
19. The value of a share of stock was $30 on Sunday. The profile of the value in the following week was as follows: The value appreciated by $1.2 on Monday. It appreciated by $3.1 on Tuesday. It depreciated by $4 on Wednesday. It appreciated by $2 on Thursday and it depreciated by $0.2 on Friday. On Friday, the stock market closed for the weekend. By what percentage did the value of the share increase in the five days?

(A) 3.2%
(B) 4%
(C) 5.6%
(D) 7%
(E) 10%

[Multiple-choice Question – Select One Answer Choice Only]
20. A project has three test cases. Three teams are formed to study the three different test cases. James is assigned to all three teams. Except for James, each researcher is assigned to exactly one team. If each team has exactly 6 members, then what is the exact number of researchers required?

(A) 12
(B) 14
(C) 15
(D) 16
(E) 18

[Multiple-choice Question – Select One Answer Choice Only]
21. If the nth term in a sequence of numbers $a_0, a_1, a_2, ..., a_n$ is defined to equal $2n + 1$, then what is the numerical difference between the 5th and 6th terms in the sequence?

(A) 1
(B) 2
(C) 4
(D) 5
(E) 6

22.   Column A $x > 0$ Column B

The number of marbles in x jars, each containing 15 marbles, plus the number of marbles in $3x$ jars, each containing 20 marbles

The number of marbles in x jars, each containing 25 marbles, plus the number of marbles in $2x$ jars, each containing 35 marbles

[Multiple-choice Question – Select One Answer Choice Only]

23. A web press prints 5 pages every 2 seconds. At this rate, how many pages will the press print in 7 minutes?

(A) 350
(B) 540
(C) 700
(D) 950
(E) 1050

[Multiple-choice Question – Select One Answer Choice Only]

24. A rancher is constructing a fence by stringing wire between posts 20 feet apart. If the fence is 400 feet long, how many posts must the rancher use?

(A) 18
(B) 19
(C) 20
(D) 21
(E) 22

Answers and Solutions Test 19:

Question	Answer
1.	A
2.	B
3.	5
4.	D
5.	B
6.	E
7.	B
8.	A
9.	A, C, E
10.	A
11.	C
12.	B
13.	B
14.	B, E
15.	A, C, E
16.	A, B
17.	C
18.	C
19.	D
20.	D
21.	B
22.	A
23.	E
24.	D

If you got 18/24 correct on this test, you are likely to get 750+ on the actual GRE **by the time you complete all the tests in the book.**

1. Prime factoring 48 yields $2 \cdot 2 \cdot 2 \cdot 2 \cdot 3$. The even factors of 48 are

$$2$$
$$2 \cdot 2 \; (= 4)$$
$$2 \cdot 2 \cdot 2 \; (= 8)$$
$$2 \cdot 2 \cdot 2 \cdot 2 \; (= 16)$$
$$3 \cdot 2 \; (= 6)$$
$$3 \cdot 2 \cdot 2 \; (= 12)$$
$$3 \cdot 2 \cdot 2 \cdot 2 \; (= 24)$$
$$3 \cdot 2 \cdot 2 \cdot 2 \cdot 2 \; (= 48)$$

Not counting the last factor (48 itself), the total number of factors is 7. Hence, Column A equals 7.

Now, $122 = 2 \cdot 61$ (61 is a prime). So, 2 is the only even factor. Hence, *(122) = 1, which is less than 7 (= Column A). Hence, the answer is (A).

2. We are given that $x = 2 \cdot 3 \cdot 7 \cdot a = 42a$ and $y = 2 \cdot 2 \cdot 8 \cdot b = 32b$.

We are given that the values of both x and y lie between 120 and 130 (not including the two).

The only multiple of 42 in this range is $42 \times 3 = 126$. Hence, $x = 126$ and $a = 3$. The only multiple of 32 in this range is $32 \times 4 = 128$. Hence, $y = 128$ and $b = 4$. Hence, $a - b = 3 - 4 = -1$. The answer is (B) only.

---

3. We are given that the numbers m and n, when divided by 6, leave remainders of 2 and 3, respectively. Hence, we can represent the numbers m and n as $6p + 2$ and $6q + 3$, respectively, where p and q are suitable integers.

Now, $m - n = (6p + 2) - (6q + 3) = 6p - 6q - 1 = 6(p - q) - 1$. A remainder must be positive, so let's add 6 to this expression and compensate by subtracting 6:

$$6(p - q) - 1 =$$

$$6(p - q) - 6 + 6 - 1 =$$

$$6(p - q) - 6 + 5 =$$

$$6(p - q - 1) + 5$$

Thus, the remainder is 5. Enter 5 in the grid.

---

4. In triangle ABC, $\angle A$ is 20°. Hence, the right angle in $\triangle ABC$ is either $\angle ABC$ or $\angle BCA$. If $\angle ABC$ is the right angle, then $\angle ABD$, of which $\angle ABC$ is a part, would be greater than the right angle (90°) and ABD would be an obtuse angle, not a right triangle. So, $\angle ABC$ is not 90° and therefore $\angle BCA$ must be a right angle. Since AD is a line, $\angle ACB + \angle BCD = 180°$. Solving for $\angle BCD$ yields $\angle BCD = 180 - 90 = 90$. Hence, $\triangle BCD$ is also a right triangle, with right angle at C. Since there can be only one right angle in a triangle, $\angle D$ is not a right angle. But, we are given that $\triangle ABD$ is right angled, and from the figure $\angle A$ equals 20°, which is not a right angle. Hence, the remaining angle $\angle ABD$ is right angled. Now, since the sum of the angles in a triangle is 180°, in $\triangle ABC$, we have $20 + 90 + x = 180$. Solving for x yields $x = 70$. The answer is (D).

5. In the figure, triangle *P* is equilateral, with each side measuring 10 units. So, as in any equilateral triangle, the altitude (*x* here) is shorter than any of the other sides of the triangle. Hence, *x* is less than 10. Now,

$$\begin{aligned} \text{I} &= \text{Area of Triangle } P \\ &= 1/2 \times base \times height \\ &= 1/2 \times 10 \times x \\ &= 5x \text{ and this is less than 50, since } x \text{ is less than 10} \end{aligned}$$

Triangle *Q* has both base and altitude measuring 10 units, and the area of the triangle is $1/2 \times base \times height = 1/2 \times 10 \times 10 = 50$. So, II = 50. We have one more detail to pick up: *y*, being the hypotenuse in the right triangle in figure *Q*, is greater than any other side of the triangle. Hence, *y* is greater than 10, the measure of one leg of the right triangle.

Triangle *R* has both the base and the altitude measuring *y* units. Hence,

$$\begin{aligned} \text{II} &= \text{Area of the Triangle } R \\ &= 1/2 \times base \times height \\ &= 1/2 \times y \times y \\ &= (1/2)y^2, \text{ and this is greater than } 1/2 \times 10^2 \text{ (since } y > 10\text{), which equals 50} \end{aligned}$$

Summarizing, the three results I < 50, II = 50, and III > 50 into a single inequality yields I < II < III. The answer is (B).

---

6. *PQRS* is a rectangle inscribed in the circle. Hence, its diagonal *PR* must pass through the center of the circle. So, *PR* is a diameter of the circle.

Similarly, *AC* is a diagonal of the rectangle *ABCD*, which is also inscribed in the same circle. Hence, *AC* must also be a diameter of the circle. Since the diameters of a circle are equal, *PR* = *AC*.

Applying The Pythagorean Theorem to $\triangle ABC$ yields $AC^2 = AB^2 + BC^2 = 5^2 + 3^2 = 25 + 9 = 34$. Hence, PR^2 also equals 34. Now, applying The Pythagorean Theorem to $\triangle PQR$ yields

$$PQ^2 + QR^2 = PR^2$$
$$l^2 + 4^2 = 34$$
$$l^2 + 16 = 34$$
$$l^2 = 18$$
$$l = \sqrt{18}$$
$$l = 3\sqrt{2}$$

The answer is (E).

---

7. Draw sample pictures of the clock at 12:15pm and 12:45pm. The figures look like

From the figures, the clockwise angle between the hands at 12:15pm is less than 90°, and at 12:45pm the counterclockwise angle is more than 90°. Hence, Column B is greater. The answer is (B).

8. The formula for the slope of a line passing through two points (x_1, y_1) and (x_2, y_2) is $\frac{y_2 - y_1}{x_2 - x_1}$. Using this formula to calculate the slope in each column yields

Column A: Slope of the line through $(-3, -4)$ and the origin $(0, 0)$ is $\frac{0 - (-4)}{0 - (-3)} = \frac{4}{3}$.

Column B: Slope of the line through $(-5, -6)$ and the origin $(0, 0)$ is $\frac{0 - (-6)}{0 - (-5)} = \frac{6}{5}$.

Since $4/3 > 6/5$, Column A is greater than Column B. The answer is (A).

9. We are given that $a = 3 + b$. This equation indicates that a is 3 units larger than b, so a is greater than $b + 2.5$. Hence, (A) is true, and (B) is false.

Similarly, a is greater than $2 + b$. Hence, (C) is true.

Choice (D) includes Choice (B), which we already know is false, reject.

Choice (E) includes (A) and (C), which we already know are true, accept.

Hence, the answer is (A), (C), and (E).

10. Method I (substitution): First factor the left side of the equation $x^2 - y^2 = 16$:

$$(x + y)(x - y) = 16$$

Now, replace $x - y$ in this equation with each of the answer-choices until we find the one that works. Let's start with Choice (E), $x - y = 7$:

$$(x + y)(7) = 16$$

We are given that $x + y > x - y$, so $x + y$ must be greater than 7. Now, multiplying 7 by a number that is greater than 7 gives a number much larger than 16, so this equation is impossible. A similar analysis shows that choices (B), (C), and (D) are impossible. For Choice (A), the equation becomes

$$(x + y)(3) = 16$$

This equation is possible:

$$\left(\frac{16}{3}\right)(3) = 16$$

And $16/3 > 3$, satisfying the inequality $x + y > x - y$. The answer is (A).

Method II: The choices given are positive. Multiplying both sides of the given inequality $x + y > x - y$ by the positive value $x - y$ yields

$$(x + y)(x - y) > (x - y)^2$$
$$x^2 - y^2 > (x - y)^2$$
$$16 > (x - y)^2$$
$$\sqrt{16} > x - y$$
$$4 > x - y$$

Since choice (A) is one such suitable choice, the answer is (A).

Test 19 — Solutions

11. Substituting $1/y$ for x in Column A yields

$$\frac{\left(\frac{1}{y}\right)^2 + 1}{\frac{1}{y}} =$$

$$\frac{\frac{1}{y^2} + 1}{\frac{1}{y}} =$$

$$\frac{\frac{1+y^2}{y^2}}{\frac{1}{y}} =$$

$$\frac{1+y^2}{y^2} \cdot \frac{y}{1} =$$

$$\frac{1+y^2}{y} =$$

Column B

The answer is (C).

12. We are given three equations $a^x = b$, $b^y = c$, and $c^z = a$. From the first equation, we have $b = a^x$. Substituting this in the second equation gives $(a^x)^y = c$. We can replace a in this equation with c^z (according to the third equation $c^z = a$):

$$\left(\left(c^z\right)^x\right)^y = c$$

$c^{xyz} = c^1$ By multiplying the exponents and writing c as c^1
$xyz = 1$ By equating the exponents of c on both sides

The answer is (B).

13. Let the number of players on team A be a and the number of players on team B be b. Since the average age of the players on team A is 20, the sum of the ages of the players on the team is $20a$. Similarly, since the average age of the players on team B is b, the sum of the ages of the players on the team is $30b$.

Now, the average age of the players of the two teams together is

$$\frac{\text{Sum of the ages of the players on each team}}{\text{Total number of players on the teams}} =$$

$$\frac{20a + 30b}{a+b}$$

We are given that this average is 26. Hence, we have

$\frac{20a + 30b}{a+b} = 26$
$20a + 30b = 26a + 26b$ by multiplying both sides by $a+b$
$4b = 6a$ by subtracting $20a + 26b$ from both sides
$b/a = 6/4 = 3/2$

Since a and b are both positive (being the team strengths) and since b/a equals 3/2, which is greater than 1, b must be greater than a. Hence, Column B is greater than Column A, and the answer is (B).

14. Since the equation $(x - 3)(x + 2) = 0$ is of degree 2, the solutions are $x = 3$ and $x = -2$. Both yield a zero on the left-hand side. Expanding the above equation yields $x^2 - x - 6 = 0$. Since the degree is 2, there are two, one, or no solutions. Hence, just choose (B) and (E) and you are safe.

---

15. We are given that $a : b : c : d : e = -6 : -2 : 1 : 3 : 5$.

Let $a = -6k$, $b = -2k$, $c = k$, $d = 3k$, and $e = 5k$.

Now, the only three choices that sum to 0 are (A), (C), and (E). Select them.

---

16. Suppose Steve spends $5 + n$ dollars. Then, Kelvin spends $3 + n$ dollars. Hence, the ratio of the expenditures of Kelvin to Steve would be $(3 + n)/(5 + n)$.

Now, let's test which choices are compatible with the expression $(3 + n)/(5 + n)$. Since Kelvin and Steve spend in positive integer number of dollars, we have to select the answer-choices that yield a positive integer for n. The rest of the ratios never occur.

- (A) $3/5 = (3 + n)/(5 + n)$; $15 + 3n = 15 + 5n$; $2n = 0$, a positive integer. Accept.
- (B) $51/52 = (3 + n)/(5 + n)$; $255 + 51n = 156 + 52n$; $n = 255 - 156 = 99$, a positive integer. Accept.
- (C) $92/155 = (3 + n)/(5 + n)$; $460 + 92n = 465 + 155n$; $92n - 155n = 465 - 460$; $-63n = 5$; $n = -5/63$, not a positive integer. Reject.
- (D) $44/75 = (3 + n)/(5 + n)$; $220 + 44n = 225 + 75n$; $44n - 75n = 225 - 220$; $-31n = 5$; $n = -5/31$, not a positive integer. Reject.
- (E) $156/155 = (3 + n)/(5 + n)$; $780 + 156n = 465 + 155n$; $n = 465 - 780 = -315$, not a positive integer. Reject.

The answer is (A) and (B).

---

17. We are given that $x = 10^{1.4}$ and $y = 10^{0.7}$. Substituting these values in the given equation $x^z = y^3$ yields

$(10^{1.4})^z = (10^{0.7})^3$
$10^{1.4z} = 10^{0.7 \cdot 3}$
$10^{1.4z} = 10^{2.1}$
$1.4z = 2.1$ since the bases are the same, the exponents must be equal
$z = 2.1/1.4 = 3/2$

The answer is (C).

---

18. 60% of a number is $60/100 = 3/5$ times the number. Hence, 60% of a number is 3/5 of the number. So, the columns are equal. The answer is (C).

Note that we did not need the statement "3/11 of a number is 23" to solve the problem.

Test 19 — Solutions

19. The initial price of the share is $30.

> After the $1.2 appreciation on Monday, its price was 30 + 1.2 = $31.2.
> After the $3.1 appreciation on Tuesday, its price was 31.2 + 3.1 = $34.3.
> After the $4 depreciation on Wednesday, its price was 34.3 – 4 = $30.3.
> After the $2 appreciation on Thursday, its price was 30.3 + 2 = $32.3.
> After the $0.2 depreciation on Friday, its price was 32.3 – 0.2 = $32.1.

The percentage increase in the price from the initial price is

$$(32.1 - 30)/30 \times 100 =$$
$$2.1/30 \times 100 =$$
$$2.1/3 \times 10 =$$
$$21/3 = 7$$

The answer is (D).

20. Since James is common to all three teams, he occupies one of six positions in each team. Since any member but James is with exactly one team, 5 different researchers are required for each team. Hence, apart from James, the number of researchers required is $5 \cdot 3 = 15$. Including James, there are $15 + 1 = 16$ researchers. The answer is (D).

21. We have the rule $a_n = 2n + 1$. By this rule,

$a_4 = 2(4) + 1 = 9$ (Note: The fifth term is a_4 because the sequence starts at a_0, so the first term is a_0, the second term is a_1, etc.)

$a_5 = 2(5) + 1 = 11$

Forming the difference $a_5 - a_4$ yields

$a_5 - a_4 = 11 - 9 = 2$

The answer is (B).

Note: Since the sequence is arithmetic, $a_6 - a_5$ will also equal 2. So, if you mistakenly subtracted the seventh and sixth terms, you will still get the correct answer.

22. In Column A, the x jars have $15x$ marbles, and $3x$ jars have $20 \cdot 3x = 60x$ marbles. Hence, Column A has a total of $15x + 60x = 75x$ marbles. Now, in Column B, the x jars have $25x$ marbles, and $2x$ jars have $35 \cdot 2x = 70x$ marbles. Hence, Column B has a total of $25x + 70x = 95x$ marbles. Thus, Column B is larger, and the answer is (B).

23. Since there are 60 seconds in a minute and the press prints 5 pages every 2 seconds, the press prints $5 \cdot 30 = 150$ pages in one minute. Hence, in 7 minutes, the press will print

$$7 \cdot 150 = 1050 \text{ pages}$$

The answer is (E).

GRE Math Tests

24. Since the fence is 400 feet long and the posts are 20 feet apart, there are 400/20 = 20 sections in the fence. Now, if we ignore the first post and associate the post at the end of each section with that section, then there are 20 posts (one for each of the twenty sections). Counting the first post gives a total of 21 posts. The answer is (D).

Test 20

GRE Math Tests

Questions: 24
Time: 45 minutes

[Quantitative Comparison Question]
1.   Column A $-1 < x < 0$ Column B
     x $1/x$

[Multiple-choice Question – Select One Answer Choice Only]
2. Define x^* by the equation $x^* = \pi/x$. Then $((-\pi)^*)^* =$

   (A) $-1/\pi$
   (B) $-1/2$
   (C) $-\pi$
   (D) $1/\pi$
   (E) π

[Numeric Entry Question]
3. If the least common multiple of m and n is 24, then what is the first integer larger than 3070 that is divisible by both m and n?

[Multiple-choice Question – Select One or More Answer Choices]
4. If *p* and *q* are two even integers and $0.05p$ and $0.07q$ lie between 0.69 and 0.78. Which of the following could $0.05p + .07q$ equal?

   (A) $0.69 + 0.78$
   (B) $0.7 + 0.77$
   (C) $0.7 + 0.78$
   (D) $0.7 + 0.7$
   (E) $0.7 + 0.78$

USE THIS SPACE FOR SCRATCHWORK.

Quantitative Comparison Question]
5.   Column A

Everyone who passes the test will be awarded a degree. The probability that Tom passes the test is 0.5, and the probability that John passes the test is 0.4. The two events are independent of each other.

Column B

The probability that both Tom and John get the degree

The probability that at least one of them gets the degree

USE THIS SPACE FOR SCRATCHWORK.

Quantitative Comparison Question]
6.   Column A   Column B
       Perimeter of a rectangle   Perimeter of a triangle
       with an area of 10   with an area of 10

USE THIS SPACE FOR SCRATCHWORK.

[Multiple-choice Question – Select One or More Answer Choices]
7. In the figure, *AD* and *BC* are lines intersecting at *O*. *a* is equal to?

    (A)   15
    (B)   30
    (C)   45
    (D) $x/2$
    (E) $y/2 - x$

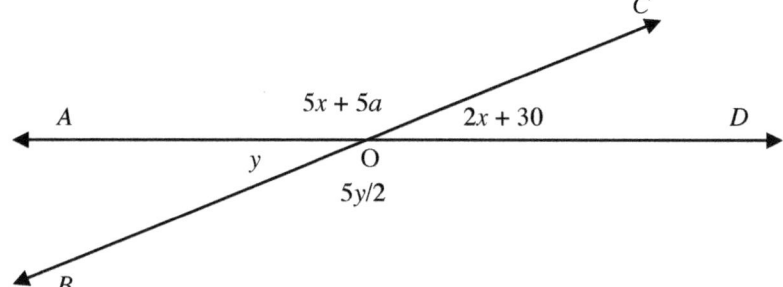

[Numeric Entry Question]
8. In the figure, AB and CD are the diameters of the circle. What is the value of x?

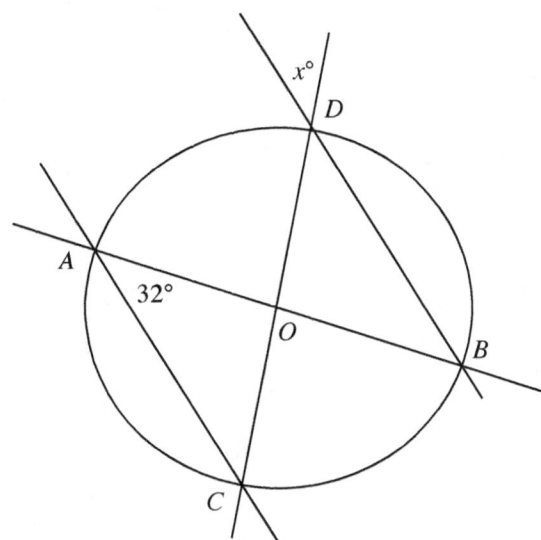

[Numeric Entry Question]
9. In the figure, what is the value of x?

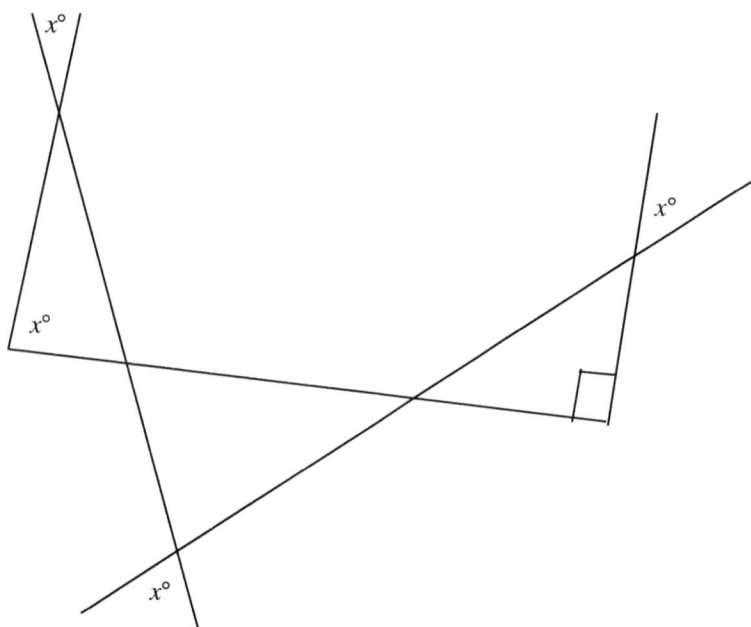

[Multiple-choice Question – Select One or More Answer Choices]
10. If exactly two of the choices below must be true about a triangle, which of the following must they be?

    (A) The angles of $\triangle ABC$ are in the ratio is $1 : 2 : 3$.
    (B) The sides of $\triangle ABC$ are in the ratio $1 : 2 : 3$.
    (C) One of the angles of $\triangle ABC$ equals the sum of the other two angles.
    (D) One of the sides of $\triangle ABC$ equals the sum of the other two sides.
    (E) The triangle is right angled.

Quantitative Comparison Question]
11. Column A $a > b > 0$ Column B
     $\dfrac{a-b}{a+b}$ $\dfrac{a^2-b^2}{a^2+b^2}$

Quantitative Comparison Question]
12. Column A $x/a > 4$ and $y/a < -6$ Column B
                   $a^2 = 9$
                   $ab^2 = -8$
     x y

GRE Math Tests

[Quantitative Comparison Question]
13. Column A

$\dfrac{2}{10} + \dfrac{3}{100} + \dfrac{4}{1000} + \dfrac{5}{10000}$

Column B

$\dfrac{4}{10} + \dfrac{3}{100} + \dfrac{2}{1000}$

[Multiple-choice Question – Select One Answer Choice Only]
14. In a country, 60% of the male citizen and 70% of the female citizen are eligible to vote. 70% of male citizens eligible to vote voted, and 60% of female citizens eligible to vote voted. What fraction of the citizens voted during the election?

(A) 0.42
(B) 0.48
(C) 0.49
(D) 0.54
(E) 0.60

[Multiple-choice Question – Select One Answer Choice Only]
15. a, b, and c are three different numbers, none of which equals the average of the other two. If $\dfrac{x}{b-c} = \dfrac{y}{c-a} = \dfrac{z}{a-b}$, then $x + y + z =$

(A) 0
(B) 1/2
(C) 1/3
(D) 2/3
(E) 3/4

Test 20 − Questions

[Quantitative Comparison Question]

16. Column A
The average of five consecutive integers starting from *m*

Column B
The average of six consecutive integers starting from *m*

[Multiple-choice Question – Select One or More Answer Choices]

17. 40% of employees in a factory earn a daily wage of $385. Some employees earn a daily wage of $395, and the remaining employees earn a daily wage of $405. Based on the given data, which of the following could be the average daily wage of the employees at the factory?

(A) 385
(B) 390
(C) 391
(D) 394
(E) 395
(F) 397
(G) 400
(H) 405
(I) 410

[Quantitative Comparison Question]

18. Column A

$\dfrac{a+10}{b+10}$

a and *b* are positive.
$(a + 6) : (b + 6) = 5 : 6$

Column B

1

391

GRE Math Tests

---

19. If two workers can assemble a car in 8 hours and a third worker can assemble the same car in 12 hours, then how long would it take the three workers together to assemble the car?

    (A) 5/12 hrs.
    (B) 2 2/5 hrs.
    (C) 2 4/5 hrs.
    (D) 3 1/2 hrs.
    (E) 4 4/5 hrs.

USE THIS SPACE FOR SCRATCHWORK.

Quantitative Comparison Question]

20. Column A
$$\dfrac{(2x-11)(2x+11)}{4}$$

Column B
$(x-11)(x+11)$

USE THIS SPACE FOR SCRATCHWORK.

Quantitative Comparison Question]

21. Column A

Miller sold apples at 125% of what it cost him.

Column B

The profit made by selling 100 apples

The profit made by selling 200 apples at a further 10% discount

USE THIS SPACE FOR SCRATCHWORK.

[Multiple-choice Question – Select One Answer Choice Only]

22. John had $42. He purchased fifty mangoes and thirty oranges with the whole amount. He then chose to return six mangoes for nine oranges as both quantities are equally priced. What is the price of each Mango?

   (A) 0.4
   (B) 0.45
   (C) 0.5
   (D) 0.55
   (E) 0.6

[Multiple-choice Question – Select One Answer Choice Only]

23. For how many positive integers n is it true that the sum of $13/n$, $18/n$, and $29/n$ is an integer?

   (A) 6
   (B) 60
   (C) Greatest common factor of 13, 18, and 29
   (D) Least common multiple of 13, 18, and 29
   (E) 12

[Multiple-choice Question – Select One Answer Choice Only]

24. Each Employee at a certain bank is either a clerk or an agent or both. Of every three agents, one is also a clerk. Of every two clerks, one is also an agent. What is the probability that an employee randomly selected from the bank is both an agent and a clerk?

   (A) 1/2
   (B) 1/3
   (C) 1/4
   (D) 1/5
   (E) 2/5

Answers and Solutions Test 20:

Question	Answer
1.	A
2.	C
3.	3072
4.	D
5.	B
6.	D
7.	A, E
8.	32
9.	45
10.	B, D
11.	B
12.	B
13.	B
14.	A
15.	A
16.	B
17.	D, E
18.	B
19.	A
20.	A
21.	C
22.	E
23.	E
24.	C

If you got 18/24 correct on this test, you are likely to get 750+ on the actual GRE by the time you complete all the tests in the book.

1. There is only one type of number between -1 and 0—negative fractions. So we need only choose one number, say, $x = -1/2$. Then $\dfrac{1}{x} = \dfrac{1}{-1/2} = -2$. Now, $-1/2$ is larger than -2 (since $-1/2$ is to the right of -2 on the number line). Hence, Column A is larger, and the answer is (A).

2. Working from the inner parentheses out, we get

$$((-\pi)^*)^* =$$

$$(\pi/(-\pi))^* =$$

$$(-1)^* =$$

$$\pi/(-1) =$$

$$-\pi$$

The answer is (C).

Method II:
We can rewrite this problem using ordinary function notation. Replacing the odd symbol $x*$ with $f(x)$ gives $f(x) = \pi/x$. Now, the expression $((-\pi)*)*$ becomes the ordinary composite function

$$f(f(-\pi)) =$$

$$f(\pi/(-\pi)) =$$

$$f(-1) =$$

$$\pi/(-1) =$$

$$-\pi$$

---

3. Any number divisible by both m and n must be a multiple of the least common multiple of the two numbers, which is given to be 24. The first multiple of 24 greater than 3070 is 3072. Hence, enter 3072 in the grid.

---

4. Multiplying the statement "$0.05p$ and $0.07q$ lie between 0.69 and 0.78" by 100 indicates that $5p$ and $7q$ lie between 69 and 78.

The only multiples of 5 between 69 and 78 are 70 and 75. Hence, $5p =$ either 70 or 75; $p =$ either $70/5 = 14$, is even (hence acceptable) or $75/5 = 15$, not even (Reject).
Since $p = 14$, $0.05p = 0.70$.

The only multiples of 7 between 69 and 78 are 70 and 77. Hence, $7q = 70$ or 77. Now, p could equal even value of $70/7 = 10$ or $77/7 = 11$. Reject 11, for it is not even.
Since $q = 10$, $0.07q = 0.70$.

The answer is $0.05p + .07q = 0.7 + 0.7 = 1.4$. The answer is (D) only.

---

5. The case of both Tom and John getting a degree is just one of the cases in which at least one of them gets the degree (Column A is one of the cases of Column B). Hence, the probability of the former is less than the probability of the later (Column B is greater). Also, the probability of the remaining case (exactly one of the two passing) is not zero. So, Column A cannot equal Column B. The answer is (B).

---

6. The eye-catcher is Column A since one expects the perimeter of a rectangle to be longer than that of a triangle of similar size. However, by making the base of the triangle progressively longer, we can make the perimeter of the triangle as long as we want. The following diagram displays a rectangle and a triangle with the same area, yet the triangle's perimeter is longer than the rectangle's:

The answer is (D).

7. Equating vertical angles ∠AOB and ∠COD in the figure yields $y = 2x + 30$. Also, equating vertical angles ∠AOC and ∠BOD yields $5y/2 = 5x + 5a$. Multiplying this equation by 2/5 yields $y = 2x + 2a$. Subtracting this equation from the equation $y = 2x + 30$ yields $2a = 30$. Hence, $a = 30/2 = 15$, and the answer is (A). Subtracting $2x$ from both sides of the equation $y = 2x + 2a$ yields $y - 2x = 2a$; $y/2 - 2x/2 = a$; $a = y/2 - x$. Also, choose (E). The choices to choose are (A) and (E).

8. *OA* and *OC* are radii of the circle and therefore equal. Hence, the angles opposite the two sides in △AOC are equal: ∠C = ∠A = 32° (from the figure). Now, summing the angles of the triangle to 180° yields ∠A + ∠C + ∠AOC = 180 or 32 + 32 + ∠AOC = 180. Solving the equation for ∠AOC, we have ∠AOC = 180 − (32 + 32) = 180 − 64 = 114.

Since ∠BOD and ∠AOC vertical angles, they are equal. Hence, we have ∠BOD = ∠AOC = 114.

Now, *OD* and *OB* are radii of the circle and therefore equal. Hence, the angles opposite the two sides in △BOD are equal: ∠B = ∠D. Now, summing the angles of the triangle to 180° yields ∠B + ∠D + ∠BOD = 180 or ∠D + ∠D + ∠BOD = 180 or 2∠D + 114 = 180. Solving the equation yields ∠D = 32. Since ∠D and angle $x°$ are vertical angles, x also equals 32.

Enter the value 32 in the grid.

9. Let's name the vertices as shown in the figure

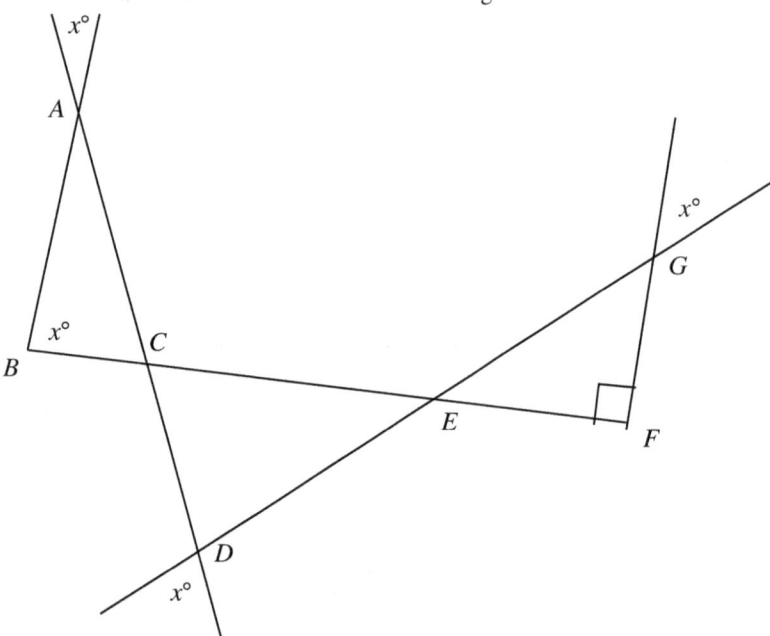

Let's start evaluating the unknown angles of the triangles in the figure from left most located triangle through the right most located triangle. Then the value of x can be derived by summing the angles of right most located triangle to 180°. This is done as follows:

In the first triangle from the left: △ABC, we have from the figure that ∠B = $x°$ and ∠A = $x°$ (vertical angles are equal). Summing the angles of the triangle to 180° yields $x + x + ∠C = 180$. Solving the equation for ∠C yields ∠C = $180 - 2x$.
In the second triangle from the left: △CED, we have from the figure that ∠D = $x°$ (vertical angles are equal), and we have ∠C [in △CED] = ∠C in △ACB (They are vertical angles) = $180 - 2x$ (Known result).

Test 20 — Solutions

Now, summing the angles of the triangle to 180° yields $(180 - 2x) + x + \angle E = 180$. Solving the equation for $\angle E$ yields $\angle E = x$.

In the third triangle from the left: $\triangle GFE$, we have from the figure that $\angle G = x°$ (vertical angles are equal) and $\angle E$ [in $\triangle GFE$] = $\angle E$ in $\triangle CED$ (Vertical angles are equal) = $x°$ (Known result). Now, we also have $\angle F = 90°$ (From the figure). Now, summing the three angles of the triangle to 180° yields $x + x + 90 = 180$. Solving for x yields $2x = 90$ or $x = 45$. Enter 45 in the grid.

---

10. Choice (A): The angles of $\triangle ABC$ are in the ratio is 1 : 2 : 3.
This indicates that $\triangle ABC$ is a right triangle.

Choice (B): The sides of $\triangle ABC$ are in the ratio 1 : 2 : 3.
This indicates that $\triangle ABC$ is not a right triangle.

Choice (C): One of the angles of $\triangle ABC$ equals the sum of the other two angles.
This indicates that $\triangle ABC$ is a right triangle.

Choice (D): One of the sides of $\triangle ABC$ equals the sum of the other two sides.
This indicates that $\triangle ABC$ is not a right triangle.

Choice (E): The triangle is right angled.

The Choices (A), (C), and (E) are mutually compatible and the choices (B) and (D) are not mutually compatible.

But the first set [Choices (A), (C), and (E)], and the second set [Choices (B), and (D)] are not mutually incompatible.

Hence, either the choices (A), (C), and (E) are true or the choices (B), and (D) are true. But since we are given that exactly two of the choices (A), (B), (C), (D), and (E) are true, the set to choose would be the second one.

The choices to select are (B) and (D).

11. Factoring Column B yields

Column A		Column B
$\dfrac{a-b}{a+b}$	$a > b > 0$	$\dfrac{(a-b)(a+b)}{a^2+b^2}$

Since $a > b$, $a - b > 0$. Hence, we can cancel $a - b$, a positive value, from both columns:

Column A		Column B
$\dfrac{1}{a+b}$	$a > b > 0$	$\dfrac{a+b}{a^2+b^2}$

Since $a > b > 0$, both a and b are positive. Hence, the Least Common Denominator of the fractions, $(a + b)(a^2 + b^2)$, is positive. So, we can multiply both columns by the LCD to clear the fractions:

Column A		Column B
$a^2 + b^2$	$a > b > 0$	$(a + b)(a + b)$

Performing the multiplication in Column B yields

Column A		Column B
$a^2 + b^2$	$a > b > 0$	$a^2 + b^2 + 2ab$

Subtracting $a^2 + b^2$ from both columns yields

Column A		Column B
0	$a > b > 0$	$2ab$

Since both a and b are positive, $2ab$ is positive. Hence, Column B is greater than Column A, and the answer is (B).

---

12. The possible solutions of the equation $a^2 = 9$ are $a = 3$ and $a = -3$.

Since the square of any nonzero number is positive, b^2 must be positive. We are given that $ab^2 = -8$, a negative number. Since the product of a and b^2 is negative, a must be negative. The negative solution of a is $a = -3$.

Substituting this value of a in the given inequality $x/a > 4$ yields

$\dfrac{x}{-3} > 4$

$x < -12$ by multiplying both sides by the negative number -3 and flipping the direction of the inequality

Hence, Column A is less than -12.

Now, replacing a with -3 in the inequality $y/a < -6$ yields

$\dfrac{y}{-3} < -6$

$y > 18$ by multiplying both sides by -3 and flipping the direction of the inequality

Hence, y is positive and therefore greater than x. Hence, Column B is greater, and the answer is (B).

Test 20 — Solutions

---

13. First, cancel (subtract) the common term 3/100 from both columns:

$\dfrac{2}{10} + \dfrac{4}{1000} + \dfrac{5}{10000}$	$\dfrac{4}{10} + \dfrac{2}{1000}$

Next, multiply both columns by 10000 to clear the fractions:

$2000 + 40 + 5$	$4000 + 20$

Finally, add the numbers:

2045	4020

The answer is (B).

Method II
Column A equals $2/10 + 3/100 + 4/1000 + 5/10000 = 0.2345$.

Column B equals $4/10 + 3/100 + 2/1000 = 0.432$.

Since 0.432 is greater than 0.2345, Column B is greater. The answer is (B).

---

14. Let the number of male and female citizens in the country be m and f, respectively.

Now, 60% of the male citizens are eligible to vote, and 60% of m is $60m/100$. 70% of female citizens are eligible to vote, and 70% of f is $70f/100$.

We are given that 70% of male citizens eligible to vote voted:

$$70\% \text{ of } 60m/100 \text{ is } \frac{70}{100} \times \frac{60m}{100} = \frac{70 \times 60m}{10{,}000} = 0.42m$$

We are also given that 60% of female citizens eligible to vote voted:

$$60\% \text{ of } 70f/100 \text{ is } \frac{60}{100} \times \frac{70f}{100} = \frac{60 \times 70f}{10{,}000} = 0.42f$$

So, out of the total $m + f$ citizens, the total number of voters who voted is

$$0.42m + 0.42f = 0.42(m + f)$$

Hence, the required fraction is

$$\frac{0.42(m+f)}{m+f} = 0.42$$

The answer is (A).

GRE Math Tests

15. Let each expression in the equation $\dfrac{x}{b-c} = \dfrac{y}{c-a} = \dfrac{z}{a-b}$ equal k. Then we have

$$\dfrac{x}{b-c} = \dfrac{y}{c-a} = \dfrac{z}{a-b} = k$$

Or

$$\dfrac{x}{b-c} = k$$

$$\dfrac{y}{c-a} = k$$

$$\dfrac{z}{a-b} = k$$

Simplifying yields

$$x = k(b - c) = kb - kc$$
$$y = k(c - a) = kc - ka$$
$$z = k(a - b) = ka - kb$$

Hence, $x + y + z = (kb - kc) + (kc - ka) + (ka - kb) = 0$. The answer is (A).

16. Column A: The five consecutive integers starting from m are $m, m + 1, m + 2, m + 3$, and $m + 4$. The average of the five numbers equals

$$\dfrac{\text{The sum of the five numbers}}{5} =$$
$$\dfrac{m + (m+1) + (m+2) + (m+3) + (m+4)}{5} =$$
$$\dfrac{5m + 10}{5} =$$
$$m + 2$$

Column B: The six consecutive integers starting from m are $m, m + 1, m + 2, m + 3, m + 4$, and $m + 5$. The average of the six numbers equals

$$\dfrac{\text{The sum of the six numbers}}{6} =$$
$$\dfrac{m + (m+1) + (m+2) + (m+3) + (m+4) + (m+5)}{6} =$$
$$\dfrac{6m + 15}{6} =$$
$$m + \dfrac{5}{2} =$$
$$m + 2 + \dfrac{1}{2} =$$
$$(m + 2) + \dfrac{1}{2} =$$
$$\text{Column A} + \dfrac{1}{2}$$

Since Column B is 1/2 units greater than Column A, the answer is (B).

Test 20 — Solutions

Method II:
Choose any five consecutive integers, say, −2, −1, 0, 1 and 2. (We chose these particular numbers to make the calculation as easy as possible. But any five consecutive integers will do. For example, 1, 2, 3, 4, and 5.) Forming the average yields $\frac{-1+(-2)+0+1+2}{5} = \frac{0}{5} = 0$. Now, add 3 to the set to form 6 consecutive integers: −2, −1, 0, 1, 2, and 3. Forming the average yields

$\frac{-1+(-2)+0+1+2+3}{6} =$

$\frac{[-1+(-2)+0+1+2]+3}{6} =$

$\frac{[0]+3}{6} =$ since the average of −1 + (−2) + 0 + 1 + 2 is zero, their sum must be zero

$\frac{3}{6} =$

$\frac{1}{2}$

Since 1/2 > 0, Column B is greater than Column A and the answer is (B).

---

17. We are given that 40% of employees in a factory earn $385 a day as daily wage.

Assuming all the remaining 60% of employees earn $395 a day, the average is 391.
Assuming all the remaining 60% of the employees earn $405 a day, the wage is 397.

So, the range of possible values is 391 through 397, not either number (not all the remaining employees earn $391 only or not all the remaining employees earn $397 only).

Select the answer-choices falling in the region (greater than 391 and less than 397).

The answer is (D) and (E).

---

18. Since a and b are positive, $a + 6$ and $b + 6$ are positive. From the ratio $(a + 6) : (b + 6) = 5 : 6$, we get $\frac{a+6}{b+6} = \frac{5}{6}$. Since 5/6 < 1, $\frac{a+6}{b+6} < 1$. Multiplying both sides of this inequality by $b + 6$, which is positive, yields $a + 6 < b + 6$. Adding 4 to both sides yields $a + 10 < b + 10$. Since b is positive, $b + 10$ is positive. Dividing the inequality by $b + 10$ yields $\frac{a+10}{b+10} < 1$. Hence, Column A < Column B, and the answer is (B).

Method II: Since b is positive, $b + 10$ is positive. So, we can multiply both columns by $b + 10$, which yields

$a + 10$ $b + 10$

Subtracting 10 from both columns yields

a b

We have reduced the problem to comparing the sizes of a and b. Let's solve the equation $\frac{a+6}{b+6} = \frac{5}{6}$ for b.

Multiplying both sides by the LCD $6(b + 6)$ yields $6(a + 6) = 5(b + 6)$, or $6a + 36 = 5b + 30$, or $5b = 6a + 6$, or $5b = 6(a + 1)$, or $b = (6/5)(a + 1)$. This equation says that to make a as large as b you must add 1 to it and then multiply it by 6/5, a number bigger than 1. Hence, b is greater than a. So, Column B is greater than Column A, and the answer is (B).

GRE Math Tests

19. The fraction of work done in 1 hour by the first two people working together is 1/8. The fraction of work done in 1 hour by the third person is 1/12. When the three people work together, the total amount of work done in 1 hour is 1/8 + 1/12 = 5/24. The time taken by the people working together to complete the job is

$$\frac{1}{\text{fraction of work done per unit time}} =$$

$$\frac{1}{5/24} =$$

$$\frac{24}{5} =$$

$$4\frac{4}{5}$$

The answer is (E).

20. Applying the Difference of Squares formula $(a + b)(a - b) = a^2 - b^2$ yields

Column A	Column B
$\dfrac{(2x)^2 - 11^2}{4}$	$x^2 - 11^2$

Column A	Column B
$\dfrac{4x^2 - 121}{4}$	$x^2 - 121$

Column A	Column B
$x^2 - \dfrac{121}{4}$	$x^2 - 121$

Subtracting x^2 from both columns yields

Column A	Column B
$-121/4$	-121

Since $-121/4 > -121$, Column A is greater than Column B and the answer is (A).

Test 20 — Solutions

21. Let each apple cost Miller x dollars. Since he sold the apples at 125% of the cost, the profit made is

$$\text{Selling price} - \text{Cost} =$$
$$(125/100)x - x =$$
$$5x/4 - x =$$
$$x/4$$

The profit on 100 apples is $100 \cdot x/4 = 25x$. Hence, Column A equals $25x$.

Now, after a 10% discount on the selling price, Mr. Miller must be selling the apples at a price equal to

$$(\text{actual selling price}) \cdot \left(1 - \frac{\text{discount percent}}{100}\right) = \left(\frac{5x}{4}\right) \cdot \left(1 - \frac{10}{100}\right) = \left(\frac{5x}{4}\right) \cdot \left(\frac{9}{10}\right) = \frac{9x}{8}$$

Hence, the profit made on each apple equals Selling price − Cost = $9x/8 - x = x/8$. The profit on 200 apples is $200 \cdot x/8 = 25x$ = Column B.

Since the columns are equal, the answer is (C).

22. Since 6 mangoes are returnable for 9 oranges, if each mango costs m and each orange costs n, then $6m = 9n$, or $2m = 3n$. Solving for n yields, $n = 2m/3$. Now, since 50 mangoes and 30 oranges together cost 42 dollars,

$$50m + 30n = 42$$
$$50m + 30(2m/3) = 42$$
$$m(50 + 30 \cdot 2/3) = 42$$
$$m(50 + 20) = 42$$
$$70m = 42$$
$$m = 42/70 = 6/10 = 0.6$$

The answer is (E).

23. The sum of $13/n$, $18/n$, and $29/n$ is $\dfrac{13+18+29}{n} = \dfrac{60}{n}$. Now, if $60/n$ is to be an integer, n must be a factor of 60. Since the factors of 60 are 1, 2, 3, 4, 5, 6, 10, 12, 15, 20, 30, and 60, there are 12 possible values for n. The answer is (E).

24. The employees of the bank can be categorized into three groups:

   1) Employees who are only Clerks. Let c be the count.
   2) Employees who are only Agents. Let a be the count.
   3) Employees who are both Clerks and Agents. Let x be the count.

Hence, the total number of employees is $c + a + x$.
The total number of clerks is $c + x$.
The total number of agents is $a + x$.

We are given that of every three agents one is also a clerk. Hence, we have that one of every three agents is also a clerk (both agent and clerk). Forming the ratio yields $\frac{x}{a+x} = \frac{1}{3}$. Solving for a yields $a = 2x$.

We are given that of every two clerks, one is also an agent. Hence, we have that one of every two clerks is also an agent (both clerk and agent). Forming the ratio yields $\frac{x}{c+x} = \frac{1}{2}$. Solving for c yields $c = x$.

Now, the probability of selecting an employee who is both an agent and a clerk from the bank is

$$\frac{x}{c+a+x} = \frac{x}{x+2x+x} = \frac{x}{4x} = \frac{1}{4}$$

The answer is (C).

Test 21

Questions: 24
Time: 45 minutes

[Multiple-choice Question – Select One Answer Choice Only]
1. If $42.42 = k(14 + m/50)$, where k and m are positive integers and $m < 50$, then what is the value of $k + m$?

    (A) 6
    (B) 7
    (C) 8
    (D) 9
    (E) 10

[Multiple-choice Question – Select One or More Answer Choices]
2. If c and d are two integers and @ is defined in a and b as $@(a, b) = (a + 1)(b + 2)$, and $@(c, d)$ equals the product of 3 and 5, then what could be the value $c + d$?

    (A) –11
    (B) 0
    (C) 5
    (D) 6
    (E) 11

Quantitative Comparison Question]
3. | Column A | | Column B |
    |---|---|---|
    | | X is a 3-digit number and Y is a 4-digit number. All the digits of X are greater than 4, and all the digits of Y are less than 5. | |
    | The sum of the digits of X | | The sum of the digits of Y |

[Multiple-choice Question – Select One Answer Choice Only]
4. How many positive integers less than 500 can be formed using the numbers 1, 2, 3, and 5 for the digits?

(A) 48
(B) 52
(C) 66
(D) 68
(E) 84

Quantitative Comparison Question]
5.   Column A $x > 0$ Column B
           $1/2x$ $2x$

[Multiple-choice Question – Select One or More Answer Choices]
6. The following are the measures of the sides of five different triangles. Which of them represents a right triangle?

(A) $\sqrt{3}, \sqrt{4}, \sqrt{5}$
(B) 1, 5, 4
(C) 3, 4, 5
(D) $\sqrt{3}, \sqrt{7}, \sqrt{4}$
(E) 4, 8, 10

[Multiple-choice Question – Select One Answer Choice Only]

7. In the figure, ABCD is a rectangle and E is a point on the side AB. If AB = 10 and AD = 5, what is the area of the shaded region in the figure?

   (A) 25
   (B) 30
   (C) 35
   (D) 40
   (E) 45

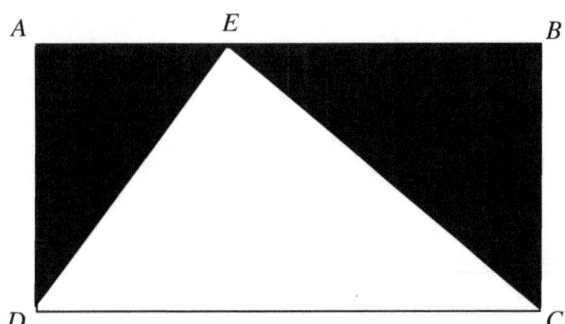

[Multiple-choice Question – Select One Answer Choice Only]

8. In the figure, lines *l* and *k* are parallel. If *a* is an acute angle, then which one of the following must be true?

   (A) $b > 10$
   (B) $b > 15$
   (C) $b < 20$
   (D) $b < 30$
   (E) $b > 45$

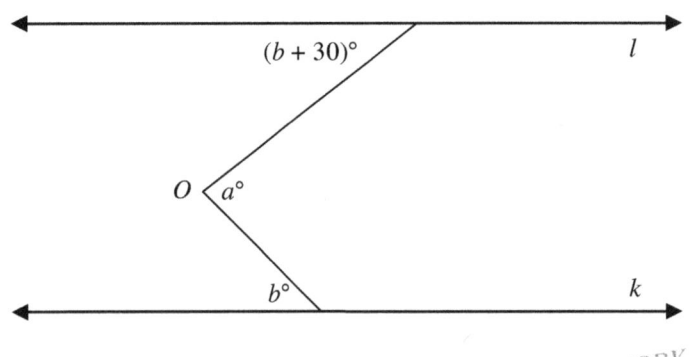

[Multiple-choice Question – Select One Answer Choice Only]
9. In the figure, △ABC is inscribed in the circle. The triangle does not contain the center of the circle O. Which one of the following could be the value of x in degrees?

   (A) 35
   (B) 70
   (C) 85
   (D) 90
   (E) 105

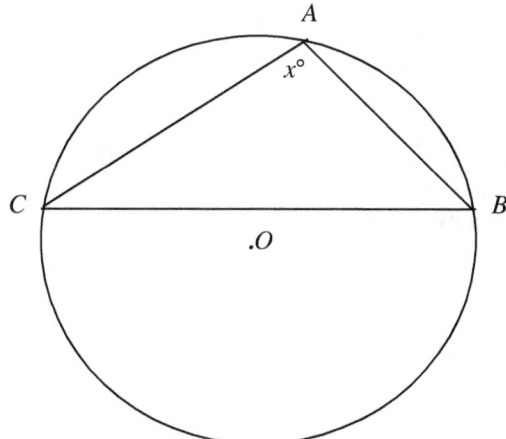

[Multiple-choice Question – Select One Answer Choice Only]
10. In the figure, which one of the following is the measure of angle θ?

   (A) $\theta < 45°$
   (B) $\theta > 45°$
   (C) $\theta = 45°$
   (D) $\theta \le 45°$
   (E) It cannot be determined from the information given

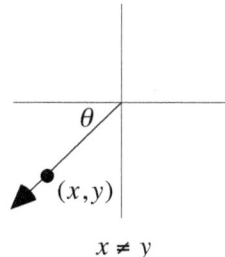

$x \ne y$

GRE Math Tests

Quantitative Comparison Question]
11.   Column A $1 < p < 3$ Column B
            p^2 $2p$

USE THIS SPACE FOR SCRATCHWORK.

Quantitative Comparison Question]
12.   Column A   Kate ate 1/3 of a cake, Fritz ate   Column B
                                1/5 of the remaining cake, and
                                what was left was eaten by Emily.
     Fraction of the cake eaten by   5/7
     Emily

USE THIS SPACE FOR SCRATCHWORK.

[Multiple-choice Question – Select One or More Answer Choices]
13. If *x* and *y* constrained by the following two different patterns:

$$3x + y = x + 3y \text{ (for all } x + y \geq 0\text{)},$$
$$\text{and } x + 3y = 2x + y \text{ (for all } x + y < 0\text{)}$$

then what is the value of $2x - y$ when $x = 2$?

(A) 0
(B) 1
(C) 2
(D) 3
(E) 4

USE THIS SPACE FOR SCRATCHWORK.

[Numeric-Entry]
14. The average of $x+2$, $x+3$, and $2x+5$ is $x+\dfrac{x}{3}+\dfrac{[\,]}{[\,]}$?

[Multiple-choice Question – Select One Answer Choice Only]
15. Kelvin takes 3 minutes to inspect a car, and John takes 4 minutes to inspect a car. If they both start inspecting different cars at 8:30 AM, what would be the ratio of the number of cars inspected by Kelvin and John by 8:54 AM of the same day?

   (A) 1 : 3
   (B) 1 : 4
   (C) 3 : 4
   (D) 4 : 3
   (E) 7 : 4

Quantitative Comparison Question]
16. Column A   Column B
     $x(x^2)^4$ $(x^3)^3$

[Multiple-choice Question – Select One Answer Choice Only]

17. If $|3x| \neq 2$, what is the value of $\dfrac{9x^2 - 4}{3x + 2} - \dfrac{9x^2 - 4}{3x - 2}$?

    (A) −9
    (B) −4
    (C) 0
    (D) 4
    (E) 9

[Multiple-choice Question – Select One Answer Choice Only]

18. A cyclist travels at 12 miles per hour. How many minutes will it take him to travel 24 miles?

    (A) 1
    (B) 2
    (C) 30
    (D) 60
    (E) 120

[Multiple-choice Question – Select One Answer Choice Only]

19. The combined salaries of three brothers is $90,000. Mr. Big earns twice what Mr. Small earns, and Mr. Middle earns 1 1/2 times what Mr. Small earns. What is the smallest salary of the three brothers?

    (A) 20,000
    (B) 22,000
    (C) 25,000
    (D) 30,000
    (E) 40,000

[Multiple-choice Question – Select One or More Answer Choices]
20. Mark and Steve sell items to each other. The seller offers a round figured number of items of the value of the items given by the buyer. For example, if Mark buys toffees from Steve by offering 3.8 toffees worth of goods, then Steve gives him the round figure 4 toffees. In this mode, once they could buy 2 Indian Mangoes for 3 Fuji Apples but could not buy 3 Fuji Apples for 2 Indian Mangoes. Which of the following could be the ratio of the prices of Indian Mangos and Fuji Apples?

(A)   2 : 3
(B)   3 : 2
(C)   3 : 4
(D)   4 : 3
(E)   3 : 7
(F)   7 : 3

[Multiple-choice Question – Select One Answer Choice Only]
21. Set *S* is the set of all numbers from 1 through 100, inclusive. What is the probability that a number randomly selected from the set is divisible by 3?

(A)   1/9
(B)   33/100
(C)   34/100
(D)   1/3
(E)   66/100

Quantitative Comparison Question]
22.   Column A   The number of distinct   Column B
                                        elements in set *A* is 8, and the
                                        number of distinct elements in
                                        set *B* is 3.

The number of distinct elements   The number of distinct elements
common to set *A* and set *B*   in set *A* that are not in set *B*

[Multiple-choice Question – Select One Answer Choice Only]

23. The sum of the first *n* even, positive integers is $2 + 4 + 6 + \cdots + 2n$ is $n(n + 1)$. What is the sum of the first 20 even, positive integers?

    (A) 120
    (B) 188
    (C) 362
    (D) 406
    (E) 420

[Multiple-choice Question – Select One Answer Choice Only]

24. A ship is sinking and 120 more tons of water would suffice to sink it. Water seeps in at a constant rate of 2 tons a minute while pumps remove it at a rate of 1.75 tons a minute. How much time in minutes has the ship to reach the shore before is sinks?

    (A) 480
    (B) 560
    (C) 620
    (D) 680
    (E) 720

Test 21 — Solutions

Answers and Solutions Test 21:

Question	Answer
1.	E
2.	A, C
3.	D
4.	D
5.	D
6.	C, D
7.	A
8.	D
9.	E
10.	E
11.	D
12.	B
13.	C, D
14.	10/3
15.	D
16.	C
17.	B
18.	E
19.	A
20.	B, D
21.	B
22.	B
23.	E
24.	A

If you got 18/24 correct on this test, you are likely to get 750+ on the actual GRE by the time you complete all the tests in the book.

---

1. We are given that k is a positive integer and m is a positive integer less than 50. We are also given that $42.42 = k(14 + m/50)$.

Suppose $k = 1$. Then $k(14 + m/50) = 14 + m/50 = 42.42$. Solving for m yields $m = 50(42.42 - 14) = 50 \times 28.42$, which is not less than 50. Hence, $k \neq 1$.

Now, suppose $k = 2$. Then $k(14 + m/50) = 2(14 + m/50) = 42.42$, or $(14 + m/50) = 21.21$. Solving for m yields $m = 50(21.21 - 14) = 50 \times 7.21$, which is not less than 50. Hence, $k \neq 2$.

Now, suppose $k = 3$. Then $k(14 + m/50) = 3(14 + m/50) = 42.42$, or $(14 + m/50) = 14.14$. Solving for m yields $m = 50(14.14 - 14) = 50 \times 0.14 = 7$, which is less than 50. Hence, $k = 3$ and $m = 7$ and $k + m = 3 + 7 = 10$.

The answer is (E).

2. According to the definition,

$$@(c, d) = (c + 1)(d + 2)$$

We are given that

$@(c, d)$ = the product of 3 and 5, which is 15
$(c + 1)(d + 2) = 15$

Since c and d are integers, so are $c + 1$ and $d + 2$. Now, the product of two integers is 15 only when they are either 3 and 5 or when they are –3 and –5.

$c + 1 = 3$ & $d + 2 = 5$	$c + 1 = -3$ & $d + 2 = -5$
$c = 2$ & $d = 3$	$c = -4$ & $d = -7$
$c + d = 5$	**$c + d = -11$**
Select choice (C).	Select choice (A).
$c + 1 = 5$ & $d + 2 = 3$	$c + 1 = -5$ & $d + 2 = -3$
$c = 4$ & $d = 1$	$c = -6$ & $d = -5$
$c + d = 5$	**$c + d = -11$**

The answer is (A) and (C).

3. Column A: Since all the digits of the 3-digit number X are greater than 4, each digit must be greater than or equal to 5. Hence, the sum of the three digits must be greater than or equal to $3 \times 5 = 15$. Also, since the maximum value of each digit is 9, the maximum possible value of the sum of its digits is $3 \times 9 = 27$. Hence, we have the inequality $15 \leq$ Column A ≤ 27.

Column B: Since all the digits of the 4-digit number Y are less than 5, each digit must be less than or equal to 4. Hence, the sum of the four digits must be less than or equal to $4 \times 4 = 16$. Also, the minimum value of the sum of the digits of a 4-digit number is 1 (for example, for 1000, the sum of the digits is 1). Hence, we have the inequality $1 \leq$ Column B ≤ 16.

Since the inequality for Column A, $15 \leq$ Column A ≤ 27, and the inequality for Column B, $1 \leq$ Column B ≤ 16, have a common range (numbers between 15 and 16, inclusive, satisfy both inequalities), an inequality between the two columns cannot be derived. Hence, we cannot know which column is greater. The answer is (D).

Method II:
Let's construct the smallest and largest numbers possible for X and Y. Since all the digits of X are greater than 4, each digit must be greater than or equal to 5. So, X will be as small as possible when it consists of only 5's and will be as large as possible when it consists of only 9's:

Smallest X

5	5	5

Sum: $5 + 5 + 5 = 15$

Largest X

9	9	9

Sum: $9 + 9 + 9 = 27$

Since all the digits of Y are less than 5, each digit must be less than or equal to 4. So, Y will be as small as possible when it consists of a leading 1 following by three zeros and will be as large as possible when it consists of only 4's:

Smallest Y

1	0	0	0

Sum: $1 + 0 + 0 + 0 = 1$

Largest Y

4	4	4	4

Sum: $4 + 4 + 4 + 4 = 16$

If we chose the smallest sum for X (15) and the largest sum for Y (16), then Column B is greater. And if we chose the largest sum for X (27) and the smallest sum for Y (1), then Column A is greater. This is a double case and therefore the answer is (D).

4. A number less than 500 will be 1) a single-digit number, or 2) a double-digit number, or a 3) triple-digit number with left-most digit less than 5.

Let the compartments shown below represent the single, double and three digit numbers.

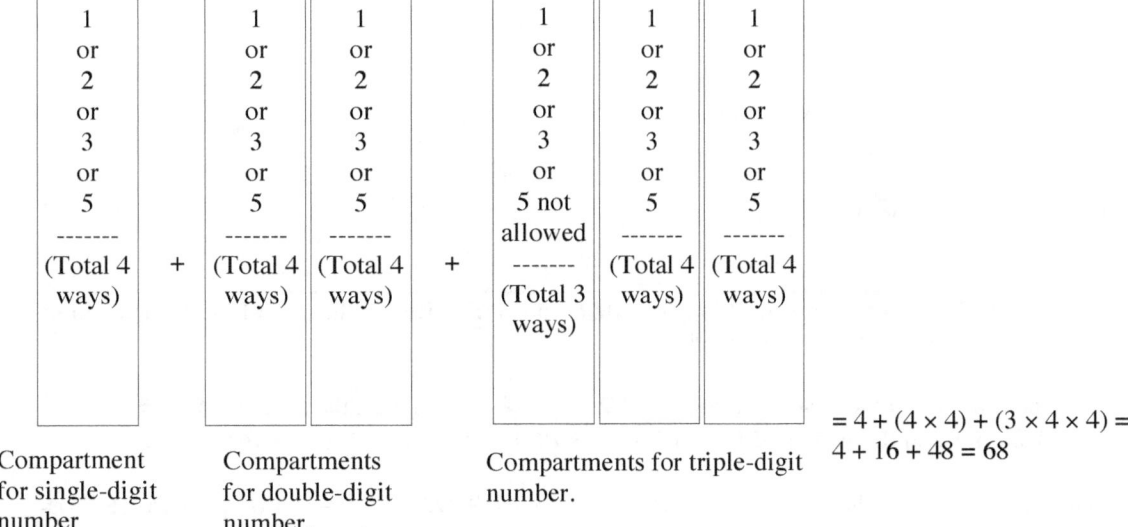

Compartment for single-digit number

Compartments for double-digit number.

Compartments for triple-digit number.

$= 4 + (4 \times 4) + (3 \times 4 \times 4) = 4 + 16 + 48 = 68$

The compartment for the single-digit number can be filled in 4 ways (with any one of the numbers 1, 2, 3, and 5).

Each of the two compartments for the double-digit number can be filled in 4 ways (with any one of the 4 numbers 1, 2, 3, and 5) each. Hence, the two-digit number can be made in $4 \times 4 = 16$ ways.

Regarding the three-digit number, the left most compartment can be filled in 3 ways (with any one of the numbers 1, 2, and 3). Each of the remaining two compartments can be filled in 4 ways (with any one of the numbers 1, 2, 3, and 5) each. Hence, total number of ways of forming the three-digit number equals

$$3 \times 4 \times 4 = 48$$

Hence, the total number of ways of forming the number is $4 + 16 + 48 = 68$. The answer is (D).

---

5. Intuitively, one expects $2x$ to be larger than the fraction $1/2x$. But that would be too easy to be the answer to a hard problem. Now, clearly, $1/2x$ cannot always be greater than $2x$, nor can it always be equal to $2x$. Hence, the answer is (D).

Let's also solve this problem by substitution. If $x = 1$, then $1/2x = 1/2$ and $2x = 2$. In this case, Column B is greater. But if $x = 1/2$, then $\dfrac{1}{2x} = \dfrac{1}{2 \cdot (1/2)} = \dfrac{1}{1} = 1$ and $2x = 2 \cdot \dfrac{1}{2} = 1$. In this case, the columns are equal.

This is a double case and the answer is (D).

6. A right triangle must satisfy The Pythagorean Theorem: the square of the longest side of the triangle is equal to the sum of the squares of the other two sides. Hence, we look for the answer-choice that satisfies this theorem:

Choice (A): $\left(\sqrt{5}\right)^2 \neq \left(\sqrt{3}\right)^2 + \left(\sqrt{4}\right)^2$. Reject.
Choice (B): $5^2 \neq 1^2 + 4^2$. Reject.
Choice (C): $5^2 = 3^2 + 4^2$. Correct.
Choice (D): $\left(\sqrt{7}\right)^2 = \left(\sqrt{3}\right)^2 + \left(\sqrt{4}\right)^2$. Correct.
Choice (E): $10^2 \neq 8^2 + 4^2$. Reject.

The answer is (C) and (D).

7. The shaded region contains two triangles $\triangle AED$ and $\triangle EBC$. Hence, the area of the shaded region equals the sum of the areas of these two triangles.

Now, the formula for the area of a triangle is $1/2 \times base \times height$. Hence, the area of $\triangle AED = 1/2 \times AE \times AD$, and the area of $\triangle EBC = 1/2 \times EB \times BC = 1/2 \times EB \times AD$ (since $BC = AD$).

Hence, the area of the shaded region equals $1/2 \times AE \times AD + 1/2 \times EB \times AD = 1/2 \times AD(AE + EB) = 1/2 \times AD \times AB$ [since $AB = AE + EB$] $= 1/2 \times 5 \times 10 = 25$.

The answer is (A).

8. Draw a line parallel to both of the lines l and k and passing through O.

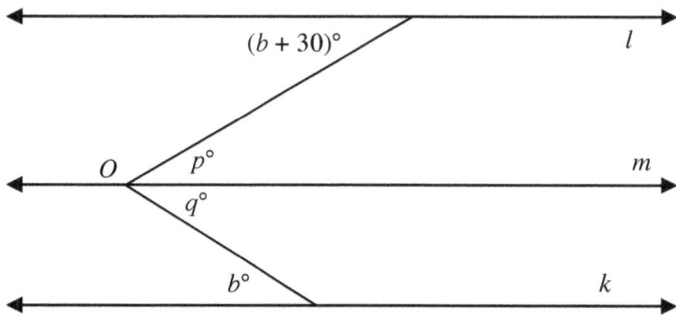

We are given that a is an acute angle. Hence, $a < 90$. Since angles p and $b + 30$ are alternate interior angles, they are equal. Hence, $p = b + 30$. Similarly, angles q and b are alternate interior angles. Hence, $q = b$. Since angle a is the sum of its sub-angles p and q, $a = p + q = (b + 30) + b = 2b + 30$. Solving this equation for b yields $b = (a - 30)/2 = a/2 - 15$. Now, dividing both sides of the inequality $a < 90$ by 2 yields $a/2 < 45$. Also, subtracting 15 from both sides of the inequality yields $a/2 - 15 < 30$. Since $a/2 - 15 = b$, we have $b < 30$. The answer is (D).

9. A chord makes an acute angle on the circle to the side containing the center of the circle and makes an obtuse angle to the other side. In the figure, BC is a chord and does not have a center to the side of point A. Hence, BC makes an obtuse angle at point A on the circle. Hence, $\angle A$, which equals $x°$, is obtuse and therefore is greater than $90°$. Since 105 is the only obtuse angle offered, the answer is (E).

10. There are two possible drawings:

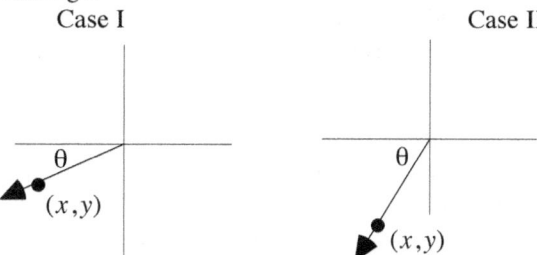

In Case I, $\theta < 45°$. Whereas, in Case II, $\theta > 45°$. This is a double case, and the answer therefore is (E).

---

11. If $p = 2$, then both columns equal 4. If $p \neq 2$, then the columns are unequal. This is a double case, and therefore the answer is (D).

---

12. We are given that Kate ate 1/3 of the cake. So, the uneaten part of the cake is $1 - 1/3 = 2/3$. Hence, regardless of how much Fritz ate, Emily could not have eaten more than 2/3 of the cake. Hence, Column A is less than 2/3; and since $2/3 < 5/7$, Column B is larger. The answer is (B).

---

13. Pattern I: $x + y \geq 0$.

Subtracting x and y from both sides of the given equation $3x + y = x + 3y$ yields

$$3x + y - x - y = x + 2y - x - y$$
$$2x = 2y$$
$$x = y$$

Now, $2x - y = 2x - x = x = 2$ (we are given that $x = 2$).

Hence, select (C).

Pattern II: $x + y < 0$.

Subtracting y and x from both sides of the given equation $x + 3y = 2x + y$ yields

$$x + 3y - x - y = 2x + y - x - y$$
$$2y = x$$
$$y = x/2$$

Now, $2x - y = 2x - x/2 = 3x/2 = 3(2)/2 = 3$ (we are given that $x = 2$).

Hence, select (D).

The answer is (C) and (D).

---

14. The sum of $x + 2$, $x + 3$, and $2x + 5$ is

$$(x + 2) + (x + 3) + (2x + 5) = 4x + 10$$

Now,

$$\text{Average} = \text{Sum}/3 = (4x + 10)/3 = 4/3\, x + 10/3 = x + x/3 + 10/3$$

Hence, grid in the fraction 10/3.

15. Kelvin takes 3 minutes to inspect a car, and John takes 4 minutes to inspect a car. Hence, after t minutes, Kelvin inspects $t/3$ cars and John inspects $t/4$ cars. Hence, the ratio of the number of cars inspected by them is $t/3 : t/4 = 1/3 : 1/4 = 4 : 3$. The answer is (D).

Method II
There are 24 minutes between 8:30 AM and 8:54 AM. Since Kelvin takes 3 minutes to inspect a car, he can inspect 8 (= 24/3) cars in 24 minutes. Since John takes 4 minutes to inspect a car, he can inspect 6 (= 24/4) cars in 24 minutes. Forming the ratio of Kelvin to John yields $8/6 = 4/3$. The answer is (D).

---

16. Column A: $x(x^2)^4 = x \cdot x^{2 \cdot 4} = x^1 \cdot x^8 = x^{1+8} = x^9$.

Column B: $(x^3)^3 = x^{3 \cdot 3} = x^9$.

The answer is (C).

17.
$$\frac{9x^2 - 4}{3x + 2} - \frac{9x^2 - 4}{3x - 2}$$

$$= (9x^2 - 4)\left(\frac{1}{3x+2} - \frac{1}{3x-2}\right) \quad \text{by factoring out the common term } 9x^2 - 4$$

$$= (9x^2 - 4)\frac{(3x-2) - (3x+2)}{(3x+2)(3x-2)}$$

$$= (9x^2 - 4)\frac{3x - 2 - 3x - 2}{(3x)^2 - 2^2}$$

$$= (9x^2 - 4)\frac{-4}{9x^2 - 4} \quad \text{Since } |3x| \neq 2, (3x)^2 \neq 4, \text{ and therefore } 9x^2 - 4 \neq 0.$$

Hence, we can safely cancel $9x^2 - 4$ from numerator and denominator.

$$= -4$$

The answer is (B).

---

18. Since the answer is in minutes, let's convert the cyclist's speed (12 miles per hour) into miles per minute. Since there are 60 minutes in an hour, his speed is $12/60 = 1/5$ miles per minute.

Remember that *Distance* = *Rate* × *Time*. Hence,

$$24 = \frac{1}{5} \times t$$

Solving for t yields $t = 5 \times 24 = 120$. The answer is (E). [If you forgot to convert hours to minutes, you may have mistakenly answered (B).]

Test 21 — Solutions

---

19. Let s be the salary of Mr. Small. Since Mr. Big earns twice what Mr. Small earns, the salary of Mr. Big is $2s$; and since Mr. Middle earns 1 1/2 times what Mr. Small earns, the salary of Mr. Middle equals $(1\ 1/2)s = 3s/2$. Since $s < 3s/2$ and $s < 2s$, Mr. Small earns the smallest salary. Summing the salaries to 90,000 (given) yields

$$2s + 3s/2 + s = 90{,}000$$

$$9s/2 = 90{,}000$$

$$s = 90{,}000 \cdot 2/9 = 20{,}000$$

The answer is (A).

---

20. Let i and a be the prices of the Indian Mangoes and Fuji Apples, respectively.

There are two possibilities:

Possibility I:

The exchange method dictates that if Fuji Apples were bought for Indian Mangoes, then

$$2.5a < 2m < 3.5a$$

Dividing the inequality by 2 yields

$$2.5a/2 < 2m/2 < 3.5a/2$$
$$1.25a < m < 1.75a$$

So, the value of a mango could be in the range 1.25 times to 1.75 times an apple.

Possibility II:

The exchange method dictates that if Indian Mangoes were bought for Fuji Apples, then

$$1.5m < 3a < 2.5m$$

Splitting the inequality yields $1.5m < 3a$. Dividing by 1.5 yields $m < 2a$.

Splitting the inequality yields $3a < 2.5m$. Dividing by 2.5 yields $1.2a < m$.

Here, the range of the price of mango is between $1.2a$ and $2a$.

The wider range of the price of mango is from $1.2a$ to $2a$. That is, m/a must be between 1.2 and 2.

Choice (A): 2 : 3; $m/a = 2/3 = 0.666$. Reject.
Choice (B): 3 : 2; $m/a = 3/2 = 1.5$. Accept.
Choice (C): 3 : 4; $m/a = 3/4 = 0.7$. Reject.
Choice (D): 4 : 3; $m/a = 4/3 = 1.333$. Accept.
Choice (E): 3 : 7; $m/a = 3/7 < 1.2$. Reject.
Choice (F): 7 : 3; $m/a = 7/3 = 2.33 > 2$. Reject.

The answer is (B) and (D).

21. The count of the numbers 1 through 100, inclusive, is 100.

Now, let $3n$ represent a number divisible by 3, where n is an integer.

Since we have the numbers from 1 through 100, we have $1 \leq 3n \leq 100$. Dividing the inequality by 3 yields $1/3 \leq n \leq 100/3$. The possible values of n are the integer values between $1/3$ (≈ 0.33) and $100/3$ (≈ 33.33). The possible numbers are 1 through 33, inclusive. The count of these numbers is 33.

Hence, the probability of randomly selecting a number divisible by 3 is 33/100. The answer is (B).

22. We are given that set A contains 8 elements and set B contains just 3 elements, so the greatest possible number of elements common to set A and set B is 3. There are 5 more elements in set A than in set B, so there are at least 5 elements in set A that cannot be in set B. Hence, Column B is always larger than Column A. The answer is (B).

23. We are given a formula for the sum of the first n even, positive integers. Plugging $n = 20$ into this formula yields

$$n(n + 1) = 20(20 + 1) = 20(21) = 420$$

The answer is (E).

24. We have that water enters the ship at 2 tons per minute and the pumps remove the water at 1.75 tons per minute. Hence, the effective rate at which water is entering the ship is $2 - 1.75 = 0.25$ tons per minute. Since it takes an additional 120 tons of water to sink the ship, the time left is (120 tons)/(0.25 tons per minute) = 120/0.25 = 480 minutes. The answer is (A).

Test 22

GRE Math Tests

Questions: 24
Time: 45 minutes

[Multiple-choice Question – Select One Answer Choice Only]
1. By how much is the greatest of five consecutive even integers greater than the smallest among them?

 (A) 1
 (B) 2
 (C) 4
 (D) 8
 (E) 10

USE THIS SPACE FOR SCRATCHWORK.

Quantitative Comparison Question]
2.   Column A   Column B
   The number of prime numbers   The number of prime numbers
   divisible by 2   divisible by 3.

USE THIS SPACE FOR SCRATCHWORK.

[Numeric Entry Question]
3. The number 3 divides a with a result of b and a remainder of 2. The number 3 divides b with a result of 2 and a remainder of 1. What is the value of a ?

USE THIS SPACE FOR SCRATCHWORK.

424

Quantitative Comparison Question]

4.	Column A	Column B
	$a + \dfrac{b}{10} > 0$ and $b + \dfrac{a}{10} < 0$	
	a	b

USE THIS SPACE FOR SCRATCHWORK.

Quantitative Comparison Question]

5. Column A	Column B
The largest power of 3 that is a factor of $5 \cdot 3^2 + 3^2 \cdot 2$	The largest power of 3 that is a factor of $3 \cdot 2 + 7 \cdot 3$

USE THIS SPACE FOR SCRATCHWORK.

GRE Math Tests

[Numeric Entry Question]
6. In the figure, O is the center of the circle. What is average of the numbers $a, b, c,$ and d ?

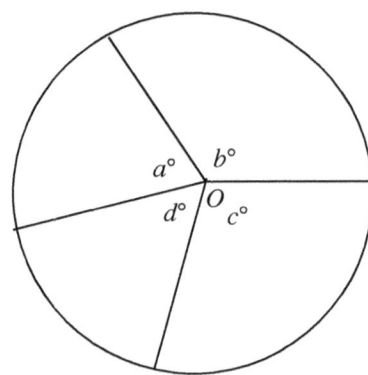

USE THIS SPACE FOR SCRATCHWORK.

[Multiple-choice Question – Select One Answer Choice Only]
7. Which of the following could be the four angles of a parallelogram?

(I) $50°, 130°, 50°, 130°$
(II) $125°, 50°, 125°, 60°$
(III) $60°, 110°, 60°, 110°$

(A) I only
(B) II only
(C) I and II only
(D) I and III only
(E) I, II and III

USE THIS SPACE FOR SCRATCHWORK.

[Multiple-choice Question – Select One or More Answer Choices]
8. In the figure, both triangles are right triangles. The area of the shaded region is

   (A) 1/2
   (B) 2/3
   (C) 7/8
   (D) 3/2
   (E) 5/2

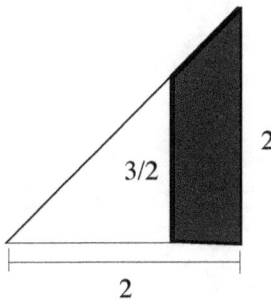

USE THIS SPACE FOR SCRATCHWORK.

[Multiple-choice Question – Select One Answer Choice Only]
9. *AC*, a diagonal of the rectangle *ABCD*, measures 5 units. The area of the rectangle is 12 sq. units. What is the perimeter of the rectangle?

   (A) 7
   (B) 14
   (C) 17
   (D) 20
   (E) 28

USE THIS SPACE FOR SCRATCHWORK.

[Multiple-choice Question – Select One Answer Choice Only]
10. In the figure, lines l_1, l_2, and l_3 are parallel to one another. Line-segments AC and DF cut the three lines. If AB = 3, BC = 4, and DE = 5, then which one of the following equals DF ?

(A) 3/30
(B) 15/7
(C) 20/3
(D) 6
(E) 35/3

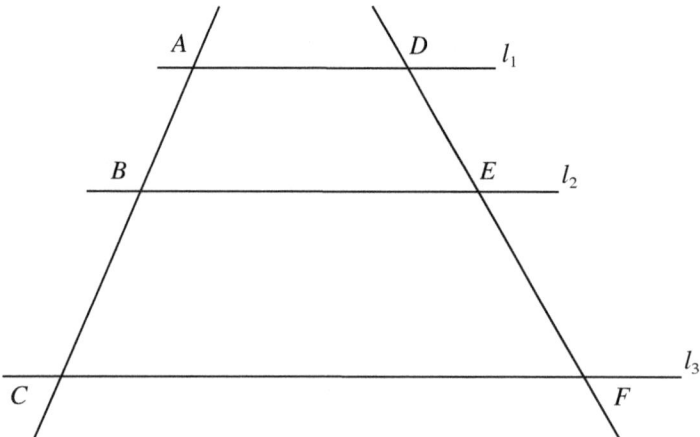

USE THIS SPACE FOR SCRATCHWORK.

[Multiple-choice Question – Select One Answer Choice Only]
11. Let P stand for the product of the first 5 positive integers. What is the greatest possible value of m if $\dfrac{P}{10^m}$ is an integer?

(A) 1
(B) 2
(C) 3
(D) 5
(E) 10

USE THIS SPACE FOR SCRATCHWORK.

[Quantitative Comparison Question]
12.   Column A $3x + 7y > 7x + 3y$ Column B
            $x + 2y$ $y + 2x$

USE THIS SPACE FOR SCRATCHWORK.

[Multiple-choice Question – Select One Answer Choice Only]
13.   If $x - 4y = 1$ and $y = x/2 + 1$, then what is the value of x ?

      (A)   –5
      (B)   –2
      (C)   2
      (D)   5
      (E)   8

USE THIS SPACE FOR SCRATCHWORK.

[Multiple-choice Question – Select One Answer Choice Only]
14.   If $mn = 3$ and $1/m + 1/n = 4/3$, then what is the value of $0.1 + 0.1^{1/m} + 0.1^{1/n}$?

      (A) $0.2 + 0.1^{1/3}$
      (B) $0.1 + 0.1^{1/3} + 0.1^{1/2}$
      (C) $0.1 + 0.1^{4/3} + 0.1^{1/2}$
      (D) $0.1 + 0.1^{1/3} + 0.1^{3/2}$
      (E) $0.1 + 0.1^{1/4} + 0.1^{1/2}$

USE THIS SPACE FOR SCRATCHWORK.

[Multiple-choice Question – Select One or More Answer Choices]
15. If *a*3 and *b*5 are two double-digit numbers, the average of which is *cd*, then *d* could be which of the following?

   (A) 2
   (B) 4
   (C) 6
   (D) 8
   (E) 9

[Multiple-choice Question – Select One Answer Choice Only]
16. 40% of the employees in a factory are workers. All the remaining employees are executives. The annual income of each worker is $390. The annual income of each executive is $420. What is the average annual income of all the employees in the factory together?

   (A) 390
   (B) 405
   (C) 408
   (D) 415
   (E) 420

[Multiple-choice Question – Select One or More Answer Choices]
17. Kelvin takes 2 minutes to inspect a car and John takes 3 minutes to inspect a car. If they both start inspecting different cars at 8:30AM, which of the following is the ratio of the number of cars inspected by Kelvin to John after they start the work?

   (A) 2 : 3
   (B) 3 : 2
   (C) 7 : 8
   (D) 8 : 7
   (E) 15 : 13

Quantitative Comparison Question]

18.	Column A	Column B
	$7^{\frac{1}{x}-\frac{1}{y}}$	7^{x-y}

$0 < x < y$

USE THIS SPACE FOR SCRATCHWORK.

Quantitative Comparison Question]

19.	Column A	Column B
	b	c

$a/2$ is $b\%$ of 30, and a is $c\%$ of 50. b is positive.

USE THIS SPACE FOR SCRATCHWORK.

Quantitative Comparison Question]

20.	Column A	Column B
	$x - y$	y

Williams has x eggs. He sells y of them at a profit of 10 percent and the remaining eggs at a loss of 10 percent. He made a profit overall.

USE THIS SPACE FOR SCRATCHWORK.

[Multiple-choice Question – Select One or More Answer Choices]
21. A man walks at a rate of 10 mph. After every ten miles, he rests for 6 minutes. After how much time would he be available at the 50 miles milestone?

    (A) 300
    (B) 318
    (C) 322
    (D) 324
    (E) 329

USE THIS SPACE FOR SCRATCHWORK.

[Multiple-choice Question – Select One Answer Choice Only]
22. According to the stock policy of a company, each employee in the technical division only is given 15 shares of the company and each employee in the recruitment division only is given 10 shares. Employees belonging to both communities get 25 shares each. There are 20 employees in the company, and each one belongs to at least one division. The cost of each share is $10. If the technical division has 15 employees and the recruitment division has 10 employees, then what is the total cost of the shares given by the company?

    (A) 2,250
    (B) 2,650
    (C) 3,120
    (D) 3,180
    (E) 3,250

USE THIS SPACE FOR SCRATCHWORK.

[Multiple-choice Question – Select One Answer Choice Only]
23. In a multi-voting system, voters can vote for more than one candidate. Two candidates A and B are contesting the election. 100 voters voted for A. Fifty out of 250 voters voted for both candidates. If each voter voted for at least one of the two candidates, then how many candidates voted only for B?

    (A) 50
    (B) 100
    (C) 150
    (D) 200
    (E) 250

USE THIS SPACE FOR SCRATCHWORK.

[Multiple-choice Question – Select One Answer Choice Only]
24. There are 58 balls in a jar. Each ball is painted with at least one of two colors, red or green. It is observed that 2/7 of the balls that have red color also have green color, while 3/7 of the balls that have green color also have red color. What is the probability that a ball randomly picked from the jar will have both red and green colors?

(A) 6/14
(B) 2/7
(C) 6/35
(D) 6/29
(E) 6/42

USE THIS SPACE FOR SCRATCHWORK.

Answers and Solutions Test 22:

Question	Answer
1.	D
2.	C
3.	23
4.	A
5.	B
6.	90
7.	A
8.	C
9.	B, E
10.	E
11.	A
12.	A
13.	A
14.	A
15.	B, E
16.	C
17.	B, E
18.	A
19.	B
20.	B
21.	D, E
22.	E
23.	C
24.	D

If you got 18/24 correct on this test, you are likely to get 750+ on the actual GRE **by the time you complete all the tests in the book.**

1. Choose any 5 consecutive even integers—say—2, 4, 6, 8, 10. The largest in this group is 10, and the smallest is 2. Their difference is $10 - 2 = 8$. The answer is (D).

2. A prime number is divisible by no other numbers, but itself and 1. Hence, the only prime number divisible by 2 is 2 itself; and the only prime number divisible by 3 is 3 itself. Since the number of primes in each column is 1, the answer is (C).

3. Since 3 divides b with a result of 2 and a remainder of 1, $b = 3 \cdot 2 + 1 = 7$. Since number 3 divides a with a result of b (which we now know equals 7) and a remainder of 2, $a = 3 \cdot b + 2 = 3 \cdot 7 + 2 = 23$. Hence, enter 23 in the grid.

4. Multiplying the given inequalities by 10 yields

$$10a + b > 0$$
$$10b + a < 0$$

Multiplying the second inequality by –1 and flipping the direction of the inequality yields

$$10a + b > 0$$
$$-a - 10b > 0$$

Adding these inequalities yields

$$9a - 9b > 0$$

Adding $9b$ to both sides of the inequality yields

$$9a > 9b$$

Finally, dividing both sides of the inequality by 9 yields

$$a > b$$

Hence, Column A is greater than Column B. The answer is (A).

---

5. At first glance, Column A appears larger than Column B since it has more 3's. But this is a hard problem, so that could not be the answer. Now, if we multiply out each expression, Column A becomes $63 = 3^2 \cdot 7$ and Column B becomes $27 = 3^3$. The power of 3^3 is larger than the power of 3^2. Hence, Column B is larger. The answer is (B).

---

6. Since the angle around a point has 360°, the sum of the four angles $a, b, c,$ and d is 360 and their average is 360/4 = 90. Enter 90 in the grid.

---

7. A quadrilateral is a parallelogram if it satisfies two conditions:

   1) The opposite angles are equal.
   2) The angles sum to 360°.

Now, in (I), opposite angles are equal (one pair of opposite angles equals 50°, and the other pair of opposite angles equals 130°). Also, all the angles sum to 360° (= 50° + 130° + 50° + 130° = 360°). Hence, (I) is true.

In (II), not all opposite angles are equal (50° ≠ 60°). Hence, (II) is not a parallelogram.

In (III), the angle sum is not equal to 360° (60° + 110° + 60° + 110° = 340° ≠ 360°). Hence, (III) does not represent a quadrilateral.

Hence, only (I) is true, and the answer is (A).

8. Since the height and base of the larger triangle are the same, the slope of the hypotenuse is 45°. Hence, the base of the smaller triangle is the same as its height, 3/2. Thus, the area of the shaded region equals

(area of the larger triangle) – (area of the smaller triangle) =

$$\left(\frac{1}{2} \cdot 2 \cdot 2\right) - \left(\frac{1}{2} \cdot \frac{3}{2} \cdot \frac{3}{2}\right) = 2 - \frac{9}{8} = \frac{7}{8}$$

The answer is (C).

---

9. If l and w are the length and width of the rectangle, respectively, then we have

The perimeter = $2(l + w)$.
The length of a diagonal = $\sqrt{l^2 + w^2}$ = 5 (given).
The area of the rectangle = lw = 12 (given).

Squaring both sides of the equation $\sqrt{l^2 + w^2}$ = 5 yields $l^2 + w^2 = 5^2 = 25$.

Multiplying both sides of the equation $lw = 12$ by 2 yields $2lw = 24$.

Adding the equations $l^2 + w^2 = 25$ and $2lw = 24$ yields $l^2 + w^2 + 2lw = 49$.

Applying the Perfect Square Trinomial formula, $(a + b)^2 = a^2 + b^2 + 2ab$, to the left-hand side yields

$$(l + w)^2 = 7^2$$

Square rooting yields

$$l + w = 7$$
(positive since side lengths and their sum are positive)

Hence, $2(l + w) = 2(7) = 14$. The answer is (B).

10. In the figure, AC and DF are transversals cutting the parallel lines l_1, l_2, and l_3. Let's move the line-segment DF horizontally until point D touches point A. The new figure looks like this:

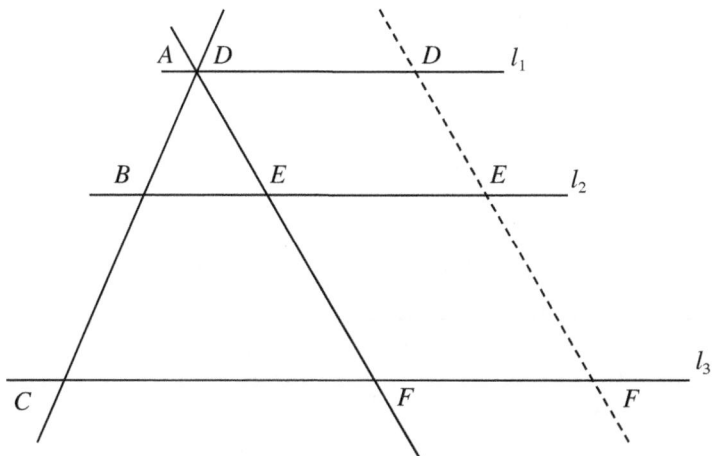

Now, in triangles ABE and ACF, $\angle B$ equals $\angle C$ and $\angle E$ equals $\angle F$ because corresponding angles of parallel lines (here l_2 and l_3) are equal. Also, $\angle A$ is a common angle of the two triangles. Hence, the three angles of triangle ABE equal the three corresponding angles of the triangle ACF. Hence, the two triangles are similar. Since the ratios of the corresponding sides of two similar triangles are equal, we have

$$\frac{AB}{AC} = \frac{AE}{AF}$$

$$\frac{AB}{AB+BC} = \frac{DE}{DF}$$ From the figure, $AC = AB + BC$. Also, from the new figure, point A is the same as point D. Hence, AE is the same as DE and AF is the same as DF.

$$\frac{3}{3+4} = \frac{5}{DF}$$ Substituting the given values

$$DF = \frac{35}{3}$$ By multiplying both sides by 7/3 ×DF

The answer is (E).

---

11. Since we are to find the greatest value of m, we eliminate (E)—the greatest. Also, eliminate 5 because it is repeated from the problem. Now, since we are looking for the largest number, start with the greatest number remaining and work toward the smallest number. The first number that works will be the answer. To this end, let $m = 3$. Then $\frac{P}{10^m} = \frac{1 \cdot 2 \cdot 3 \cdot 4 \cdot 5}{10^3} = \frac{120}{1000} = \frac{3}{25}$. This is not an integer, so eliminate (C).

Next, let $m = 2$. Then $\frac{P}{10^m} = \frac{1 \cdot 2 \cdot 3 \cdot 4 \cdot 5}{10^2} = \frac{120}{100} = \frac{6}{5}$. This still is not an integer, so eliminate (B). Hence, by process of elimination, the answer is (A).

12. We are given the inequality $3x + 7y > 7x + 3y$. Subtracting $3x + 3y$ from both sides of the inequality yields $4y > 4x$, and dividing both sides of this inequality by 4 yields $y > x$.

Column A	$y > x$	Column B
$x + 2y$		$y + 2x$

Now, subtracting $x + y$ from both columns yields

Column A	$y > x$	Column B
y		x

Since $y > x$, Column A is greater than Column B and the answer is (A).

13. We have the system of equations

$$x - 4y = 1$$
$$y = x/2 + 1$$

Substituting the bottom equation in to the top yields

$$x - 4(x/2 + 1) = 1$$
$$x - 2x - 4 = 1$$
$$-x = 4 + 1$$
$$x = -5$$

The answer is (A).

14. We are given the two equations $1/m + 1/n = 4/3$ and $mn = 3$. From the second equation, we have $n = 3/m$. Substituting this in the equation $1/m + 1/n = 4/3$ yields $1/m + m/3 = 4/3$. Multiplying the equation by $3m$ yields $m^2 - 4m + 3 = 0$. The two possible solutions of this equation are 1 and 3.

When $m = 1$, $n = 3/m = 3/1 = 3$ and the expression $0.1 + 0.1^{1/m} + 0.1^{1/n}$ equals $0.1 + 0.1^{1/1} + 0.1^{1/3}$; and when $m = 3$, $n = 3/m = 3/3 = 1$ and the expression $0.1 + 0.1^{1/m} + 0.1^{1/n}$ equals $0.1 + 0.1^{1/3} + 0.1^{1/1}$.

In either case, the expressions equal $0.1 + 0.1^{1/3} + 0.1^{1/1} = 0.2 + 0.1^{1/3}$. Hence, the answer is (A).

15. A 2-digit number xy can be represented as $x \times 10 + y$. For example, $53 = 5 \times 10 + 3$.

We know that the value of $a3$ is equal to $10a + 3$ and the value of $b5$ is $10b + 5$. Hence, $a3 + b5 = (10a + 3) + (10b + 5) = 10a + 3 + 10b + 5 = 10(a + b) + 3 + 5 = 10(a + b) + 8 = 10c + 8$, letting c equal $a + b$.

Therefore, the average of $a3$ and $b5$ equals $(10c + 8)/2 = (10/2)c + 8/2 = 5c + 4$.

Now, there are two possible cases:

If c is even, $5c$ ends with 0. For example if $c = 4$, then $5c = 20$. This term contributes zero to the units place. Therefore the last digit becomes $0 + 4 = 4$ itself. In the example, $5c + 4 = 20 + 4 = 24$, last digit is 4. Choose (B).

If c is odd, $5c$ ends with 5. For example, if $c = 3$, then $5c = 15$ and contributes 5 to the units place. Therefore the last digit becomes $5 + 4 = 9$. Choose (E).

The answers are (B) and (E).

16. Let e be the number of employees.

We are given that 40% of the employees are workers. Now, 40% of e is $40/100 \times e = 0.4e$. Hence, the number of workers is $2e/5$.

All the remaining employees are executives, so the number of executives equals

(The number of Employees) − (The number of Workers) =

$e - 2e/5 =$

$3e/5$

The annual income of each worker is $390. Hence, the total annual income of all the workers together is $2e/5 \times 390 = 156e$.

Also, the annual income of each executive is $420. Hence, the total income of all the executives together is $3e/5 \times 420 = 252e$.

Hence, the total income of the employees is $156e + 252e = 408e$.

The average income of all the employees together equals

(The total income of all the employees) ÷ (The number of employees) =

$408e/e =$

408

The answer is (C).

17. The speed of Kelvin to John is 2 minutes to 3 minutes. Hence, after 6 minutes they would have completed 3 cars and 2 cars, respectively. So, the ratio (B) is encountered. The ratio (A) 2 : 3 is never encountered since it implies that Kelvin is slower than John, which is not as given. Hence, reject (A). Similarly, reject (C).

After say $6n$ minutes (where n is positive integer), the number of cars inspected by Kelvin would be $3n$ and the number of cars inspected by John will be $2n$.

Now,

At $t = 6n$ minus 1 minute, Kelvin inspected $3n - 1$ cars, and John inspected $2n - 1$ cars.
At $t = 6n$ minus 2 minute, Kelvin inspected $3n - 1$ cars, and John inspected $2n - 1$ cars.
At $t = 6n$ minus 3 minute, Kelvin inspected $3n - 2$ cars, and John inspected $2n - 1$ cars.
At $t = 6n$ minus 4 minute, Kelvin inspected $3n - 2$ cars, and John inspected $2n - 2$ cars.
At $t = 6n$ minus 5 minute, Kelvin inspected $3n - 3$ cars, and John inspected $2n - 2$ cars.
At $t = 6n$ minus 6 minute, Kelvin inspected $3n - 3$ cars, and John inspected $2n - 2$ cars.

The cycle repeats.

Therefore, the ratios encountered are

$3n - 1 : 2n - 1$
$3n - 2 : 2n - 1$
$3n - 2 : 2n - 2$
$3n - 3 : 2n - 2 = 3 : 2$

Now, we compare the choices with these available ratios.

Choice (D): 8 : 7.

$(3n - 1)/(2n - 1) = 8/7$
$21n - 7 = 16n - 8$
$5n = 7 - 8 = -1$
$n = -1/5$, not a positive integer.

$(3n - 2)/(2n - 1) = 8/7$
$21n - 14 = 16n - 8$
$5n = 14 - 8 = 6$
$n = 6/5$, not an integer.

$(3n - 2)/(2n - 2) = 8/7$
$21n - 14 = 16n - 16$
$5n = 14 - 16$
$5n = -2$
$n = -2/5$, not a positive integer.

Choice (E): 10 : 7.

$(3n - 1)/(2n - 1) = 10/7$
$21n - 7 = 20n - 10$
$n = 7 - 10 = -3$, not positive integer.

$(3n - 2)/(2n - 1) = 10/7$
$21n - 14 = 20n - 10$
$n = 14 - 10 = 4$. Accept.

The answers are (B) and (E).

Test 22 — Solutions

---

18. From the inequality $0 < x < y$, x and y are positive and $x < y$. Hence, we may reciprocate both sides of the inequality $x < y$ and reverse the direction of the inequality. Thus, $1/x > 1/y$.

Subtracting y from both sides of the inequality $x < y$ yields $x - y < 0$.

Finally, subtracting $1/y$ from both sides of the inequality $1/x > 1/y$ yields $1/x - 1/y > 0$.

Thus, Column A has 7 raised to a positive number, while Column B has 7 raised to a negative number. Hence, Column A is greater than 1, and Column B is less than 1. The answer is (A).

---

19. We are given that $a/2$ is $b\%$ of 30. Now, $b\%$ of 30 is $\frac{30}{100}b$. Hence, $\frac{a}{2} = \frac{30}{100}b$. Solving for a yields $a = \frac{3}{5}b$. We are also given that a is $c\%$ of 50. Now, $c\%$ of 50 is $\frac{c}{100} \cdot 50 = \frac{c}{2}$. Hence, $a = c/2$. Plugging this into the equation $a = \frac{3}{5}b$ yields $\frac{c}{2} = \frac{3}{5}b$. Multiplying both sides by 2 yields $c = \frac{6}{5}b$. Since b is positive, c is also positive; and since $6/5 > 1$, $c > b$. Hence, the answer is (B).

---

20. Let a dollars be the cost of each egg to Williams. Hence, the net cost of the x eggs is ax dollars.

Now, the selling price of the eggs when selling at 10% profit is $a(1 + 10/100) = 11a/10$.

The selling price of the eggs when selling at 10% loss is $a(1 - 10/100) = 9a/10$.

We are given that Williams has x eggs and he sold y of them at 10 percent profit (at a selling price of $11a/10$) and the rest, $x - y$, at 10 percent loss (at a selling price of $9a/10$). Hence, the net selling price is $y(11a/10) + (x - y)(9a/10) = a(0.2y + 0.9x)$.

Since overall he made a profit, the net selling price must be greater than the net cost. Hence, we have the inequality $a(0.2y + 0.9x) > ax$.

Canceling a from both sides of the inequality yields $0.2y + 0.9x > x$. Subtracting $0.9x$ from both sides of the inequality yields $0.2y > 0.1x$. Multiplying both sides by 10 yields $2y > x$. Now, subtracting y from both sides yields $y > x - y$.

Hence, Column B is greater than Column A. The answer is (B).

---

21. Remember that *Time = Distance ÷ Speed*. Hence, the time taken by the man to walk 10 miles is 10 miles/10 mph = 1 hour.

Since the man walks 50 miles in five installments of 10 miles each, each installment should take him 1 hour. Hence, the total time for which he walked equals $5 \cdot 1$ hr. = 5 hr. = $5 \cdot 60$ = 300 mins.

Since he takes a break after each installment (until reaching the 50 mile point; one after 10 miles; one after 20 miles; one after 30 miles; final one after 40 miles. The 50th mile is his destination.), he takes four breaks; and since each break lasts 6 minutes, the net time spent in the breaks is $4 \cdot 6$ mins = 24 mins.

Hence, the total time taken to reach the destination is 300 + 24 = 324 mins. The answer is (D).

He will then stay for 6 minutes there. Therefore, at 324 + 5 = 329 mins, he is still there. Select choice (E) as well.

22. Since each person in both the technical and recruitment divisions is given 25 (= 15 + 10) shares, which is the same as giving that person 15 shares for being in the technical division and 10 for being in the recruitment division, the allotment of shares amounts to merely two independent allotments: 15 shares to each technical person and 10 shares to each recruitment person.

We have that the number of employees in the technical division is 15 and the number of employees in the recruitment division is 10. Hence, the total shares given equals $15 \cdot 15 + 10 \cdot 10 = 225 + 100 = 325$. Each share is worth 10 dollars, so the net worth of the shares is $325 \cdot 10 = 3{,}250$. The answer is (E).

23. There are three kinds of voters:

       1) Voters who voted for A only. Let the count of such voters be a.
       2) Voters who voted for B only. Let the count of such voters be b.
       3) Voters who voted for both A and B. The count of such voters is 50 (given).

Since the total number of voters is 250, we have

$a + b + 50 = 250$
$a + b = 200$ (1)   By subtracting 50 from both sides

Now, we have that 100 voters voted for A. Hence, we have

(Voters who voted for A only) + (Voters who voted for both A and B) = 100

Forming this as an equation yields

$$a + 50 = 100$$
$$a = 50$$

Substituting this in equation (1) yields $50 + b = 200$. Solving for b yields $b = 150$.

The answer is (C).

24. Let T be the total number of balls, R the number of balls having red color, G the number having green color, and B the number having both colors.

Hence, the number of balls having only red is $R - B$, the number having only green is $G - B$, and the number having both is B. Now, the total number of balls is

$$T = (R - B) + (G - B) + B = R + G - B$$

We are given that 2/7 of the balls having red color have green also. This implies that $B = 2R/7$. Also, we are given that 3/7 of the green balls have red color. This implies that $B = 3G/7$. Solving for R and G in these two equations yields $R = 7B/2$ and $G = 7B/3$. Substituting this into the equation $T = R + G - B$ yields

$$T = 7B/2 + 7B/3 - B$$

Solving for B yields $B = 6T/29$. Hence, the probability of selecting such a ball is the fraction

$$(6T/29)/T = 6/29$$

The answer is (D).

Test 23

Questions: 25
Time: 45 minutes

[Numeric Entry Question]
1. (The average of five consecutive integers starting from *m*) − (the average of six consecutive integers starting from *m*) =

USE THIS SPACE FOR SCRATCHWORK.

[Quantitative Comparison Question]
2. Column A | Column B
    The last digit in the number 252^{56} | The last digit in the number 152^{56}

USE THIS SPACE FOR SCRATCHWORK.

[Multiple-choice Question – Select One or More Answer Choices]
3. The remainder when the positive integer *m* is divided by 7 is *x*. The remainder when *m* is divided by 14 is *x* + 7. Which of the following could *m* equal?

(A) 45
(B) 53
(C) 72
(D) 85
(E) 92

USE THIS SPACE FOR SCRATCHWORK.

444

Multiple-choice Question – Select One Answer Choice Only]
4. a, b, c, d, and e are five consecutive numbers in increasing order of size. Deleting one of the five numbers from the set decreased the sum of the remaining numbers in the set by 20%. Which one of the following numbers was deleted?
    (A) a
    (B) b
    (C) c
    (D) d
    (E) e

[Multiple-choice Question – Select One or More Answer Choices]
5. What is the value of y in the figure?

    (A) a
    (B) 20
    (C) 30
    (D) 35
    (E) 45

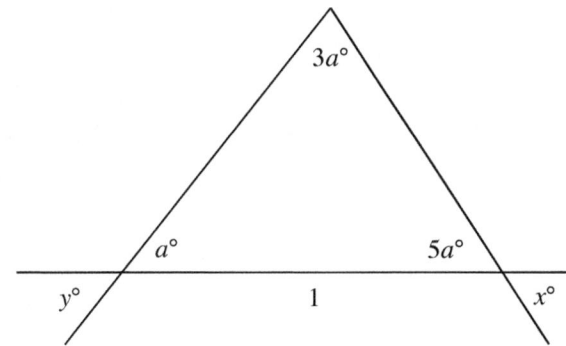

GRE Math Tests

[Multiple-choice Question – Select One Answer Choice Only]

6. If $A, B, C, D,$ and E are points in a plane such that line CD bisects $\angle ACB$ and line CB bisects right angle $\angle ACE$, then $\angle DCE =$

   (A) 22.5°
   (B) 45°
   (C) 57.5°
   (D) 67.5°
   (E) 72.5°

[Multiple-choice Question – Select One Answer Choice Only]

7. In the figure shown, if $\angle A = 60°$, $\angle B = \angle C$, and $BC = 20$, then $AB =$

   (A) 20
   (B) $10\sqrt{2}$
   (C) $10\sqrt{3}$
   (D) $20\sqrt{2}$
   (E) $20\sqrt{3}$

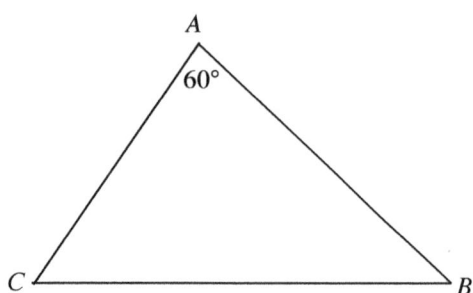

[Quantitative Comparison Question]

8. Column A — A regular polygon of 24 sides is inscribed in a circle. — Column B

    The perimeter of the polygon   The circumference of the circle

[Multiple-choice Question – Select One Answer Choice Only]
9. In the figure, if *AB* = 10, what is the length of the side *CD* ?

    (A)   5
    (B) $5\sqrt{3}$
    (C) $\dfrac{10}{\sqrt{3}}$
    (D)   10
    (E) $10\sqrt{3}$

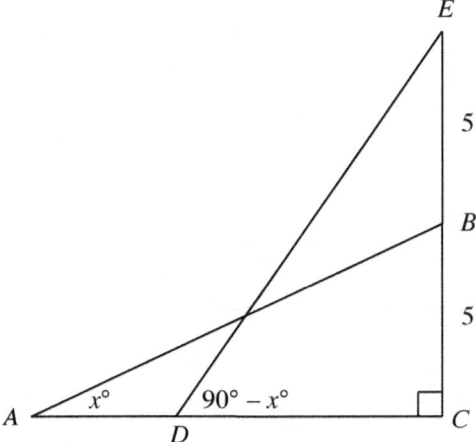

[Multiple-choice Question – Select One Answer Choice Only]
10. Which one of the following statements is true about the line segment with endpoints (–1, 1) and (1, –1)?

    (A)   Crosses the *x*-axis only.
    (B)   Crosses the *y*-axis only.
    (C)   Crosses the *y*-axis on its positive side.
    (D)   Passes through the origin (0, 0).
    (E)   Crosses the *x*- and *y*-axes on their negative sides.

[Multiple-choice Question – Select One Answer Choice Only]
11. If **w** is 10 percent less than **x**, and **y** is 30 percent less than **z**, then **wy** is what percent less than **xz**?

   (A) 10%
   (B) 20%
   (C) 37%
   (D) 40%
   (E) 100%

Quantitative Comparison Question]
12.   Column A   Column B
   $2 \times 10^1 + 3 \times 10^0 + 4 \times 10^{-1} + 5 \times 10^{-2}$ $1 \times 10^{-3} + 2 \times 10^{-2} + 3 \times 10^{-1} + 4 \times 10^0 + 5 \times 10^1$

[Multiple-choice Question – Select One Answer Choice Only]
13. If $|2x - 4|$ is equal to 2 and $(x - 3)^2$ is equal to 4, then what is the value of x?

   (A) 1
   (B) 2
   (C) 3
   (D) 4
   (E) 5

[Multiple-choice Question – Select One or More Answer Choices]
14. Which two of the following numbers can be removed (without replacement) from the set $S = \{1, 2, 3, 4, 5, 6, 7\}$ without changing the average of set S?

   (A) 1
   (B) 2
   (C) 3
   (D) 4
   (E) 6

[Multiple-choice Question – Select One Answer Choice Only]
15. In Figure 1, $y = \sqrt{3}x$ and $z = 2x$. What is the ratio $p : q : r$ in Figure 2?

   (A) $1 : 2 : 3$
   (B) $\sqrt{3} : 1 : 2$
   (C) $1 : \sqrt{3/2} : 1$
   (D) $2 : \sqrt{3} : 1$
   (E) $3 : 2 : 1$

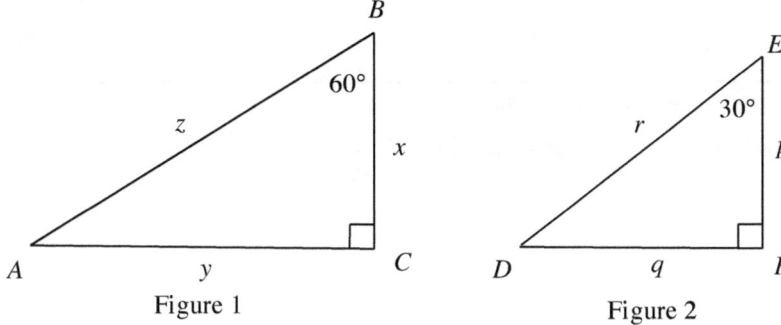

Figure 1   Figure 2

Note: The figures are not drawn to scale.

[Quantitative Comparison Question]

16.
Column A	Five years ago, in a zoo, the ratio of the number of cheetahs to the number of pandas was 1 : 3. The ratio is now 1 : 2.	Column B
The increase in the number of cheetahs in the zoo in the last five years		The increase in the number of pandas in the zoo in the last five years

USE THIS SPACE FOR SCRATCHWORK.

[Multiple-choice Question – Select One Answer Choice Only]

17. The sum of three consecutive positive integers must be divisible by which one of the following?

(A) 2
(B) 3
(C) 4
(D) 5
(E) 6

USE THIS SPACE FOR SCRATCHWORK.

[Multiple-choice Question – Select One Answer Choice Only]

18. If a is positive and b is one-fourth of a, then what is the value of $\dfrac{a+b}{\sqrt{ab}}$?

(A) 1/5
(B) 1/3
(C) 1/2
(D) $1\dfrac{1}{2}$
(E) $2\dfrac{1}{2}$

USE THIS SPACE FOR SCRATCHWORK.

[Multiple-choice Question – Select One Answer Choice Only]

19. The price of a car was *m* dollars. It then depreciated by *x*%. Later, it appreciated by *y*% to *n* dollars. If there are no other changes in the price and if $y = \dfrac{x}{1 - \dfrac{x}{100}}$, then which one of the following must *n* equal?

　　(A)　3*m*/4
　　(B)　*m*
　　(C)　4*m*/3
　　(D)　3*m*/2
　　(E)　2*m*

USE THIS SPACE FOR SCRATCHWORK.

The next two questions refer to the discussion below:
Mike and Fritz ran a 30-mile Marathon. Mike ran 10 miles at 10 miles per hour and then ran at 5 miles per hour for the remaining 20 miles. Fritz ran for the first one-third of the time of the run at 10 miles per hour, and for the remaining two-thirds of the time of the run at 5 miles per hour.

[Multiple-choice Question – Select One Answer Choice Only]

20. How much time in hours did Mike take to complete the Marathon?

　　(A)　3
　　(B)　3.5
　　(C)　4
　　(D)　4.5
　　(E)　5

USE THIS SPACE FOR SCRATCHWORK.

[Multiple-choice Question – Select One Answer Choice Only]

21. How much time in hours did Fritz take to complete the Marathon?

　　(A)　3
　　(B)　3.5
　　(C)　4
　　(D)　4.5
　　(E)　5

USE THIS SPACE FOR SCRATCHWORK.

[Multiple-choice Question – Select One Answer Choice Only]

22. A sequence of positive integers $a_1, a_2, a_3, ..., a_n$ is given by the rule $a_{n+1} = 2a_n + 1$. The only even number in the sequence is 38. What is the value of a_2?

    (A) 11
    (B) 25
    (C) 38
    (D) 45
    (E) 77

USE THIS SPACE FOR SCRATCHWORK.

[Multiple-choice Question – Select One Answer Choice Only]

23. The ratio of the number of red balls, to yellow balls, to green balls in a urn is 2 : 3 : 4. What is the probability that a ball chosen at random from the urn is a red ball?

    (A) 2/9
    (B) 3/9
    (C) 4/9
    (D) 5/9
    (E) 7/9

USE THIS SPACE FOR SCRATCHWORK.

[Multiple-choice Question – Select One Answer Choice Only]

24. There are 5 doors to a lecture room. Two are red and the others are green. In how many ways can a lecturer enter the room and leave the room from different colored doors?

    (A) 1
    (B) 3
    (C) 6
    (D) 9
    (E) 12

USE THIS SPACE FOR SCRATCHWORK.

[Multiple-choice Question – Select One Answer Choice Only]

25. A menu offers 2 entrees, 3 main courses, and 3 desserts. How many different combinations of dinner can be made? (A dinner must contain an entrée, a main course, and a dessert.)

    (A) 12
    (B) 15
    (C) 18
    (D) 21
    (E) 24

USE THIS SPACE FOR SCRATCHWORK.

Test 23 — Solutions

Answers and Solutions Test 23:

Question	Answer
1.	–1/2 **or** –0.5
2.	C
3.	B, E
4.	C
5.	A, B
6.	D
7.	A
8.	B
9.	C
10.	D
11.	C
12.	B
13.	A
14.	B, E
15.	B
16.	D
17.	B
18.	E
19.	B
20.	E
21.	D
22.	E
23.	A
24.	E
25.	C

If you got 18/25 correct on this test, you are likely to get 750+ on the actual GRE **by the time you complete all the tests in the book.**

---

1. Choose any five consecutive integers, say, –2, –1, 0, 1 and 2. (We chose these particular numbers to make the calculation as easy as possible. But any five consecutive integers will do. For example, 1, 2, 3, 4, and 5.) Forming the average yields (–1 + (–2) + 0 + 1 + 2)/5 = 0/5 = 0. Now, add 3 to the set to form 6 consecutive integers: –2, –1, 0, 1, 2, and 3. Forming the average yields

$$\frac{-1+(-2)+0+1+2+3}{6} =$$

$$\frac{[-1+(-2)+0+1+2]+3}{6} =$$

$$\frac{[0]+3}{6} = \qquad \text{since the average of } -1 + (-2) + 0 + 1 + 2 \text{ is zero, their sum must be zero}$$

$$3/6 =$$

$$1/2$$

(The average of five consecutive integers starting from *m*) – (The average of six consecutive integers starting from *m*) = (0) – (1/2) = –1/2. Enter in the grid.

453

Method II (without substitution):
The five consecutive integers starting from m are $m, m + 1, m + 2, m + 3$, and $m + 4$. The average of the five numbers equals

$$\frac{\text{the sum of the five numbers}}{5} =$$

$$\frac{m + (m + 1) + (m + 2) + (m + 3) + (m + 4)}{5} =$$

$$\frac{5m + 10}{5} =$$

$$m + 2$$

The six consecutive integers starting from m are $m, m + 1, m + 2, m + 3, m + 4$, and $m + 5$. The average of the six numbers equals

$$\frac{\text{the sum of the six numbers}}{6} =$$

$$\frac{m + (m + 1) + (m + 2) + (m + 3) + (m + 4) + (m + 5)}{6} =$$

$$\frac{6m + 15}{6} =$$

$$m + 5/2 =$$

$$m + 2 + 1/2 =$$

$$(m + 2) + 1/2$$

(The average of five consecutive integers starting from m) – (The average of six consecutive integers starting from m) = $(m + 2) - [(m + 2) + 1/2] = -1/2$. Enter in the grid.

---

2. The last digit of the number 252 (in Column A) is 2, and the last digit of the number 152 (in Column B) is also 2. Hence, both numbers raised to the same power (here 56) should end with the same digit. So, 152^{56} should end with the same digit as 252^{56}. The answer is (C). A small example: the last digit of 6^2 (= 36) is 6 and the last digit of 16^2 (= 256) is also 6.

3. Choice (A): 45/7 = 6 + 3/7, so x = 3. Now, 45/14 = 3 + 3/14. The remainder is 3, not $x + 7$ (= 10). Reject.

Choice (B): 53/7 = 7 + 4/7, so x = 4. Now, 53/14 = 3 + 11/14. The remainder is 11, and equals $x + 7$ (= 11). Accept the choice.

Choice (C): 72/7 = 10 + 2/7, so x = 2. Now, 72/14 = 5 + 2/14. The remainder is 2, not $x + 7$ (= 9). Reject.

Choice (D): 85/7 = 12 + 1/7, so x = 1. Now, 85/14 = 6 + 1/14. The remainder is 1, not $x + 7$ (= 8). Reject.

Choice (E): 92/7 = 13 + 1/7, so x = 1. Now, 92/14 = 6 + 8/14. The remainder is 8, not $x + 7$ (= 8). Accept the choice.

The answer is (B) and (E).

---

4. Since a, b, c, d, and e are consecutive numbers in the increasing order, we have $b = a + 1, c = a + 2, d = a + 3$ and $e = a + 4$. The sum of the five numbers is $a + (a + 1) + (a + 2) + (a + 3) + (a + 4) = 5a + 10$.

Now, we are given that the sum decreased by 20% when one number was deleted. Hence, the new sum should be $(5a + 10)(1 - 20/100) = (5a + 10)(1 - 1/5) = (5a + 10)(4/5) = 4a + 8$. Now, since New Sum = Old Sum − Dropped Number, we have $(5a + 10) = (4a + 8) +$ (Dropped Number). Hence, the number dropped is $(5a + 10) - (4a + 8) = a + 2$. Since $c = a + 2$, the answer is (C).

---

5. Summing the angles of the triangle in the figure to 180° yields $a + 3a + 5a = 180$. Solving this equation for a yields $a = 180/9 = 20$. Angles y and a in the figure are vertical and therefore are equal. So, $y = a = 20$. The answer is (A) and (B).

---

6. Drawing the figure given in the question yields

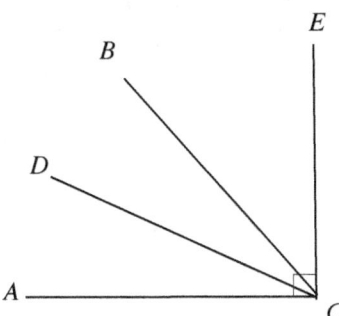

Figure not drawn to scale.

CD bisects ∠ACB
CB bisects ∠ACE

We are given that CB bisects the right-angle ∠ACE. Hence, ∠ACB = ∠BCE = ∠ACE/2 = 90°/2 = 45°. Also, since CD bisects ∠ACB, ∠ACD = ∠DCB = ∠ACB/2 = 45°/2 = 22.5°. Now, ∠DCE = ∠DCB + ∠BCE = 22.5° + 45° = 67.5°. The answer is (D).

7. Summing the angles of △ABC to 180° yields

$\angle A + \angle B + \angle C = 180$
$60 + \angle B + \angle B = 180$ since $\angle A = 60°$ and $\angle B = \angle C$
$2\angle B = 120$ by subtracting 60 from both sides
$\angle B = 60$

Hence, $\angle A = \angle B = \angle C = 60°$. Since the three angles of △ABC are equal, the three sides of the triangle must also be equal. Hence, $AB = BC = 20$. The answer is (A).

8. Each side of the regular polygon represents a chord of the circle, and each chord subtends a unique arc on the circle. The length of a chord is always less than the length of the arc it subtends. Hence, the sum of the lengths of the 24 sides (which are the 24 chords of the circle) is less than the sum of the lengths of the arcs that each one of the chords subtends. Hence, the perimeter of the polygon is less than the circumference of the circle, and therefore Column A is less than Column B. The answer is (B).

9. Applying The Pythagorean Theorem to the right triangle ABC yields

$$BC^2 + AC^2 = AB^2$$

$$5^2 + AC^2 = 10^2 \text{ given that } AB = 10 \text{ and } BC = 5 \text{ (from the figure)}$$

$$AC^2 = 10^2 - 5^2 = 100 - 25 = 75$$

Square rooting yields $AC = \sqrt{75} = \sqrt{25 \cdot 3} = \sqrt{25} \times \sqrt{3} = 5\sqrt{3}$.

Hence, the sides opposite angles measuring $x°$ (A in △ABC) and $90° - x°$ (B in △ABC) are in the ratio $5 : 5\sqrt{3} = 1 : \sqrt{3}$.

Similarly, in △ECD, the ratio of the sides opposite the angles E (measuring $x°$) and D (measuring $90° - x°$) must also be $1 : \sqrt{3}$.

Hence, we have

$$\frac{CD}{EC} = 1 : \sqrt{3}$$

$$\frac{CD}{5+5} = 1 : \sqrt{3}$$

$$CD = \frac{10}{\sqrt{3}}$$

The answer is (C).

10. Locating the points $(-1, 1)$ and $(1, -1)$ on the xy-plane gives

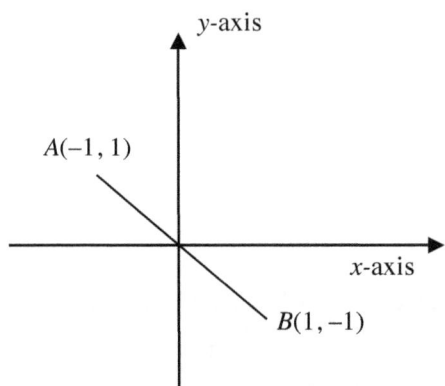

The midpoint of two points is given by

(Half the sum of the x-coordinates of the two points, Half the sum of the y-coordinates of the two points)

Hence, the midpoint of A and B is

$$\left(\frac{-1+1}{2}, \frac{1-1}{2}\right) = (0, 0)$$

Hence, the line-segment passes through the origin. The answer is (D).

---

11. We eliminate (A) since it repeats the number 10 from the problem. We can also eliminate choices (B), (D), and (E) since they are derivable from elementary operations:

$$20 = 30 - 10$$

$$40 = 30 + 10$$

$$100 = 10 \cdot 10$$

This leaves choice (C) as the answer.
    Let's also solve this problem directly. The clause

w is 10 percent less than x

translates into

$$w = x - .10x$$

Simplifying yields

1) $\quad w = .9x$

Next, the clause

y is 30 percent less than **z**

translates into

$$y = z - .30z$$

Simplifying yields

$$2)\ y = .7z$$

Multiplying 1) and 2) gives

$$wy = (.9x)(.7z) = .63xz = xz - .37xz$$

Hence, **wy** is 37 percent less than **xz**. The answer is (C).

---

12. The dominant term 10^1 appears in both columns, but has more weight (5) in Column B. Hence, Column B is greater. Let's still evaluate the expressions:

Column A = $2 \times 10^1 + 3 \times 10^0 + 4 \times 10^{-1} + 5 \times 10^{-2} = 20 + 3 + 0.4 + 0.05 = 23.45$.
Column B = $1 \times 10^{-3} + 2 \times 10^{-2} + 3 \times 10^{-1} + 4 \times 10^0 + 5 \times 10^1 = 0.001 + 0.02 + 0.3 + 4 + 50 = 54.321$.

Hence, Column B is greater than Column A. The answer is (B).

---

13. We have that $|2x - 4| = 2$. Since $|2x - 4|$ is only the positive value of $2x - 4$, the expression $2x - 4$ could equal 2 or –2. If $2x - 4$ equals 2, x equals 3; and if $2x - 4$ equals –2, x equals 1. We also have that $(x - 3)^2$ is equal to 4. By square rooting, we have that $x - 3$ may equal 2 (Here, $x = 3 + 2 = 5$), or $x - 3$ equals –2 (Here, $x = 3 - 2 = 1$). The common solution is $x = 1$. Hence, the answer is (A).

---

14. The average of the numbers is $(1 + 2 + 3 + 4 + 5 + 6 + 7)/7 = 28/7 = 4$. So, we have to remove a number on either side of the average. We can remove either 1 and 7 or 2 and 6 or 3 and 5. The pair that is available in the answer-choices is 2 and 6. Choose (B) and (E).

---

15. Angles $\angle B$ and $\angle C$ in triangle ABC equal 60° and 90°, respectively. Since the sum of the angles in a triangle is 180°, the third angle of the triangle, $\angle A$, must equal 180° – (60° + 90°) = 30°.

So, in the two triangles, ABC and EDF, we have $\angle A = \angle E = 30°$ and $\angle C = \angle F = 90°$ (showing that at least two corresponding angles are equal). So, the two triangles are similar. Hence, the ratios of the corresponding sides in the two triangles are equal. Hence, we have

$EF : DF : DE = AC : BC : AB$
$p : q : r = y : x : z$ after substitutions from the figure
$= \sqrt{3}x : x : 2x = \sqrt{3} : 1 : 2$ $y = \sqrt{3}x$ and $z = 2x$, given

The answer is (B).

Test 23 — Solutions

---

16. Let k and $3k$ be the number of cheetahs and pandas five years ago. Let d and $2d$ be the corresponding numbers now.

The increase in the number of cheetahs is $d - k$, and the increase in the number of pandas is $2d - 3k$.

Now, suppose the increase in the number of cheetahs equals the increase in the number of pandas. Then we have the equation $d - k = 2d - 3k$. Solving the equation for d yields $d = 2k$. Hence, we have a case here. Suppose $k = 3$. Then $d = 2k = 6$. The case supposes there were 3 cheetahs and 9 pandas 5 years ago, and now, there are 6 cheetahs and 12 pandas. Hence, the increase is the same.

Now, suppose the increase in the number of cheetahs is less than that of the pandas. Then we have the inequality $d - k < 2d - 3k$. Solving the inequality yields $d > 2k$. Hence, suppose $k = 3$ and $d = 7$. The case refers to when there are 3 cheetahs and 9 pandas five years ago and now there are 7 cheetahs and 14 pandas.

Now, suppose the increase in the number of cheetahs is greater than that of the pandas. Then we have the inequality $d - k > 2d - 3k$. Solving the inequality yields $d < 2k$. Hence, suppose $k = 3$ and $d = 5$. The case refers to when there are 3 cheetahs and 9 pandas five years ago, and now, there are 5 cheetahs and just 10 pandas. Here, the increase in cheetahs is greater than the increase in pandas.

Hence, we have a double case, and the answer is (D).

---

17. Let the three consecutive positive integers be n, $n + 1$, and $n + 2$. The sum of these three positive integers is

$$n + (n + 1) + (n + 2) =$$
$$3n + 3 =$$
$$3(n + 1)$$

Since we have written the sum as a multiple of 3, it is divisible by 3. The answer is (B).

GRE Math Tests

18. We are given that b is 1/4 of a. Hence, we have the equation $b = a/4$. Multiplying both sides of this equation by $4/b$ yields $4 = a/b$.

Now,

$$\frac{a+b}{\sqrt{ab}} =$$

$$\frac{a}{\sqrt{ab}} + \frac{b}{\sqrt{ab}} =$$

$$\frac{\sqrt{a^2}}{\sqrt{ab}} + \frac{\sqrt{b^2}}{\sqrt{ab}} =$$

$$\sqrt{\frac{a^2}{ab}} + \sqrt{\frac{b^2}{ab}} =$$

$$\sqrt{\frac{a}{b}} + \sqrt{\frac{b}{a}} =$$

$$\sqrt{4} + \sqrt{\frac{1}{4}} =$$

$$2 + \frac{1}{2} =$$

$$2\frac{1}{2}$$

The answer is (E).

19. After a depreciation of $x\%$ on the m dollars, the depreciated price of the car is $m(1 - x/100)$.

After an appreciation of $y\%$ on this price, the appreciated price, n, is $m(1 - x/100)(1 + y/100) = (m/100)(100 - x)(1 + y/100)$. Hence, $n = (m/100)(100 - x)(1 + y/100)$.

We are given that $y = \dfrac{x}{1 - \dfrac{x}{100}} = \dfrac{x}{\dfrac{100-x}{100}} = \dfrac{100x}{100-x}$. Substituting this in the equation $n = (m/100)(100 - x)$ $(1 + y/100)$ yields

$$\frac{m}{100}(100-x)\left(1 + \frac{\frac{100x}{100-x}}{100}\right) =$$

$$\frac{m}{100}(100-x)\left(1 + \frac{x}{100-x}\right) =$$

$$\frac{m}{100}(100-x)\left(\frac{100-x+x}{100-x}\right) =$$

$$\frac{m}{100}(100-x)\left(\frac{100}{100-x}\right) =$$

$$m$$

Hence, $n = m$, and the answer is (B).

20. W13. Mike ran 10 miles at 10 miles per hour (*Time = Distance/Rate* = 10 miles/10 miles per hour = 1 hour). He ran at 5 miles per hour for the remaining 20 miles (*Time = Distance/Rate* = 20 miles/5 miles per hour = 4 hrs). The total length of the Marathon track is 30 miles, and the total time taken to cover the track is 5 hours. Hence, the answer is (E).

21. Suppose Fritz took t hours to complete the 30-mile Marathon. Then as given, Fritz ran at 10 miles per hour for $t/3$ hours and 5 miles per hour for the remaining $2t/3$ hours. Now, by the formula, *Distance = Rate · Time*, the total distance covered would be (10 miles per hour) · $t/3$ + (5 miles per hour) · $2t/3$ = $(10/3 + 10/3)t$ = 30 miles. Solving the equation for t yields $t = 90/20$ hours = 4.5 hours. The answer is (D).

22. 2(an integer) + 1 is always odd. The rule $a_{n+1} = 2a_n + 1$ indicates that each term in the series, except possibly the first one, must be odd. The first term may be even. Hence, assign the even number 38 to the only possible even term in the sequence. By the rule $a_{n+1} = 2a_n + 1$, we have $a_2 = 2a_1 + 1 = 2(38) + 1 = 77$. The answer is (E).

23. Let the number of red balls in the urn be $2k$, the number of yellow balls $3k$, and the number of green balls $4k$, where k is a common factor of the three. Now, the total number of balls in the urn is $2k + 3k + 4k = 9k$. Hence, the fraction of red balls from all the balls is $2k/9k = 2/9$. This also equals the probability that a ball chosen at random from the urn is a red ball. The answer is (A).

24. There are 2 red and 3 green doors. We have two cases:

> The room can be entered from a red door (2 red doors, so 2 ways) and can be left from a green door (3 green doors, so 3 ways): $2 \cdot 3 = 6$.

> The room can be entered from a green door (3 green doors, so 3 ways) and can be left from a red door (2 red doors, so 2 ways): $3 \cdot 2 = 6$.

Hence, the total number of ways is

$$2 \cdot 3 + 3 \cdot 2 = 6 + 6 = 12$$

The answer is (E).

25. The problem is a mix of 3 combinational problems. The goal is to choose 1 of 2 entrees, then 1 of 3 main courses, then 1 of 3 desserts. The choices can be made in 2, 3, and 3 ways, respectively. Hence, the total number of ways of selecting the combinations is $2 \cdot 3 \cdot 3 = 18$. The answer is (C).

We can also count the combinations by the Fundamental Principle of Counting:

	Main Course 1	Dessert 1 Dessert 2 Dessert 3
Entrée 1	Main Course 2	Dessert 1 Dessert 2 Dessert 3
	Main Course 3	Dessert 1 Dessert 2 Dessert 3
	Main Course 1	Dessert 1 Dessert 2 Dessert 3
Entrée 2	Main Course 2	Dessert 1 Dessert 2 Dessert 3
	Main Course 3	Dessert 1 Dessert 2 Dessert 3
		Total 18

The Fundamental Principle of Counting states:

The total number of possible outcomes of a series of decisions, making selections from various categories, is found by multiplying the number of choices for each decision.

Counting the number of choices in the final column above yields 18.

Part Two
Summary of Math Properties

Arithmetic

1. A *prime number* is an integer that is divisible only by itself and 1.
2. An even number is divisible by 2, and can be written as $2x$.
3. An odd number is not divisible by 2, and can be written as $2x + 1$.
4. Division by zero is undefined.
5. Perfect squares: 1, 4, 9, 16, 25, 36, 49, 64, 81, ...
6. Perfect cubes: 1, 8, 27, 64, 125, ...
7. If the last digit of a integer is 0, 2, 4, 6, or 8, then it is divisible by 2.
8. An integer is divisible by 3 if the sum of its digits is divisible by 3.
9. If the last digit of a integer is 0 or 5, then it is divisible by 5.
10. Miscellaneous Properties of Positive and Negative Numbers:

    A. The product (quotient) of positive numbers is positive.
    B. The product (quotient) of a positive number and a negative number is negative.
    C. The product (quotient) of an even number of negative numbers is positive.
    D. The product (quotient) of an odd number of negative numbers is negative.
    E. The sum of negative numbers is negative.
    F. A number raised to an even exponent is greater than or equal to zero.

    $$even \times even = even$$
    $$odd \times odd = odd$$
    $$even \times odd = even$$

    $$even + even = even$$
    $$odd + odd = even$$
    $$even + odd = odd$$

11. Consecutive integers are written as $x, x + 1, x + 2, \ldots$
12. Consecutive even or odd integers are written as $x, x + 2, x + 4, \ldots$
13. The integer zero is neither positive nor negative, but it is even: $0 = 2 \cdot 0$.
14. Commutative property: $x + y = y + x$. Example: $5 + 4 = 4 + 5$.
15. Associative property: $(x + y) + z = x + (y + z)$. Example: $(1 + 2) + 3 = 1 + (2 + 3)$.
16. Order of operations: Parentheses, Exponents, Multiplication, Division, Addition, Subtraction.
17. $-\dfrac{x}{y} = \dfrac{-x}{y} = \dfrac{x}{-y}$. Example: $-\dfrac{2}{3} = \dfrac{-2}{3} = \dfrac{2}{-3}$
18.
    $33\dfrac{1}{3}\% = \dfrac{1}{3}$ $\quad 20\% = \dfrac{1}{5}$
    $66\dfrac{2}{3}\% = \dfrac{2}{3}$ $\quad 40\% = \dfrac{2}{5}$
    $25\% = \dfrac{1}{4}$ $\quad 60\% = \dfrac{3}{5}$
    $50\% = \dfrac{1}{2}$ $\quad 80\% = \dfrac{4}{5}$

GRE Math Tests

19.
$\dfrac{1}{100} = .01$ $\dfrac{1}{10} = .1$ $\dfrac{2}{5} = .4$

$\dfrac{1}{50} = .02$ $\dfrac{1}{5} = .2$ $\dfrac{1}{2} = .5$

$\dfrac{1}{25} = .04$ $\dfrac{1}{4} = .25$ $\dfrac{2}{3} = .666\ldots$

$\dfrac{1}{20} = .05$ $\dfrac{1}{3} = .333\ldots$ $\dfrac{3}{4} = .75$

20. Common measurements:
1 foot = 12 inches
1 yard = 3 feet
1 quart = 2 pints
1 gallon = 4 quarts
1 pound = 16 ounces

21. Important approximations: $\sqrt{2} \approx 1.4$ $\sqrt{3} \approx 1.7$ $\pi \approx 3.14$

22. "The remainder is r when p is divided by q" means $p = qz + r$; the integer z is called the quotient. For instance, "The remainder is 1 when 7 is divided by 3" means $7 = 3 \cdot 2 + 1$.

23. Probability = $\dfrac{\text{number of outcomes}}{\text{total number of possible outcomes}}$

Algebra

24. Multiplying or dividing both sides of an inequality by a negative number reverses the inequality. That is, if $x > y$ and $c < 0$, then $cx < cy$.

25. Transitive Property: If $x < y$ and $y < z$, then $x < z$.

26. Like Inequalities Can Be Added: If $x < y$ and $w < z$, then $x + w < y + z$.

27. Rules for exponents:

$x^a \cdot x^b = x^{a+b}$ Caution, $x^a + x^b \neq x^{a+b}$

$(x^a)^b = x^{ab}$

$(xy)^a = x^a \cdot y^a$

$\left(\dfrac{x}{y}\right)^a = \dfrac{x^a}{y^a}$

$\dfrac{x^a}{x^b} = x^{a-b}$, if $a > b$. $\dfrac{x^a}{x^b} = \dfrac{1}{x^{b-a}}$, if $b > a$.

$x^0 = 1$

28. There are only two rules for roots that you need to know for the GRE:

$\sqrt[n]{xy} = \sqrt[n]{x}\sqrt[n]{y}$ For example, $\sqrt{3x} = \sqrt{3}\sqrt{x}$.

$\sqrt[n]{\dfrac{x}{y}} = \dfrac{\sqrt[n]{x}}{\sqrt[n]{y}}$ For example, $\sqrt[3]{\dfrac{x}{8}} = \dfrac{\sqrt[3]{x}}{\sqrt[3]{8}} = \dfrac{\sqrt[3]{x}}{2}$.

Caution: $\sqrt[n]{x+y} \neq \sqrt[n]{x} + \sqrt[n]{y}$.

29. Factoring formulas:

$$x(y + z) = xy + xz$$
$$x^2 - y^2 = (x + y)(x - y)$$
$$(x - y)^2 = x^2 - 2xy + y^2$$
$$(x + y)^2 = x^2 + 2xy + y^2$$
$$-(x - y) = y - x$$

30. Adding, multiplying, and dividing fractions:

$$\frac{x}{y} + \frac{z}{y} = \frac{x+z}{y} \quad \text{and} \quad \frac{x}{y} - \frac{z}{y} = \frac{x-z}{y}$$

Example: $\frac{2}{4} + \frac{3}{4} = \frac{2+3}{4} = \frac{5}{4}$.

$$\frac{w}{x} \cdot \frac{y}{z} = \frac{wy}{xz}$$

Example: $\frac{1}{2} \cdot \frac{3}{4} = \frac{1 \cdot 3}{2 \cdot 4} = \frac{3}{8}$.

$$\frac{w}{x} \div \frac{y}{z} = \frac{w}{x} \cdot \frac{z}{y}$$

Example: $\frac{1}{2} \div \frac{3}{4} = \frac{1}{2} \cdot \frac{4}{3} = \frac{4}{6} = \frac{2}{3}$.

31. $x\% = \dfrac{x}{100}$

32. Quadratic Formula: $x = \dfrac{-b \pm \sqrt{b^2 - 4ac}}{2a}$ are the solutions of the equation $ax^2 + bx + c = 0$.

Geometry

33. There are four major types of angle measures:

    An **acute angle** has measure less than 90°:

    A **right angle** has measure 90°:

    An **obtuse angle** has measure greater than 90°:

    A **straight angle** has measure 180°: 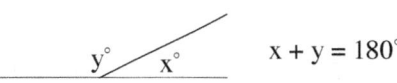 $x + y = 180°$

34. Two angles are supplementary if their angle sum is 180°: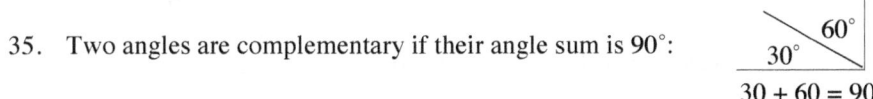

35. Two angles are complementary if their angle sum is 90°:

36. Perpendicular lines meet at right angles:

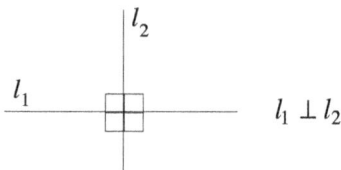

$l_1 \perp l_2$

37. When two straight lines meet at a point, they form four angles. The angles opposite each other are called vertical angles, and they are congruent (equal). In the figure, $a = b$, and $c = d$.

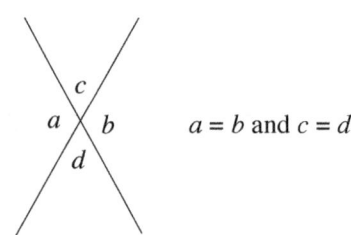

$a = b$ and $c = d$

38. When parallel lines are cut by a transversal, three important angle relationships exist:

| Alternate interior angles are equal. | Corresponding angles are equal. | Interior angles on the same side of the transversal are supplementary. |

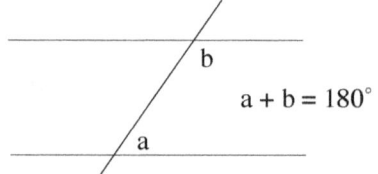

$a + b = 180°$

39. The shortest distance from a point not on a line to the line is along a perpendicular line.

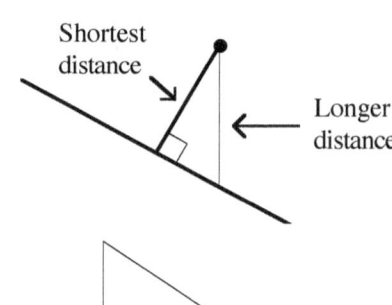

40. A triangle containing a right angle is called a *right triangle*. The right angle is denoted by a small square:

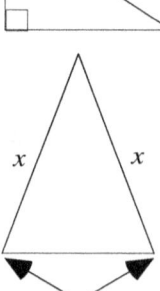

41. A triangle with two equal sides is called isosceles. The angles opposite the equal sides are called the base angles:

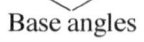

Base angles

42. In an equilateral triangle, all three sides are equal and each angle is 60°:

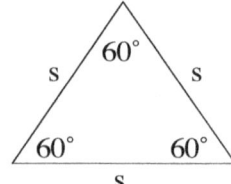

43. The altitude to the base of an isosceles or equilateral triangle bisects the base and bisects the vertex angle:

Isosceles: 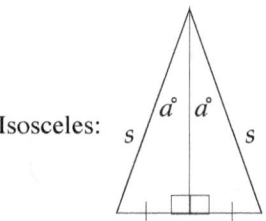   Equilateral: 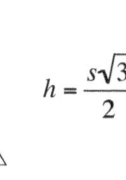 $h = \dfrac{s\sqrt{3}}{2}$

44. The angle sum of a triangle is 180°:

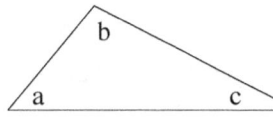 $a + b + c = 180°$

45. The area of a triangle is $\dfrac{1}{2}bh$, where b is the base and h is the height.

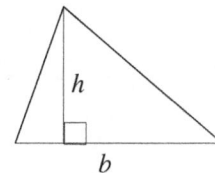 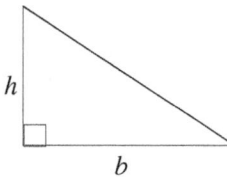 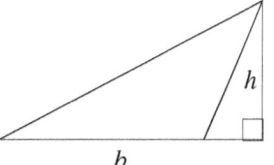 $A = \dfrac{1}{2}bh$

46. In a triangle, the longer side is opposite the larger angle, and vice versa:

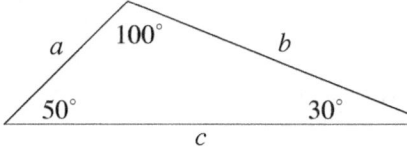   50° is larger than 30°, so side b is longer than side a.

47. Pythagorean Theorem (right triangles only): The square of the hypotenuse is equal to the sum of the squares of the legs.

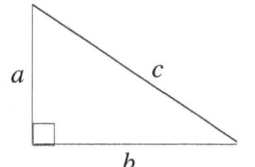 $c^2 = a^2 + b^2$

48. A Pythagorean triple: the numbers 3, 4, and 5 can always represent the sides of a right triangle and they appear very often: $5^2 = 3^2 + 4^2$.

49. Two triangles are similar (same shape and usually different size) if their corresponding angles are equal. If two triangles are similar, their corresponding sides are proportional:

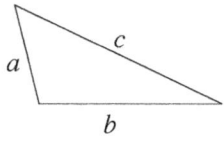 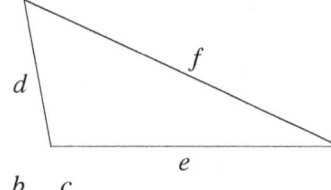

$$\dfrac{a}{d} = \dfrac{b}{e} = \dfrac{c}{f}$$

50. If two angles of a triangle are congruent to two angles of another triangle, the triangles are similar.
    In the figure, the large and small triangles are similar because both contain a right angle and they share $\angle A$.

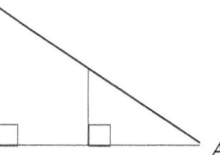

51. Two triangles are congruent (identical) if they have the same size and shape.

52. In a triangle, an exterior angle is equal to the sum of its remote interior angles and is therefore greater than either of them:

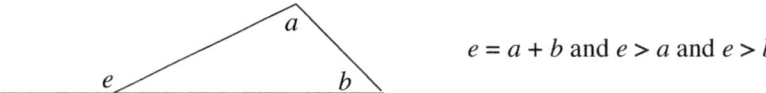

$e = a + b$ and $e > a$ and $e > b$

53. In a triangle, the sum of the lengths of any two sides is greater than the length of the remaining side:

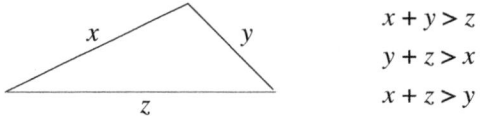

$x + y > z$
$y + z > x$
$x + z > y$

54. In a 30°–60°–90° triangle, the sides have the following relationships:

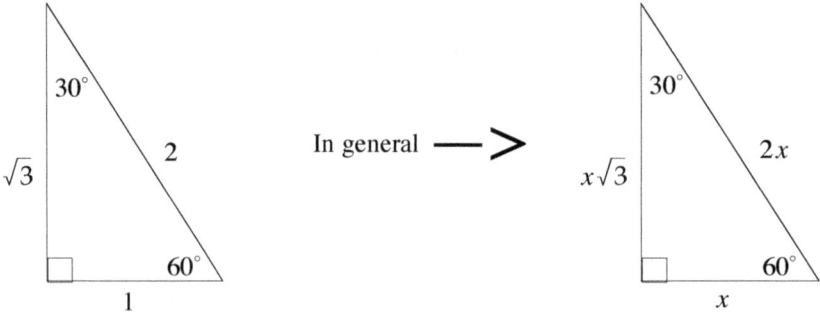

In general ⟶

55. In a 45°–45°–90° triangle, the sides have the following relationships:

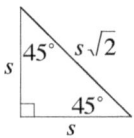

56. Opposite sides of a parallelogram are both parallel and congruent:

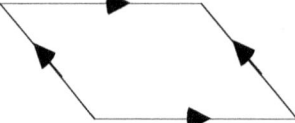

57. The diagonals of a parallelogram bisect each other:

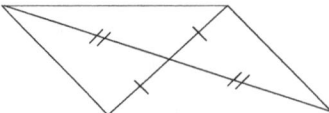

58. A parallelogram with four right angles is a *rectangle*. If w is the width and l is the length of a rectangle, then its area is $A = lw$ and its perimeter is $P = 2w + 2l$:

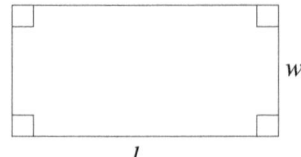

$A = l \cdot w$
$P = 2w + 2l$

59. If the opposite sides of a rectangle are equal, it is a square and its area is $A = s^2$ and its perimeter is $P = 4s$, where s is the length of a side:

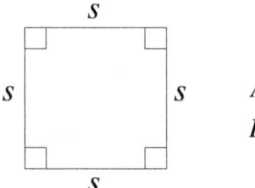

$A = s^2$
$P = 4s$

60. The diagonals of a square bisect each other and are perpendicular to each other:

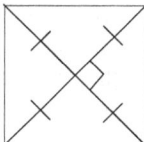

61. A quadrilateral with only one pair of parallel sides is a *trapezoid*. The parallel sides are called *bases*, and the non-parallel sides are called *legs*:

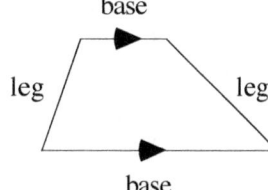

62. The area of a trapezoid is the average of the bases times the height:

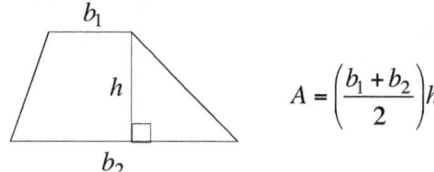

$$A = \left(\frac{b_1 + b_2}{2}\right)h$$

63. The volume of a rectangular solid (a box) is the product of the length, width, and height. The surface area is the sum of the area of the six faces:

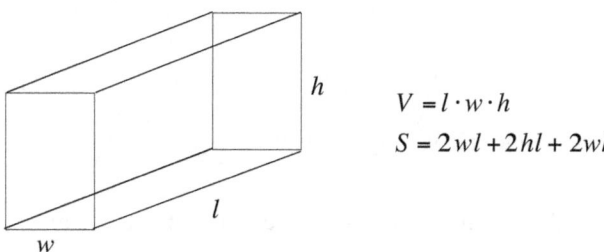

$$V = l \cdot w \cdot h$$
$$S = 2wl + 2hl + 2wh$$

64. If the length, width, and height of a rectangular solid (a box) are the same, it is a cube. Its volume is the cube of one of its sides, and its surface area is the sum of the areas of the six faces:

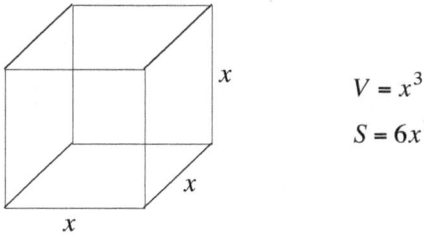

$$V = x^3$$
$$S = 6x^2$$

65. The volume of a cylinder is $V = \pi r^2 h$, and the lateral surface (excluding the top and bottom) is $S = 2\pi rh$, where r is the radius and h is the height:

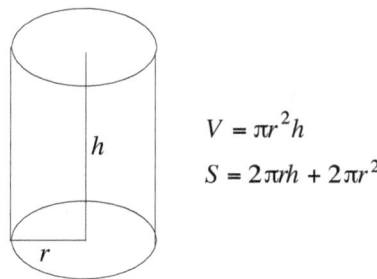

$$V = \pi r^2 h$$
$$S = 2\pi rh + 2\pi r^2$$

66. A line segment form the circle to its center is a *radius*.
A line segment with both end points on a circle is a *chord*.
A chord passing though the center of a circle is a *diameter*.
A diameter can be viewed as two radii, and hence a diameter's length is twice that of a radius.
A line passing through two points on a circle is a *secant*.
A piece of the circumference is an *arc*.
The area bounded by the circumference and an angle with vertex at the center of the circle is a *sector*.

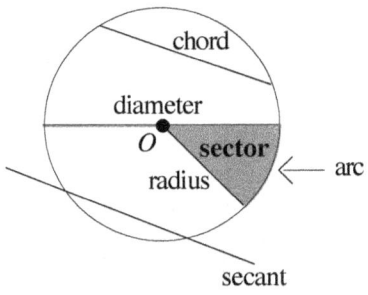

67. A tangent line to a circle intersects the circle at only one point. The radius of the circle is perpendicular to the tangent line at the point of tangency:

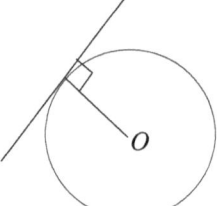

68. Two tangents to a circle from a common exterior point of the circle are congruent:

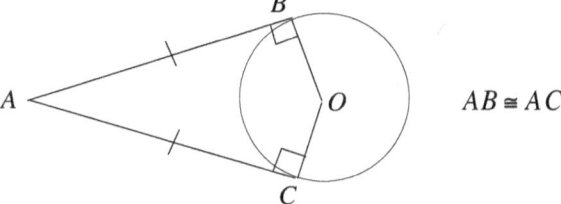

$AB \cong AC$

69. An angle inscribed in a semicircle is a right angle: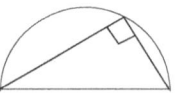

70. A central angle has by definition the same measure as its intercepted arc.

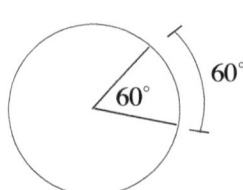

71. An inscribed angle has one-half the measure of its intercepted arc.

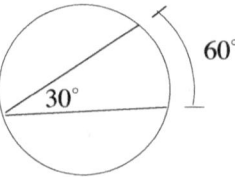

72. The area of a circle is πr^2, and its circumference (perimeter) is $2\pi r$, where r is the radius:

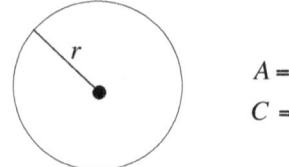

$A = \pi r^2$
$C = 2\pi r$

73. To find the area of the shaded region of a figure, subtract the area of the unshaded region from the area of the entire figure.

74. When drawing geometric figures, don't forget extreme cases.

Summary of Math Properties

Miscellaneous

75. To compare two fractions, cross-multiply. The larger product will be on the same side as the larger fraction.

76. Taking the square root of a fraction between 0 and 1 makes it larger.

    Caution: This is not true for fractions greater than 1. For example, $\sqrt{\frac{9}{4}} = \frac{3}{2}$. But $\frac{3}{2} < \frac{9}{4}$.

77. Squaring a fraction between 0 and 1 makes it smaller.

78. $ax^2 \neq (ax)^2$. In fact, $a^2 x^2 = (ax)^2$.

79. $\frac{1/a}{b} \neq \frac{1}{a/b}$. In fact, $\frac{1/a}{b} = \frac{1}{ab}$ and $\frac{1}{a/b} = \frac{b}{a}$.

80. $-(a+b) \neq -a+b$. In fact, $-(a+b) = -a-b$.

81. $\text{percentage increase} = \dfrac{\text{increase}}{\text{original amount}}$

82. Systems of simultaneous equations can most often be solved by merely adding or subtracting the equations.

83. When counting elements that are in overlapping sets, the total number will equal the number in one group plus the number in the other group minus the number common to both groups.

84. The number of integers between two integers <u>inclusive</u> is one more than their difference.

85. Elimination strategies:
    A. On hard problems, if you are asked to find the least (or greatest) number, then eliminate the least (or greatest) answer-choice.
    B. On hard problems, eliminate the answer-choice "not enough information."
    C. On hard problems, eliminate answer-choices that <u>merely</u> repeat numbers from the problem.
    D. On hard problems, eliminate answer-choices that can be derived from elementary operations.
    E. After you have eliminated as many answer-choices as you can, choose from the more complicated or more unusual answer-choices remaining.

86. To solve a fractional equation, multiply both sides by the LCD (lowest common denominator) to clear fractions.

87. You can cancel only over multiplication, not over addition or subtraction. For example, the c's in the expression $\dfrac{c+x}{c}$ cannot be canceled.

88. The average of N numbers is their sum divided by N, that is, $average = \dfrac{sum}{N}$.

89. *Weighted average:* The average between two sets of numbers is closer to the set with more numbers.

90. $Average\ Speed = \dfrac{Total\ Distance}{Total\ Time}$

91. $Distance = Rate \times Time$

92. *Work = Rate × Time*, or $W = R \times T$. The amount of work done is usually 1 unit. Hence, the formula becomes $1 = R \times T$. Solving this equation for R yields $R = \dfrac{1}{T}$.

93. *Interest = Amount × Time × Rate*

94. Principles for solving quantitative comparisons

    A. You can add or subtract the same term (number) from both sides of a quantitative comparison problem.

    B. You can multiply or divide both sides of a quantitative comparison problem by the same positive term (number). (Caution: this cannot be done if the term can ever be negative or zero.)

    C. When using substitution on quantitative comparison problems, you must plug in all five major types of numbers: positives, negatives, fractions, 0, and 1. Test 0, 1, 2, –2, and 1/2, in that order.

    D. If there are only numbers (i.e., no variables) in a quantitative comparison problem, then "not-enough-information" cannot be the answer.

95. Substitution (Special Cases):

    A. In a problem with two variables, say, x and y, you must check the case in which $x = y$. (This often gives a double case.)

    B. When you are given that $x < 0$, you must plug in negative whole numbers, negative fractions, and –1. (Choose the numbers –1, –2, and –1/2, in that order.)

    C. Sometimes you have to plug in the first three numbers (but never more than three) from a class of numbers.

www.ingramcontent.com/pod-product-compliance
Lightning Source LLC
Chambersburg PA
CBHW080722230426
43665CB00020B/2579